An Introduction to the Kinetic Theory of Gases and Magnetoplasmas

An Introduction to the Kinetic Theory of Gases and Magnetoplasmas

L. C. Woods
Balliol College, University of Oxford

OXFORD NEW YORK MELBOURNE
OXFORD UNIVERSITY PRESS
1993

Oxford University Press, Walton Street, Oxford OX2 6DP
Oxford New York Toronto
Delhi Bombay Calcutta Madras Karachi
Kuala Lumpur Singapore Hong Kong Tokyo
Nairobi Dar es Salaam Cape Town
Melbourne Auckland Madrid
and associated companies in
Berlin Ibadan

Oxford is a trade mark of Oxford University Press

Published in the United States
by Oxford University Press Inc., New York

A catalogue record for this book is available from the British Library

Library of Congress Cataloging in Publication Data

Woods, L. C.(Leslie Colin), 1922–
An introduction to the kinetic theory of gases and magnetoplasm / L. C. Woods
1. Kinetic theory of gases. 2. Plasma (Ionised gases) I. Title
QC175.W66 1993 533'.7—dc20 92–43326
ISBN 0–19–856393–0

Typeset using LaTeX by the author
Printed in Great Britain by St Edmundsbury Press,
Bury St Edmunds, Suffolk

PREFACE

The kinetic theory contained in this book is concerned mainly with the connection between the motions and interactions of microscopic particles comprising a gas and the transport of macroscopic properties like fluid momentum and energy through that gas. The oldest example relating macroscopic properties to microscopic behaviour is provided by the pressure force acting on the walls of a gas container. That it is due to the near-continuous bombardment of the walls by the vast number of neighbouring molecules, is a concept dating back to Boyle and Newton. The more subtle relationship between heat and the energy of molecular agitation required more than another century before it was revealed with increasing detail in the works of Waterston, Clausius, and Maxwell.

Most modern accounts of kinetic theory (see References) are strong on mathematical formalism, but because they too readily dismiss mean-free-path arguments, they do not convey much of the *physics* of the various processes involved. Older texts like those authored by Jeans, Kennard, and Present were more physically oriented, but since their publication, there has been a great increase in the applications of kinetic theory. Recent advances in chemical engineering, aeronautics, plasma physics and astrophysics have considerably enhanced the importance of the subject. Furthermore, the development of high-speed computers has made it possible to solve previously inaccessible problems. But computing is one stage further removed from the physics than the underlying mathematical theory. There remains a need to understand the *mechanisms* of transport, not only to judge the value of computer output, but also to check the merit of mathematical analysis, which is often too lightly underpinned by physical reasoning. It therefore appeared timely to write a new account of the kinetic theory of gases and magnetoplasmas, not only giving physical arguments priority, but also covering the more useful parts of the mathematical structure of the subject.

The aim in this book is to give a treatment of the kinetic theory of gases and magnetoplasmas, covering the standard material in as simple a way as possible, using mean-free-path arguments where appropriate and identifying problem areas where received theory has either completely failed or has fallen short of expectations. Comparisons between theory and experiment are made in the text when possible; this includes studies on ultrasonic waves and strong shock waves. There is also a wealth of observations available on tokamak phenomena, some of which is relevant to the theory developed in the book. Chapter 3 shows how mean-free-path methods can be refined to yield results almost as accurate as those derived from advanced kinetic theory. It forms the basis for the new treatment of transport given in Chapters 5, 8, and 11.

Initially the difference between convection and diffusion was not apparent, and Maxwell's earliest work on thermal diffusion contained a distortion due to convection. Indeed, failure to make precise the distinction between these transport processes has remained a source of error in kinetic theory until the present day. The usual prescription is that diffusion is due to molecular agitation as seen in a frame F_1 moving with the local fluid velocity $\mathbf{v}$. Excepting that due to a body force like gravity, the acceleration of the fluid element is ignored. But to remove all systematic or non-random particle motions, the molecular agitation—which is responsible for diffusion—must be reckoned in a frame F that not only has the fluid velocity $\mathbf{v}$, but also has the fluid acceleration $\mathbf{a}$ and (ideally) the time derivatives of $\mathbf{a}$ to all orders.

A basic, non-dimensional parameter central to transport theory is the Knudsen number. This is the ratio of the average length λ of a particle trajectory, taken between successive collisions with other particles, and the characteristic macroscopic length scale L, over which a macroscopic variable like the temperature is significantly changed. It can also be defined as a ratio of microscopic to macroscopic time scales. For what may be described as 'collision dominated' flows, $\epsilon = \lambda/L$ is much smaller than unity. Most of the transport phenomena discussed in this book apply to this regime, in which case it is sufficiently accurate to specify diffusion in F_1. This is termed 'first-order' transport, since only terms of order ϵ are retained. When ϵ is greater than about 0.2, but still clearly smaller than one—as occurs in strong shock waves—'second-order' transport, i.e. that involving $O(\epsilon^2)$ terms, provides a significant correction to first-order transport. To calculate these terms accurately, diffusion needs to be calculated in a frame F_2 that has the accelerations, including the spin, of the fluid element.

Boltzmann's equation for the evolution of the velocity distribution function f is taken as the starting point for almost all calculations of transport in neutral gases. But its formulation is correct *only* for diffusion in F_1. It follows that the $O(\epsilon^2)$ terms, rigorously derived from Boltzmann's equation and known as the 'Burnett' terms, include effects resulting from the neglected accelerations. The failure of the unmodified Burnett expressions to yield continuous shock wave structures above a Mach number of about 1.7, together with the absence of a physical interpretation of the terms, has led some kinetic theorists to doubt their value.

For the reason given above, and for a more important reason to follow, in its treatment of the second-order terms, our account of kinetic theory diverges from the traditional account of the subject. Several other related changes are also made. To help readers new to the subject, these departures from standard kinetic theory are summarized in the Appendix.

A particularly important change concerns the definition of the pressure tensor: the standard treatment defines it as being momentum *flux*, which is not the same as the fluid definition, namely that it is the *force* transmitted across unit area of a surface by particle *collisions*. For first-order transport this discrepancy in definitions is unimportant, but pressure in the role of a *force* contributes to the acceleration of F_2 and therefore cannot influence second-order diffusion. Neglect of this point results in pressure gradients appearing in the Burnett terms.

The most striking example of the failure of standard kinetic theory comes from plasma physics, where losses from tokamaks are orders of magnitude in excess of mathematical prediction. Turbulence is usually indited, but the experimental data is remarkably regular from machine to machine and under greatly varying conditions. In a previous work, **Principles of magnetoplasma dynamics** (OUP, 1987), the author gave a theory for this phenomenon that explains a wide range of observations, and achieves close agreement with experiment. It is based on second-order transport across the magnetic field which, being much less suppressed by strong fields, proves to be orders of magnitude larger than the usual first-order transport. However, the approach adopted in 1987 was based on the assumption that Boltzmann's equation was correct for *all* values of the Knudsen number. With this abandoned, it became possible to develop a better and one hopes, a more convincing treatment, albeit leading to the same transport equations. One of the aims of this work is to present this new theory.

My thanks are due to Dr G.B. Deane (now at the Scripps Institution of Oceanography, U.S.C.D., California), whose grasp of the complexities of second-order transport, made him a valued colleague. His challenges enabled me to sharpen arguments and to reduce speculations masquerading as proofs; I hope no imposters remain. I am also grateful to Richard Kennaugh and Jason Reese of the Mathematical Institute, University of Oxford, for their critical reading of the text, enabling me to remove some errors. The staff of the Oxford University Press have been most helpful in bringing the work to publication.

Balliol College
University of Oxford
November, 1992

L.C.W.

FOR SUZANNE

CONTENTS

1

BASIC CONCEPTS

1.1 Molecular models

A substance in the gaseous state consists of an assembly of a vast number of microscopic particles that, excepting when they collide with each other, move freely and independently through the region of physical space available to them. The nature of the particles depends largely on the temperature of the assembly. At low temperatures, but above the critical value at which liquefication can occur, they are molecules. At higher temperatures the molecules dissociate into atoms, and at still higher temperatures the atoms will become ions by shedding some of their electrons. The resulting assembly is termed a 'plasma'. Partially ionized plasmas consist of a mixture of neutral (i.e. uncharged) atoms, electrons, and ions, requiring at least three distinct species of microscopic particles to be included in a mathematical representation of their collective behaviour.

The simplest model of a (microscopic) particle is a small featureless sphere, possessing a spherically-symmetric force field. For neutral particles this field has a very short range, and the particles can be pictured as being almost rigid 'billiard balls', with an effective diameter equal to the range of the force field. As they have no structure, these particles have only energy of translation. The gas is assumed to be sufficiently tenuous for collisions involving more than two particles at a time to be ignored, i.e. only *binary* collisions are considered. The model is appropriate for monatomic uncharged molecules.

Diatomic and more complex molecules do not have symmetric force fields, but for many purposes they are also well represented by the billiard ball model. Their relative orientations at collisions may be assumed to be randomly distributed, so that averages taken over a large number of encounters will have values independent of orientation, just as with symmetric force fields. It is the internal energy possessed by multi-atomic particles that gives rise to the largest discrepancies between the predictions of the billiard ball model and observation.

The intermolecular force law plays a central role in the model. Classical kinetic theory, our concern here, proceeds on the assumption that this law has been separately established, either empirically or from quantum theory, except with charged particles, when the Coulomb force law applies.

1.1.1 *The mean free path*

Two microscopic parameters will play a large role in our account of kinetic theory. These are the mean free path λ, which is the average distance moved by a particle between successive encounters, and the collision interval τ, which is the average time taken by a particle to move this distance. The reciprocal of τ is known as the 'collision' frequency, $\nu = \tau^{-1}$. The terminology is particularly fitting for 'hard' molecules, i.e. those with force fields abruptly falling to zero outside a molecular diameter σ, say.

An approximate formula for λ can be found as follows. Suppose there are n molecules per unit volume, and we assume that all are stationary, save one that has a velocity v_r relative to the others. In a tenuous or dilute gas, $\lambda \gg \sigma$ and hence $\pi\sigma^2 v_r$ is a good approximation to the volume swept in one second by the sphere of influence of the moving particle. Those molecules with centres lying within this volume will experience a collision. The collision frequency per molecule is therefore $\tau^{-1} = \pi\sigma^2 n v_r$. Replacing v_r by the average molecular speed $\bar{\mathbf{c}}$ relative to the centre of mass of all the similar molecules within a macroscopic volume element, and writing $\lambda = \tau\bar{\mathbf{c}}$, we arrive at the estimate

$$\lambda \approx 1/(\pi\sigma^2 n) \qquad (\tau = \lambda/\bar{\mathbf{c}}). \tag{1.1}$$

An accurate formula will be derived in §2.8.2. The only modification to equation (1.1)* is the appearance of a numerical factor $1/\sqrt{2}$ on its right-hand side.

'Soft' molecules have extended force fields that make only slight changes in the momentum and energy of most passing molecules, so many such 'grazing' collisions are required to accumulate significant changes in these properties for a given test particle. However, by using σ to denote an 'effective' diameter, equation (1.1) can be extended to the case of soft molecules. Then λ becomes the average distance that a sequence of small-angle collisions takes to stop a test particle moving in a given direction, i.e. to give a 90° deflection, and τ is the time it takes for this change to happen. Even with hard molecules, a sequence of collisions is required to stop a particle. Another consequence of this cascade process is that momentum and energy require related but difference times to be transported in a specified direction.

The Coulomb force fields of electrons and ions have ranges extensive enough to influence great numbers of nearby particles, so that purely binary collisions are very rare. The billiard ball model and the associated concept of a mean free path are not strictly relevant, although it is usual to describe the distance required for a 90° deflection of a test particle as being a 'mean free path'. More precisely, it is the distance over which this particle loses its momentum along its initial direction of motion.

*We shall usually omit 'equation', and refer to the equation numbers only.

1.2 Macroscopic variables

By a *macroscopic point* $P(\mathbf{r}, t)$ is meant an infinitesimal volume element $d\mathbf{r}$ ($= dx\,dy\,dz$) centred at the precise point $(\mathbf{r}, t)$, and with sufficient extension so that $d\mathbf{r}$ contains a great number of particles of the species under consideration. (Note that $d\mathbf{r}$ is not a vector; when a vector element is intended, this will be clear from the context.) Let this number be $n\,d\mathbf{r}$, then, as $\mathbf{r}$ and t change continuously, fluctuations in n due solely to the discreteness of matter will be negligible. We may therefore introduce the *number density* $n(\mathbf{r}, t)$ as a continuum variable, provided the length $(d\mathbf{r})^{1/3}$ is very much larger than the interparticle distance $n^{-1/3}$. On the other hand, we must not take $d\mathbf{r}$ to be too large, otherwise significant variations in $n(\mathbf{r}, t)$ may be smothered. Thus $d\mathbf{r}$ should satisfy

$$n^{-1/3} \ll (d\mathbf{r})^{1/3} \ll L_n \qquad (L_n = (d \ln n/dx)^{-1}),$$

where L_n is a typical distance over which n changes by a significant fraction of its local value.

The *fluid density* is the macroscopic variable

$$\rho(\mathbf{r}, t) = mn(\mathbf{r}, t),$$

where m is the particle mass, and for the present we are assuming that only one species is present.

Let $\mathbf{w}$ be the velocity of a typical particle at P, measured relative to a frame L (the 'laboratory' frame), and use $\langle\ \rangle$ to denote average values taken over the particles in $d\mathbf{r}$. The *fluid velocity* at P is the macroscopic variable $\mathbf{v}(\mathbf{r}, t) = \langle \mathbf{w} \rangle$. It is the velocity of the centre of mass of the particles at P; the fluid momentum relative to L is $\rho\mathbf{v}\,d\mathbf{r}$. The curl of $\mathbf{v}$ is called the *fluid vorticity*: $\zeta \equiv \nabla \times \mathbf{v}$. Its importance in kinetic theory follows from the fact (shown in §1.13) that the fluid at P spins relative to L with the angular velocity

$$\boldsymbol{\Omega} = \tfrac{1}{2}\nabla \times \mathbf{v},$$

which we shall term the *fluid spin*. One of the principal aims of kinetic theory is finding expressions for the rates at which properties like energy, momentum and species concentration are transported through the gas. Transport is divided into two distinct types, namely *convection* and *diffusion*. Convection is the transport of a property carried along 'bodily' by the fluid motion, whereas diffusion is due to the random particle motions superimposed on this mass motion. To distinguish the transport due to molecular agitation from that due to convection, it is essential to be able to express the equations of kinetic theory in a reference frame convected with the fluid.

A macroscopic point P_c that coincides momentarily with $P(\mathbf{r}, t)$, and that moves with the fluid velocity, is called a *convected point*, and its locus is termed a *path line*. If axes are fixed in P_c and allowed to rotate with the local fluid spin $\boldsymbol{\Omega}$, then P_c thus augmented is a *convected frame*. Viewed from this frame, the fluid near P_c will appear to be almost stationary, and without spin. It is the appropriate frame in which to determine transport due to diffusion. By 'at P_c' we shall mean the convected frame at $\mathbf{r}, t$. The spin of this frame is important when time derivatives of vectors and tensors are required at the point.

At P_c the particle velocity measured relative to the convected frame is $\mathbf{c} \equiv \mathbf{w} - \mathbf{v}$; it is termed the *peculiar* velocity. By its definition, $\langle \mathbf{c} \rangle = 0$. In kinetic theory this velocity serves to classify the particles at P_c, and the thermodynamic variables can be defined in terms of it.

1.3 The thermodynamic variables

1.3.1 *Temperature*

The basic thermodynamic variable is *temperature*; it is usually introduced with the aid of the concept of thermal equilibrium between two contiguous macroscopic systems, say G and W. Such systems are said to be in thermal equilibrium when no net energy transfer occurs between them when they are in physical contact. At the microscopic level 'physical contact' means that the molecules of G and W are colliding with each other. One of the basic laws of thermodynamics— known as the 'zeroth' law—states that

Two systems in thermal equilibrium with a third are in thermal equilibrium with each other.

The third system can be regarded as being a 'thermometer', and the three systems are said to be at the same temperature.

Now choose G to be a gas and W to be the rigid boundary wall confining it. Take the line of impact at the collision to be the axis OX, not necessarily perpendicular to the wall (see Fig. 1.1). Let the velocity components of a gas particle G_1 and a wall molecule W_1 be u, v, w, and U, V, W just before the collision and u', v', w' and U', V', W' just after it, then from momentum and energy conservation, the velocity components parallel to OY and OZ will be unchanged, whereas in the OX direction

$$\begin{aligned} mu + MU &= mu' + MU', \\ \tfrac{1}{2}mu^2 + \tfrac{1}{2}MU^2 &= \tfrac{1}{2}mu'^2 + \tfrac{1}{2}MU'^2, \end{aligned}$$

or

$$\begin{aligned} m(u - u') &= -M(U - U'), \\ m(u^2 - u'^2) &= -M(U^2 - U'^2). \end{aligned}$$

Dividing the second forms we obtain $u + u' = U + U'$, whence the relative velocity $u - U$ is reversed by the collision, as required by perfect elasticity. This condition and the momentum equation gives

$$(m + M)u' = (m - M)u + 2MU, \qquad (m + M)U' = -(m - M)U + 2mu,$$

and hence the gain in the wall's kinetic energy per collision is

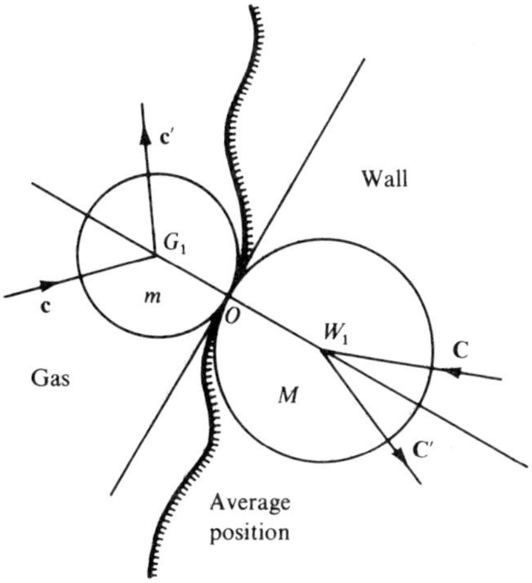

FIG. 1.1. Collision between gas and wall molecules

$$\tfrac{1}{2}M(U'^2 - U^2) = \frac{2mM}{(m+M)^2}\{mu^2 - MU^2 + (M-m)uU\}.$$

As the wall is stationary, W_1 oscillates about a mean position fixed in the wall. Since u and U are uncorrelated, over a large number of collisions along OX the product uU will have positive and negatives values with equal probability. The average of uU is zero, so the net gain of wall energy is proportional to the average of $(mu^2 - MU^2)$, or equivalently to the average of $(mu^2 + mu'^2 - MU^2 - MU'^2)$. We now extend the averaging to all directions of OX to find that the average gain of energy by the wall is proportional to Q, where

$$Q = \frac{4mM}{(m+M)^2}\left\{\langle\tfrac{1}{2}mc^2\rangle_G - \langle\tfrac{1}{2}Mc^2\rangle_W\right\};$$

here c is the molecular speed (the fluid velocities are zero) and the subscripts denote gas and wall molecules. Thermal equilibrium therefore requires that

$$\langle\tfrac{1}{2}mc^2\rangle_G = \langle\tfrac{1}{2}Mc^2\rangle_W,$$

in which case the gas and wall are at the same temperature.

If a second gas is present, also in thermal equilibrium with the wall, then

$$\langle\tfrac{1}{2}m_1c_1^2\rangle = \langle\tfrac{1}{2}Mc^2\rangle_W = \langle\tfrac{1}{2}m_2c_2^2\rangle,$$

the subscripts denoting the first and second gases. Hence

$$\langle\tfrac{1}{2}m_1c_1^2\rangle = \langle\tfrac{1}{2}m_2c_2^2\rangle,$$

and the gases are in thermal equilibrium with each other; this is the zeroth law described earlier. We have now established the result:

When two gases are mixed at the same temperature, the average kinetic energy of their molecules is the same.

The above suggests that we should *define* the absolute temperature $T(\mathbf{r}, t)$ at a point $P(\mathbf{r}, t)$ as being proportional to the average energy of translation of the particles in P_c. Hence we take*

$$\tfrac{3}{2}kT = m\langle \tfrac{1}{2}c^2 \rangle \qquad (k = 1.380 \times 10^{-23}\,\mathrm{J\,K^{-1}}). \tag{1.2}$$

where the constant of proportionality k is termed Boltzmann's constant.

1.3.2 *Equations of state*

At P_c a particle has the energy $m(\frac{1}{2}c^2 + \epsilon)$, where $m\epsilon$ is the energy due to its internal motions and its intermolecular potential. The average particle energy per unit mass will be denoted by u: thus

$$u = \langle \tfrac{1}{2}c^2 + \epsilon \rangle \tag{1.3}$$

is a macroscopic variable $u(\mathbf{r}, t)$, termed the *specific energy* (i.e. energy per unit mass of the medium). From (1.2) and (1.3),

$$u = \tfrac{3}{2}RT + \langle \epsilon \rangle \qquad (R = k/m), \tag{1.4}$$

where R is termed the *gas constant*. As $\langle \epsilon \rangle$ is found to depend on ρ and T, a more general form of (1.4) is

$$u = u(\rho, T), \tag{1.5}$$

a relation known as the *caloric equation of state*.

If the particles have no internal structure, i.e. possess energy of translation only, (1.4) gives

$$u = c_v T \qquad (c_v = \tfrac{3}{2}R), \tag{1.6}$$

where c_v is the *specific heat at constant volume* (see §1.10.2). Later in §1.5.2 we shall show that with such particles the pressure, i.e. the force per unit area, acting normal to a convected surface, has an average value of

$$p = \tfrac{1}{3}\rho\langle c^2 \rangle = \tfrac{2}{3}\rho u = R\rho T = nkT. \tag{1.7}$$

A gas to which (1.6) and (1.7) apply is said to be a *perfect gas*. More generally the pressure is related to ρ and T by a relation

$$p = p(\rho, T), \tag{1.8}$$

known as the *thermal* equation of state.

One mole is a mass of gas in grams numerically equal to its molecular weight M; e.g. one mole of O_2 is 32 grams of oxygen. From (1.7) applied to a volume V of gas containing N molecules,

$$pV = NkT = (Nm/M)\,M(k/m)\,T,$$

*Throughout this text we use SI units.

where (Nm/M) is the number of kilomoles in the volume. When this number is 10^3, the equation is written in the form

$$pV_A = N_A kT = \mathcal{R}T \qquad (\mathcal{R} \equiv kN_A; \;\; N_A \equiv 10^3\, M/m), \qquad (1.9)$$

where $\mathcal{R}$ is termed the *universal gas constant*, V_A is the *molar volume* and N_A is called *Avogadro's number*. In a perfect gas $\mathcal{R}$ and N_A are constants with the values

$$\mathcal{R} = 8.3144 \;\; \mathrm{J\,K^{-1}\,mol^{-1}}; \qquad N_A = 6.0220 \times 10^{23} \;\; \mathrm{mol^{-1}}.$$

At 'standard' conditions, (p = 1 st. atmos. = 1.0133×10^5 Pa, T = 273.16 K = 0 °C), $V_A = \mathcal{R}T/p = 2.2415 \times 10^{-2}\,\mathrm{m^3\,mole^{-1}}$.

1.4 Dynamic pressure

1.4.1 *Wall pressure*

Pressure is such a central concept in kinetic theory that it is important to have a clear physical picture of its nature. We shall start with the classical problem of determining the relationship between molecular impacts on a rigid wall separating two regions ① and ② of the same gas and the force per unit area acting on the wall. We shall generalize the problem slightly by assuming the wall to be uniformly perforated on a very small scale, so that of any macroscopic area element dS, only the fraction $\Re$ is solid. Our aim is to calculate the force exerted by the gas in region ① on that in region ②, using the wall as an intermediary. Collisions between gas molecules will be ignored for the present. Let $\mathbf{n}$ denote unit normal to the wall directed away

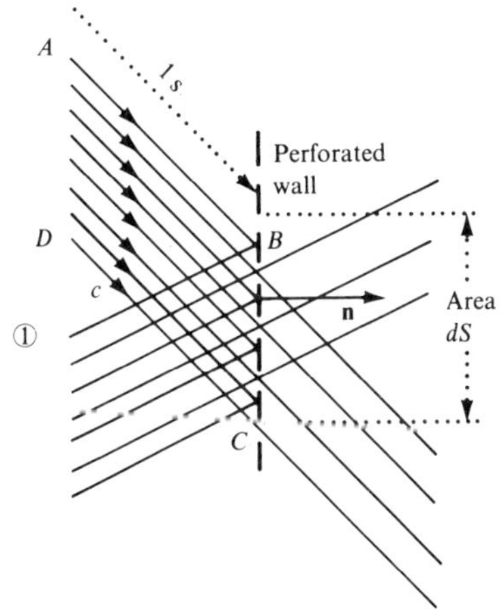

FIG. 1.2. Force on a perforated wall due to impacting molecules

from region ①, and suppose that of the n molecules per unit volume in ①, the number nf is moving towards the wall with velocity $\mathbf{c}$. In one second all the molecules lying in the cylinder $ABCD$ (see Fig. 1.2) of volume $\mathbf{n}\cdot\mathbf{c}\,dS$

and moving towards the wall will reach it and either collide with it, or pass through it. As $\Re$ is the probability of a collision, the average collision rate is $nf\mathbf{n}\cdot\mathbf{c}\,\Re\,dS$. On striking the wall and coming to rest (instantaneously) each molecule loses momentum $m\mathbf{c}$, so by Newton's second law, the force on the element dS of the wall is $mnf\mathbf{n}\cdot\mathbf{cc}\,\Re\,dS$. If the integral of $f\mathbf{cc}$ over all $\mathbf{c}$, with $\mathbf{c}\cdot\mathbf{n}\geq 0$, is denoted by $\langle\mathbf{cc}\rangle^+$, and writing $\langle\mathbf{cc}\rangle^+_{col} = \Re\langle\mathbf{cc}\rangle^+$ for the 'collided' flux, then the force per unit area on the wall due to all the approaching molecules is

$$\mathbf{F}^+ = \rho\mathbf{n}\cdot\langle\mathbf{cc}\rangle^+_{col} \qquad (\mathbf{c}\cdot\mathbf{n}\geq 0). \tag{1.10}$$

Besides providing the force to stop the approaching molecules, the wall also gives the reflected molecules their new momentum. Some of the molecules for which $\mathbf{c}\cdot\mathbf{n}\leq 0$ in region ① will have passed through the wall without change in momentum. When discounting these uncollided molecules by the factor $\Re$, we have, by an argument similar to that used to obtain (1.10), that the force on unit area of wall required to accelerate the departing molecules is

$$\mathbf{F}^- = \rho\mathbf{n}\cdot\langle\mathbf{cc}\rangle^-_{col} \qquad (\mathbf{c}\cdot\mathbf{n}\leq 0). \tag{1.11}$$

Note that by using the probability $\Re$ of a wall collision in (1.10) and (1.11), we are, in effect, averaging the force over a region containing many perforations.

1.4.2 *The pressure tensor*

By adding (1.10) and (1.11) and introducing the average collided flux $\langle\mathbf{cc}\rangle^+_{col} + \langle\mathbf{cc}\rangle^-_{col} = \langle\mathbf{cc}\rangle^*$, we obtain for the force per unit area of the *whole* the wall—including perforations—the expression

$$\mathbf{F} = \mathbf{n}\cdot\mathsf{p},$$

where

$$\mathsf{p} \equiv \rho\langle\mathbf{cc}\rangle^*. \tag{1.12}$$

The macroscopic variable p, defined in this way, is the *pressure tensor* that arises in the equations of fluid mechanics. For the perforated wall its relationship to the total momentum flux (collided plus uncollided)

$$\mathsf{M} \equiv \rho\langle\mathbf{cc}\rangle \tag{1.13}$$

is

$$\mathsf{p} = \Re\mathsf{M}. \tag{1.14}$$

Finally, we note that the net force per unit area of wall transmitted from gas ① with $\mathsf{p} = \mathsf{p}_1$ to gas ② with $\mathsf{p} = \mathsf{p}_2$, is

$$\mathbf{F}_t = \mathbf{n}\cdot(\mathsf{p}_1 - \mathsf{p}_2).$$

1.5 Pressure at points interior to the gas

1.5.1 *Role of collisions*

It follows from (1.14) that $\mathbf{p}$ and $\mathbf{M}$ are equal only if *all* the molecules contributing to the average $\langle \mathbf{cc} \rangle$ experience collisions. A 'collisionless' gas is one in which collisions between molecules can be ignored. Hence from (1.14) in such a gas $\mathbf{p}$ is zero. To enable a non-zero pressure to be defined at points interior to the gas, it is essential that the molecules collide with each other, which requires the mean free path λ to be much smaller than the dimensions of the containing vessel. Suppose that the wall in Fig. 1.2 is replaced by a gas layer of thickness ϵ extending into region ②, as indicated in Fig. 1.3. The gas is the same throughout, so the surface S

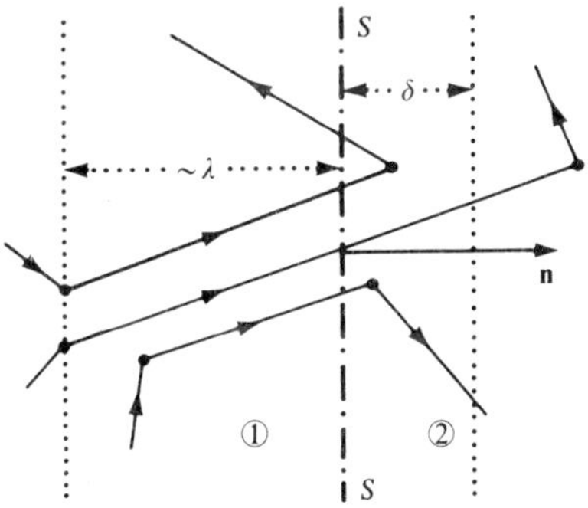

FIG. 1.3. Collisional interface ϵ

between ① and ② is a notional interface in a continuous gas field. First suppose that the layer is much smaller than the mean free path λ. Most of the molecules in ① within a distance λ of S and moving towards it, will pass through S before experiencing a collision. An estimate of the probability of a collision occurring in a distance ϵ is ϵ/λ, so it follows from the discussion in §1.4 that the force exerted on the layer by the gas in ① is $\mathbf{F}_\epsilon = \mathbf{n} \cdot \rho \langle \mathbf{cc} \rangle^* = (\epsilon/\lambda)\mathbf{n} \cdot \mathbf{M}$. Almost all the molecules that pass through the layer will sooner or later experience a collision, so to replace $\mathbf{F}_\epsilon$ by $\mathbf{F} = \mathbf{n} \cdot \mathbf{M}$, we need only to increase ϵ to a value that somewhat exceeds λ, e.g. 3λ or so. Precision is not possible here, because as we shall see in §3.3.1, the probability of a collision in the distance x is proportional to $\exp(-x/\lambda)$, so some molecules will move quite large distances undeflected.

It follows from the above that provided the interface between two regions of gas is envisaged as being a layer several mean free paths thick, the force transmitted across the interface in the direction of the unit normal $\mathbf{n}$ can be expressed as

$$\mathbf{F} = \mathbf{n} \cdot \mathbf{p},$$

where

$$\mathbf{p} \equiv \rho \langle \mathbf{cc} \rangle^* = \rho \langle \mathbf{cc} \rangle = \mathbf{M}. \tag{1.15}$$

Thus in these circumstances it is consistent to set p equal to the momentum flux at a macroscopic point $P_c(\mathbf{r}, t)$, but the role of collisions should not be forgotten. The equality between $\langle \mathbf{cc} \rangle$ and $\langle \mathbf{cc} \rangle^*$ in (1.15) requires that the molecules involved must experience collisions with like molecules within a distance of P_c less than that to the nearest (impenetrable) boundary.

1.5.2 *The pressure force per unit mass*

The normal pressure between regions ① and ② is $\mathbf{F} \cdot \mathbf{n} = \mathbf{n} \cdot \mathsf{p} \cdot \mathbf{n}$. Let $\mathbf{i}, \mathbf{j}, \mathbf{k}$ denote the unit Cartesian vectors and take $\mathbf{n}$ to be parallel to each in turn. The average value of the normal pressure at P_c is

$$\tfrac{1}{3}(\mathbf{i} \cdot \mathsf{p} \cdot \mathbf{i} + \mathbf{j} \cdot \mathsf{p} \cdot \mathbf{j} + \mathbf{k} \cdot \mathsf{p} \cdot \mathbf{k}) = \tfrac{1}{3}\mathbf{1} : \mathsf{p} = \tfrac{1}{3}\rho\langle \mathbf{c} \cdot \mathbf{1} \cdot \mathbf{c} \rangle = \tfrac{1}{3}\rho\langle c^2 \rangle,$$

where $\mathbf{1} \equiv \mathbf{ii} + \mathbf{jj} + \mathbf{kk}$ is the unit tensor, i.e. for any vector $\mathbf{a}$, $\mathbf{1} \cdot \mathbf{a} = \mathbf{a}$ and $\mathbf{a} \cdot \mathbf{1} = \mathbf{a}$. The scalar or thermodynamic pressure is defined to be the average normal pressure at a point convected with the fluid, i.e.

$$p \equiv \tfrac{1}{3}\mathbf{1} : \mathsf{p} = \tfrac{1}{3}\mathrm{trace}\, \mathsf{p} = \tfrac{1}{3}\rho\langle c^2 \rangle. \tag{1.16}$$

The perfect gas relationship, $p = nkT$, which follows from (1.16) and the definition of T given in (1.2), is valid only if the medium is 'collision dominated', i.e. if the mean free path is somewhat shorter than any macroscopic length scale of interest. A more general relationship between p and T is given in §3.6.1.

The fluid external to a volume V of gas exerts an inward force $-\mathbf{n} \cdot \mathsf{p}\, dS$ on each element of the surface ∂V of V. So by the divergence theorem the total force on the medium within V due to the medium outside V is

$$-\int_{\partial V} \mathbf{n} \cdot \mathsf{p}\, dS = -\int_V \nabla \cdot \mathsf{p}\, d\mathbf{r}.$$

Let $\mathbf{P}$ denote the pressure force per unit mass acting on the volume element $d\mathbf{r}$, then as $\rho\, d\mathbf{r}$ is the mass of the element,

$$\mathbf{P}\rho\, d\mathbf{r} = -\lim_{V \to d\mathbf{r}} \int_{\partial V} \mathbf{n} \cdot \mathsf{p}\, dS = -\nabla \cdot \mathsf{p}\, d\mathbf{r},$$

therefore

$$\mathbf{P} = -\frac{1}{\rho}\nabla \cdot \mathsf{p}. \tag{1.17}$$

1.6 Ideal fluids

1.6.1 *Hydrostatic or thermodynamic pressure*

In dyadic notation the second order tensor p maybe expanded as

$$\mathsf{p} = \begin{array}{l} p_{xx}\mathbf{ii} + p_{xy}\mathbf{ij} + p_{xz}\mathbf{ik} \\ +p_{yx}\mathbf{ji} + p_{yy}\mathbf{jj} + p_{yz}\mathbf{jk} \\ +p_{zx}\mathbf{ki} + p_{zy}\mathbf{kj} + p_{zz}\mathbf{kk}, \end{array} \tag{1.18}$$

where by (1.15),

$$\left.\begin{aligned} p_{xx} &= \rho\langle c_x^2\rangle, & p_{yy} &= \rho\langle c_y^2\rangle, & p_{zz} &= \rho\langle c_z^2\rangle, \\ p_{xy} = p_{yx} &= \rho\langle c_x c_y\rangle, & p_{yz} = p_{zy} &= \rho\langle c_y c_z\rangle, & p_{zx} = p_{xz} &= \rho\langle c_z c_x\rangle. \end{aligned}\right\} \tag{1.19}$$

Notice that $\mathbf{p}$ is a symmetric tensor. The force $\mathbf{i}\cdot\mathbf{p}$ is that acting on unit area orthogonal to OX, and p_{xx}, p_{xy}, p_{xz} are the components of this force in the directions indicated by the second subscript (see Fig. 1.4).

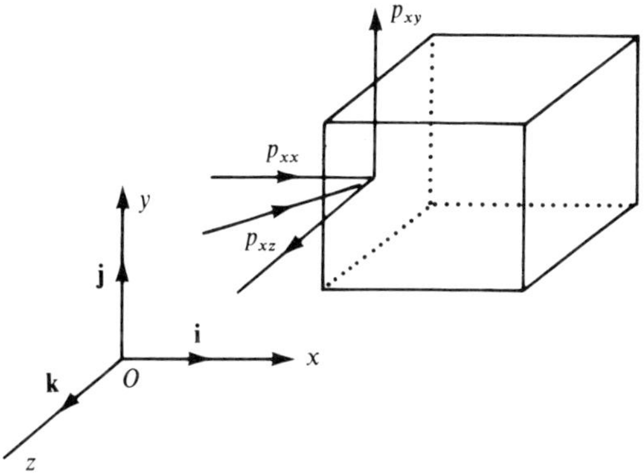

FIG. 1.4. Pressure components

By (1.18),

$$\mathbf{1}:\mathbf{p} = p_{xx} + p_{yy} + p_{zz} = 3p,$$

so that

$$\begin{aligned} \mathbf{nn}:\mathbf{p} &= p + (p_{xx} - p)n_x^2 + (p_{yy} - p)n_y^2 + (p_{zz} - p)n_z^2 \\ &\quad + 2p_{xy}n_x n_y + 2p_{yz}n_y n_z + 2p_{zx}n_z n_x, \end{aligned} \tag{1.20}$$

where $\mathbf{n} = n_x\mathbf{i} + n_y\mathbf{j} + n_z\mathbf{k}$. It follows that the normal pressure on a convected surface is independent of the orientation of that surface only if

$$p = p_{xx} = p_{yy} = p_{zz}, \quad 0 = p_{xy} = p_{yz} = p_{zx}.$$

In this case (1.18) reduces to $\mathbf{p} = p(\mathbf{ii} + \mathbf{jj} + \mathbf{kk})$, or

$$\mathbf{p} = p\,\mathbf{1}. \qquad \text{(ideal fluid)} \tag{1.21}$$

This equation certainly holds in 'hydrostatics', that is, when the fluid velocity is zero or constant. An ideal fluid is one for which (1.21) holds even when $\mathbf{v}$ is not uniform. Such a fluid does not transmit *shear* stress across a surface, and is said to be *inviscid*.

1.6.2 *Viscous stress tensor*

It is usually the case that p is dominated by its normal stress components, each of which is nearly equal to p. It is therefore convenient to divide p into its ideal and viscous components;

$$\mathsf{p} = p\mathbf{1} + \boldsymbol{\pi}, \tag{1.22}$$

where $\boldsymbol{\pi}$ is called the *viscous stress tensor*. In many problems in fluid dynamics the approximation $\boldsymbol{\pi} \approx 0$ is satisfactory, especially if there are no solid boundaries or shock waves in the neighbourhood.

Note from (1.17) that the pressure exerts a force per unit volume:

$$\rho\mathbf{P} = -\nabla\cdot\mathsf{p} = -\nabla p - \nabla\cdot\boldsymbol{\pi} \tag{1.23}$$

on the fluid, which in an ideal fluid reduces to $-\nabla p$.

When the magnetic field $\mathbf{B}$ in a magnetoplasma (see Chapter 9) is sufficiently strong, the fluid is anisotropic. Let L denote a family of planes containing the vector $\mathbf{B}$, then if the normal pressure, say $p_\perp$, is the same on all members of L, the medium is an ideal magnetoplasma. Suppose that $\mathbf{B}$ lies along OZ, then on L, n_z is zero and n_x, n_y are arbitrary. It follows from (1.20) that for an ideal fluid of this type,

$$p_{xx} = p_{yy} = p_\perp, \qquad p_{zz} = p_\parallel, \qquad 0 = p_{xy} = p_{yz} = p_{zx}, \tag{1.24}$$

where $p_\parallel$ is the normal pressure in a direction parallel to $\mathbf{B}$, that is $p_\parallel = \mathbf{b}\cdot\mathsf{p}\cdot\mathbf{b}$, where $\mathbf{b}$ is unit vector parallel to $\mathbf{B}$. In this case (1.18) gives $\mathsf{p} = p_\perp(\mathbf{ii}+\mathbf{jj}) + p_\parallel\mathbf{kk}$. More generally we have

$$\mathsf{p} = p_\perp(\mathbf{1} - \mathbf{bb}) + p_\parallel\mathbf{bb} \qquad \text{(ideal magnetoplasma).} \tag{1.25}$$

The peculiar velocity has the perpendicular and parallel components $\mathbf{c}_\perp = \mathbf{c} - \mathbf{c}_\parallel$, $\mathbf{c}_\parallel = \mathbf{bb}\cdot\mathbf{c}$. If $\mathbf{B}$ is parallel to $\mathbf{k}$, $p_\perp = \frac{1}{2}(p_{xx}+p_{yy}) = \frac{1}{2}\rho\langle c_x^2 + c_y^2\rangle$, $p_\parallel = p_{zz} = \rho\langle c_z^2\rangle$; in general

$$p_\perp = \tfrac{1}{2}\rho\langle c_\perp^2\rangle, \qquad p_\parallel = \rho\langle c_\parallel^2\rangle. \tag{1.26}$$

Corresponding tensors are defined by

$$\mathsf{p}_\perp = \rho\langle\mathbf{c}_\perp\mathbf{c}_\perp\rangle, \qquad \mathsf{p}_\parallel = \rho\langle\mathbf{c}_\parallel\mathbf{c}_\parallel\rangle, \tag{1.27}$$

and since the fluid is ideal, $\langle\mathbf{c}_\perp\mathbf{c}_\parallel\rangle = 0$, whence

$$\mathsf{p} = \rho\langle\mathbf{cc}\rangle = \mathsf{p}_\perp + \mathsf{p}_\parallel. \tag{1.28}$$

Likewise for the specific energy,

$$u_\perp = \tfrac{1}{2}\langle c_\perp^2\rangle = \frac{p_\perp}{\rho}, \quad u_\parallel = \tfrac{1}{2}\langle c_\parallel^2\rangle = \frac{p_\parallel}{2\rho}, \quad u = u_\perp + u_\parallel. \tag{1.29}$$

1.7 Convection and diffusion

1.7.1 *The diffusion vector*

As noted in §1.2, convection is the transport of a macroscopic property, such as density, momentum, energy, the concentration level of a contaminant, and so on, by the fluid motion. Let $\Phi(\mathbf{r}, t)$ be such an attribute, specified as an amount per unit mass of fluid—known as a 'specific' property—then a volume element of mass $\rho\, d\mathbf{r}$ will possess an amount $\rho\Phi\, d\mathbf{r}$ of it. As the volume of fluid crossing a stationary surface $\mathbf{n}\, dS$ in one second is $\mathbf{v} \cdot \mathbf{n}\, dS$, it follows that the transport of Φ due to convection occurs at the rate $\rho\Phi\mathbf{v} \cdot \mathbf{n}\, dS$ across this surface. Hence

$$\text{convective flux of } \Phi = \rho\Phi\mathbf{v} \cdot \mathbf{n}.$$

This description can be generalized by introducing a specific property $\phi(\mathbf{r}, \mathbf{w}, t)$, whose value may depend on the velocity $\mathbf{w} = \mathbf{v} + \mathbf{c}$ of the particles involved in its transport. Suppose that the average value of ϕ taken over particles at a macroscopic point is Φ, i.e.

$$\Phi(\mathbf{r}, t) = \langle \phi(\mathbf{r}, \mathbf{w}, t) \rangle, \tag{1.30}$$

then the local transport of Φ across $\mathbf{n}\, dS$ is the average of $\rho\phi\mathbf{w} \cdot \mathbf{n}\, dS$. This gives a total flux of $\rho\langle \phi\mathbf{w} \rangle \cdot \mathbf{n} = \rho\langle \phi(\mathbf{v} + \mathbf{c}) \rangle \cdot \mathbf{n}$, or

$$\text{total flux of } \Phi = \rho\Phi\mathbf{v} \cdot \mathbf{n} + \mathbf{J}_\phi \cdot \mathbf{n}, \tag{1.31}$$

where

$$\mathbf{J}_\phi \equiv \rho\langle \phi\mathbf{c} \rangle. \tag{1.32}$$

We term $\mathbf{J}_\phi$ the diffusion vector for ϕ. Diffusion is thus the transport of a property by the purely random component of molecular motion. It is very important to distinguish between diffusion and convection. A central task in kinetic theory is that of obtaining the diffusive flux $\mathbf{J}_\phi \cdot \mathbf{n}$ for various properties ϕ.

In some circumstances an expression for $\mathbf{J}_\phi$ of the form

$$\mathbf{J}_\phi = -\rho\kappa_\phi \nabla\Phi \tag{1.33}$$

can be found; κ_ϕ is termed the *coefficient of diffusion* for Φ. It follows from (1.33) that κ_ϕ has the dimensions: $(\text{length})^2/\text{time}$, and since it is due to particle transport, we may write

$$\kappa_\phi = \alpha\, \lambda^2/\tau, \tag{1.34}$$

where α is a constant of order unity. Sometimes there exist several 'routes' for the diffusion of Φ, i.e. a number of distinct processes each contribute to $\mathbf{J}_\phi$. If these are independent, the total flux is obtained by summation, and κ_ϕ in (1.33) becomes

$$\kappa_\phi = \sum_j \frac{\rho_j}{\rho} \kappa_{\phi j} = \sum_j \frac{\rho_j}{\rho} \alpha_j \frac{\lambda_j^2}{\tau_j}, \tag{1.35}$$

the subscript j denoting a particular process.

The convection term in (1.31), viz. $\rho\Phi\, \mathbf{v} \cdot \mathbf{n}$, is dependent on the choice of reference frame in which the velocity is measured, whereas the diffusion term is not. In fact we shall adopt 'frame-indifference' as being the essential property that distinguishes diffusion from convection. In some circumstances it is not evident from the physics where to draw the line between these two transport processes, and a mathematical definition is

Table 1.1 *Basic variables for a simple fluid*

	mass	momentum	energy	entropy
ϕ	1	$\mathbf{w}$	$\frac{1}{2}w^2 + \epsilon$	$-kf\ln f$
Φ	1	$\mathbf{v}$	$\frac{1}{2}v^2 + u$	s
$\mathbf{J}_\phi$	0	P	$\mathsf{P}\cdot\mathbf{v} + \mathbf{q}$	$\mathbf{q}/T$
S_ϕ	0	$\rho\mathbf{F}$	$\rho\mathbf{F}\cdot\mathbf{v}$	σ

useful.

The physical assumption involved in (1.33) is that ϕ is a property transported by individual particles, whose random motions are independent of the amount of ϕ that they carry. In a simplified description, if we take $\rho\phi$ to be the density of molecules labelled in a particular way, the intensity ϕ becomes the fraction of labelled molecules at a given point. If ϕ is initially uniform, variations in the density ρ cannot affect ϕ, since labelled and unlabelled molecules will have the same propensity to migrate at each point in the fluid. Thus $\mathbf{J}_\phi$ cannot depend on $\nabla\rho$, at least in a linear model of the phenomenon. The random molecular motions will disperse any nonuniformity in Φ, and by the principle of local action (see §3.4.4), the first spatial derivative of Φ is the dominant term driving the system towards equilibrium.

1.7.2 *Basic variables for a simple fluid*

Consider the case of a gas consisting of a single species (a 'simple' fluid) and subject to a body force (like gravity) $\mathbf{F}$ per unit mass. The interesting choices for ϕ are the specific values of the mass, momentum, energy, and entropy, as set out in Table 1.1. In the final column we have anticipated equations in §§2.5 and 1.11.2. For the present it can be ignored.

To calculate Φ and $\mathbf{J}_\phi$ for the other entries, we have used (1.3), viz.

$$u = \langle \tfrac{1}{2}c^2 + \epsilon \rangle, \tag{1.36}$$

equation (1.15), $\mathbf{w} = \mathbf{v} + \mathbf{c}$, where $\langle \mathbf{c} \rangle = 0$, and introduced a new diffusion vector,

$$\mathbf{q} \equiv \rho\langle \mathbf{c}(\tfrac{1}{2}c^2 + \epsilon)\rangle, \tag{1.37}$$

known as the *heat flux* if collisions are dominant and the *energy flux* otherwise.

1.8 A general balance equation in physical space

The basic equation in kinetic theory is an expression for the conservation or 'balance' of a general property $\phi(\mathbf{r}, \mathbf{w}, t)$. But before deriving this equation, we shall find it instructive to undertake the simpler task of obtaining an averaged balance equation in physical space, with $\mathbf{r}$ and t the independent variables.

Consider a volume V of the medium, with a surface ∂V across which ϕ is being convected and diffused. Let $S_\phi(\mathbf{r}, t)$ be the rate per unit volume at which ϕ is being introduced at points within V due to physical or chemical processes, the details of which will be specified later. The amount of ϕ within V is $\langle \rho\phi \rangle = \rho\Phi$ per unit volume. Its increase is due to its flux across ∂V and its creation within V. Hence by (1.31) we obtain the balance equation

$$\frac{\partial}{\partial t}\int_V \rho\Phi \, d\mathbf{r} = -\int_{\partial V} \mathbf{n} \cdot (\rho\Phi\mathbf{v} + \mathbf{J}_\phi)\, dS + \int_V S_\phi \, d\mathbf{r}.$$

By Gauss' divergence theorem, this can be expressed as

$$\int_V \left\{ \frac{\partial}{\partial t}(\rho\Phi) + \nabla \cdot (\rho\Phi\mathbf{v} + \mathbf{J}_\phi) - S_\phi \right\} d\mathbf{r} = 0.$$

Granted that this equation holds for all subdivisions of V—or equivalently that V may be chosen arbitrarily—we deduce that

$$\frac{\partial}{\partial t}(\rho\Phi) + \nabla \cdot (\rho\Phi\mathbf{v} + \mathbf{J}_\phi) = S_\phi, \tag{1.38}$$

which is the required balance equation for Φ.

There may be several gas components present, in which case it is necessary to distinguish the macroscopic variables ρ, $\mathbf{v}$, T, p, $\mathbf{q}$, etc. by a different subscript for each species. Thus for what we shall term the 'i-fluid' and denote by P_i:

$$\frac{\partial}{\partial t}(\rho_i\Phi_i) + \nabla \cdot (\rho_i\Phi_i\mathbf{v}_i + \mathbf{J}_{\phi i}) = S_{\phi i}. \tag{1.39}$$

The source term is partly due to mass transfer between the fluid components resulting from chemical or physical interactions between them. Let $\dot{\rho}_{ij}$ be the rate at which P_i gains mass from P_j, then $\dot{\rho}_{ij}\phi_j$ is the rate at which the property ϕ is received from P_j by P_i. Conversely, P_i loses ϕ at the rate $\dot{\rho}_{ji}\phi_i$ to P_j. The net gain of ϕ by P_i due to mass transfers is therefore

$$(\Delta\phi)_i = \sum_j \left(\dot{\rho}_{ij}\phi_j - \dot{\rho}_{ji}\phi_i\right).$$

There may also be transfers of ϕ without mass exchange. Let $\dot{\phi}_{ij}\rho_j$ be the rate per unit volume of such a transfer from P_j to P_i, then by conservation of ϕ during this transfer,

$$\dot{\phi}_{ij}\rho_j = -\dot{\phi}_{ji}\rho_i.$$

The total source due to internal and external transfers is therefore

$$S_{\phi i} = S_{\phi i}^{ex} + \sum_j \left(\dot{\rho}_{ij}\phi_j - \dot{\rho}_{ji}\phi_i + \dot{\phi}_{ij}\rho_j\right), \tag{1.40}$$

where $S_{\phi i}^{ex}$ is the rate at which P_i receives ϕ per unit volume from sources outside the system.

1.9 The equations of motion for a simple fluid

1.9.1 *The conservation laws*

The equations of motion for a single-component gas can be written down directly from Table 1 on p. 15 and equation (1.38). Since there are no internal transfers, the source term in (1.38) is due to external effects only, in the sense implied in (1.40). The body force $\mathbf{F}$ generates fluid momentum and energy at the rates $\rho\mathbf{F}$ and $\rho\mathbf{F}\cdot\mathbf{v}$ per unit volume. Thus the four basic fluid equations are:

mass:
$$\frac{\partial\rho}{\partial t} + \nabla\cdot\rho\mathbf{v} = 0, \tag{1.41}$$

momentum:
$$\frac{\partial}{\partial t}(\rho\mathbf{v}) + \nabla\cdot(\rho\mathbf{v}\mathbf{v} + \mathsf{p}) = \rho\mathbf{F}, \tag{1.42}$$

energy:
$$\frac{\partial}{\partial t}\left\{\rho(\tfrac{1}{2}v^2 + u)\right\} + \nabla\cdot\left\{\rho\mathbf{v}(\tfrac{1}{2}v^2 + u) + \mathsf{p}\cdot\mathbf{v} + \mathbf{q}\right\} = \rho\mathbf{F}\cdot\mathbf{v}, \tag{1.43}$$

entropy:
$$\frac{\partial}{\partial t}(\rho s) + \nabla\cdot(\rho\mathbf{v}s + \mathbf{q}/T) = \sigma. \tag{1.44}$$

It remains to explain the last of these equations. First we transform them into a frame P_c convected with the fluid.

Let $\Phi(\mathbf{r}, t)$ be a differentiable function of $\mathbf{r}$ and t, then

$$d\Phi = dt\left(\frac{\partial\Phi}{\partial t}\right)_{\mathbf{r}} + d\mathbf{r}\cdot\left(\frac{\partial\Phi}{\partial\mathbf{r}}\right)_t$$

is true in general. If we now associate $d\mathbf{r}$ and dt by requiring their ratio $d\mathbf{r}/dt$ to be the fluid velocity $\mathbf{v}$, we obtain the *material* time derivative at a convected point $P_c(\mathbf{r}, t)$:

$$\mathrm{D}\Phi \equiv \left(\frac{d\Phi}{dt}\right)_{P_c} = \left(\frac{\partial}{\partial t} + \mathbf{v}\cdot\nabla\right)\Phi, \tag{1.45}$$

where the commonly adopted notation $\mathrm{D}\Phi/\mathrm{D}t$ has been simplified to $\mathrm{D}\Phi$.

1.9.2 *Convected forms for the conservation laws*

Writing (1.41) to (1.44) in the forms $L_1 = 0$, $\mathbf{L}_2 = 0$, etc., we find that $\mathbf{L}_2 = \mathbf{v}L_1 + \ldots, L_3 = (\frac{1}{2}v^2 + u)L_1 + \mathbf{v}\cdot\mathbf{L}_2 + \ldots$, and $L_4 = s\,L_4 + \ldots$, and as the terms identified by 'L' are zero, with the aid of (1.45), we are able to reduce equations (1.41) to (1.44) to the 'convected' forms:

mass:
$$\mathrm{D}\rho + \rho\nabla\cdot\mathbf{v} = 0, \tag{1.46}$$

momentum:
$$\rho\mathrm{D}\mathbf{v} + \nabla\cdot\mathsf{p} = \rho\mathbf{F}, \tag{1.47}$$

energy:
$$\rho\mathrm{D}u + \mathsf{p}:\nabla\mathbf{v} = -\nabla\cdot\mathbf{q}, \tag{1.48}$$

entropy:
$$\rho\mathrm{D}s + \nabla\cdot(\mathbf{q}/T) = \sigma. \tag{1.49}$$

In the reduction of the energy equation, we used the symmetry of $\mathbf{p}$ (see (1.19)) and the expansion

$$\nabla \cdot (\mathbf{p} \cdot \mathbf{v}) = (\nabla \cdot \mathbf{p}) \cdot \mathbf{v} + \nabla \mathbf{v} : \mathbf{p} = \mathbf{v} \cdot (\nabla \cdot \mathbf{p}) + \mathbf{p} : \nabla \mathbf{v}.$$

By (1.22)
$$\mathbf{p} : \nabla \mathbf{v} = p\mathbf{1} : \nabla \mathbf{v} + \boldsymbol{\pi} : \nabla \mathbf{v} = p \nabla \cdot \mathbf{v} + \boldsymbol{\pi} : \nabla \mathbf{v},$$

whence from (1.46)
$$\mathbf{p} : \nabla \mathbf{v} = p \rho \mathrm{D} \rho^{-1} + \boldsymbol{\pi} : \nabla \mathbf{v}. \tag{1.50}$$

Now (1.48) can be written

$$\rho \mathrm{D} u + p \rho \mathrm{D} \rho^{-1} = -\boldsymbol{\pi} : \nabla \mathbf{v} - \nabla \cdot \mathbf{q}, \tag{1.51}$$

which is the first law of thermodynamics for a fluid element.

1.10 The first law of thermodynamics

1.10.1 *Heat and irreversible work*

We need an expression for the material time derivative of a volume integral. Let $\Psi = \Psi(\mathbf{r}, t)$ be a differentiable function, then the derivative of its volume integral is defined by

$$\mathrm{D} \int_V \Psi \, d\mathbf{r} \equiv \frac{\partial}{\partial t} \int_V \Psi \, d\mathbf{r} + \int_{\partial V} \mathbf{n} \cdot \mathbf{v} \, \Psi \, dS = \int_V \left\{ \frac{\partial \Psi}{\partial t} + \nabla \cdot (\mathbf{v} \Psi) \right\} d\mathbf{r},$$

i.e.
$$\mathrm{D} \int_V \Psi \, d\mathbf{r} = \int_V (\mathrm{D}\Psi \, d\mathbf{r} + \Psi \nabla \cdot \mathbf{v}) \, d\mathbf{r}.$$

Alternatively, one can regard $d\mathbf{r}$ as being the variable volume of a convected element:

$$\mathrm{D} \int_V \Psi \, d\mathbf{r} \equiv \int_V \mathrm{D}(\Psi \, d\mathbf{r}) = \int_V \{\mathrm{D}\Psi \, d\mathbf{r} + \Psi \, \mathrm{D}(d\mathbf{r})\}.$$

It follows that

$$\mathrm{D}(d\mathbf{r}) = \nabla \cdot \mathbf{v} \, d\mathbf{r}, \tag{1.52}$$

a relation that may be deduced directly by considering the volume changes in a parallelepiped element $d\mathbf{r} = dx \, dy \, dz$, with each face being convected with the local fluid velocity.

Conservation of mass may now be expressed as

$$\mathrm{D}(\rho \, d\mathbf{r}) = 0, \tag{1.53}$$

so that when (1.47) to (1.49) are multiplied by $d\mathbf{r}$, each leading term—say $\rho \mathrm{D}\phi \, d\mathbf{r}$—can be written as $\mathrm{D}(\rho\phi \, d\mathbf{r})$.

By writing (1.51) as

$$\mathrm{D}(\rho u \, d\mathbf{r}) + p\mathrm{D}(d\mathbf{r}) = -\boldsymbol{\pi} : \nabla \mathbf{v} \, d\mathbf{r} - \nabla \cdot \mathbf{q} \, d\mathbf{r},$$

and integrating over a constant volume V and a time interval dt, we obtain

$$dU + p \, dV = d'W_i + d'Q, \tag{1.54}$$

where U is the total energy content of the fluid system $\mathcal{S}$ within V, $d'W_i$ is the irreversible work done on $\mathcal{S}$ by viscous forces during the time dt, and $d'Q$ is the heat supplied to $\mathcal{S}$ in this time, i.e.

$$U = \int_V \rho u \, d\mathbf{r}, \tag{1.55}$$

$$d'W_i = dt \int_V -\boldsymbol{\pi} : \nabla \mathbf{v} \, d\mathbf{r}, \tag{1.56}$$

and

$$d'Q = dt \int_{\partial V} -\mathbf{n} \cdot \mathbf{q} \, dS. \tag{1.57}$$

Equation (1.54) is the familiar form of the first law of thermodynamics. The dashed differentials are used to indicate that, unlike the left-hand differentials, those on the right are not exact, i.e. their operands are not functions of state. The total work done on $\mathcal{S}$ is $-p\,dV + d'W_i$, of which the first term is reversible. Later we shall show that $\boldsymbol{\pi}$ is proportional to $\nabla\mathbf{v}$, whence $d'W_i$ is quadratic in the fluid velocity. It follows that if the changes take place slowly enough, $d'W_i$ can be made very small. In the limit, we obtain the ideal or 'quasi-static' changes favoured in classical thermodynamics (see Woods 1986, p. 11).

1.10.2 *Perfect gas*

We shall develop the theory further for the special case of a perfect gas. On a macroscopic time scale much longer than the collision interval τ defined in (1.1), many collisions occur between the particles, and it is a basic theorem in statistical mechanics that these result in an equipartition of energy between all the modes of motion available to the particles. In this state of thermal equilibrium the following theorem applies:

If the energy associated with any degree of freedom is a quadratic function of the variable specifying the motion, the equilibrium value of the corresponding energy is $\frac{1}{2}kT$.

Thus if the particle is a molecule with F degrees of freedom in its structure, the internal motions contribute $(\frac{1}{2}RT)F$ to $\langle\epsilon\rangle$, the energy defined in 1.3.2. If there is no intermolecular potential, the gas is said to be 'perfect', and (1.4) can be written

$$u = c_v T \qquad \left(c_v = (\tfrac{3}{2} + \tfrac{1}{2}F)R\right), \tag{1.58}$$

where c_v is the specific heat at constant volume. In general F is a function of temperature. In a range in which F is a constant integer, the gas is said to be 'calorifically perfect'. We shall deal only with this case below.

Like the temperature, the pressure is related only to the translatory degrees of freedom, so for a perfect gas, with F non-zero, the relation given in (1.7), viz.

$$p = R\rho T = nkT, \quad (R = k/m) \tag{1.59}$$

remains. Hence

$$u = \frac{c_v}{R}\frac{p}{\rho}. \tag{1.60}$$

From (1.58) to (1.60), we can express (1.51) as

$$\rho c_v \mathrm{D}T + p\rho\mathrm{D}\rho^{-1} = \rho c_p \mathrm{D}T - \mathrm{D}p = -\boldsymbol{\pi} : \nabla\mathbf{v} - \nabla\cdot\mathbf{q}, \tag{1.61}$$

where

$$c_p = c_v + R = (\tfrac{5}{2} + \tfrac{1}{2}F)R \tag{1.62}$$

is termed the *specific heat at constant pressure*. In slow fluid motions, when the viscous dissipation $-\boldsymbol{\pi} : \nabla\mathbf{v}$ is negligible, it follows from (1.57) and (1.61) that for a given temperature rise, the heat required is proportional to either c_v or c_p, depending on whether the specific volume ρ^{-1}, or the pressure p is held constant.

1.11 The second law of thermodynamics

1.11.1 *Specific entropy*

Another form of the energy equation (1.61) is

$$\rho T\mathrm{D}s = -\boldsymbol{\pi} : \nabla\mathbf{v} - \nabla\cdot\mathbf{q}, \tag{1.63}$$

where

$$s \equiv c_v \ln T - R\ln\rho + \text{const.} \tag{1.64}$$

is called the *specific entropy*. This formula for s is often written in the form

$$p = K\rho^\gamma \exp(s/c_v), \qquad (K = \text{const.}) \tag{1.65}$$

where

$$\gamma \equiv \frac{c_p}{c_v} = 1 + \frac{R}{c_v} = 1 + \frac{2}{3+F}. \tag{1.66}$$

If the conditions at the convected point are *adiabatic*, i.e. no heat is removed or supplied to the volume element $d\mathbf{r}$ at P_c, then $\mathbf{q}$ is zero. Furthermore, if the fluid is also ideal (see §1.6.1), $\boldsymbol{\pi}$ is zero, and (1.63) reduces to

$$\mathrm{D}s = 0 \qquad \text{(isentropic flow)}. \tag{1.67}$$

Thus s is constant at P_c and if this holds throughout the fluid, the motion is said to be *isentropic*. For such flows (1.65) is reduced to

$$p = K\rho^\gamma, \tag{1.68}$$

where K is another constant.

1.11.2 *Entropy production rate*

Equation (1.63) can be rearranged as

$$\rho\mathrm{D}s + \nabla\cdot(\mathbf{q}/T) = \sigma, \tag{1.69}$$

where

$$\sigma \equiv -\frac{\boldsymbol{\pi} : \nabla\mathbf{v}}{T} - \frac{\mathbf{q}\cdot\nabla T}{T^2} \tag{1.70}$$

is the entropy production rate per unit volume (see Table 1.1, p.15, (1.44) and (1.49)). The entropy of the system $\mathcal{S}$ within a volume V of the gas is

$$S = \int_V \rho s \, d\mathbf{r}. \tag{1.71}$$

Proceeding as in the derivation of (1.54), we deduce from (1.69) that

$$dS = -dt \int_{\partial V} \frac{\mathbf{n} \cdot \mathbf{q}}{T} \, d\Sigma + dt \int_V \sigma \, d\mathbf{r}, \tag{1.72}$$

changing the surface element to $d\Sigma$ to avoid confusion with the entropy.

If T and p are constant throughout V (including ∂V), it follows from (1.54) to (1.57) and (1.72) that

$$T \, dS = dU + p \, dV = d'W_i + d'Q_r, \tag{1.73}$$

which is the second law of thermodynamics in its most familiar form. The subscript 'r' has been added to $d'Q$ because, being independent of the temperature gradient—which is zero in the present case—it could pass either *in* or *out* of $\mathcal{S}$, i.e. it is a *reversible* heat transfer. Of course this is an idealization, because in the absence of electrical phenomena, such transfers are strictly not possible. The local form of this reversible heat transfer, say $\nabla \cdot \mathbf{q}_r$, is T times the divergence term in (1.69), i.e.

$$\nabla \cdot \mathbf{q}_r = T \nabla \cdot \left(\frac{\mathbf{q}}{T} \right).$$

The local form of the first equation in (1.73) is (cf. (1.51) and (1.63))

$$T \, ds = du + p \, d(1/\rho), \tag{1.74}$$

whence (1.69) can be expressed

$$\rho \mathrm{D}u - \{ \overset{(1)}{-p\rho \mathrm{D}\rho^{-1}} - \overset{(2)}{\nabla \cdot \mathbf{q}_r} \} = T\sigma.$$

The terms (1) and (2) in this equation represent the power supplied reversibly to the macroscopic point P_c by compression and heating. The stability of the process requires the internal energy of P_c to increase at a rate not less than that provided by the reversible power supply. Hence to secure stability it is necessary to have a non-negative entropy production rate,

$$\sigma \geq 0, \tag{1.75}$$

at all points in the gas.

1.11.3 *Law of entropy increase*

From (1.72) and (1.75) it follows that the entropy of an adiabatically enclosed system $\mathcal{S}$ cannot decrease:

$$\frac{dS}{dt} \geq 0. \tag{1.76}$$

Thus if equilibrium is achieved, i.e. the gradients vanish throughout $\mathcal{S}$, the entropy is at its maximum value.

Later, in §3.6.3, it will be shown that in a first approximation, the viscous stress tensor in a neutral monatomic gas is proportional to the deviatoric rate of strain, i.e.

$$\boldsymbol{\pi} = -2\mu \overset{\circ}{\nabla \mathbf{v}}, \tag{1.77}$$

where (see §1.13)

$$\overset{\circ}{\nabla \mathbf{v}} \equiv \tfrac{1}{2}(\nabla \mathbf{v} + \widetilde{\nabla \mathbf{v}}) - \tfrac{1}{3}\mathbf{1}\, \nabla \cdot \mathbf{v} \tag{1.78}$$

and μ is known as the *coefficient of shear viscosity*. For this case Fourier's law is applicable, i.e.

$$\mathbf{q} = -\kappa\, \nabla T, \tag{1.79}$$

where κ is the *thermal conductivity*. Then by (1.70) and (1.75)

$$\sigma = 2\mu T^{-1}\, \overset{\circ}{\nabla \mathbf{v}} : \overset{\circ}{\nabla \mathbf{v}} + \kappa T^{-2} \nabla T \cdot \nabla T \geq 0. \tag{1.80}$$

As $\overset{\circ}{\nabla \mathbf{v}}$ and ∇T are independent gradients, it follows that μ and κ are non-negative,

$$\mu \geq 0, \qquad \kappa \geq 0. \tag{1.81}$$

1.12 Specific entropy in a mixture of gases

Let the 'i-th fluid' P_i have a number density n_i, a fluid velocity $\mathbf{v}_i$, a temperature T_i, and so on; values applying to the mixture as a whole will have no subscripts. Thus for a perfect gas,

$$\rho = \sum \rho_i, \quad \rho \mathbf{v} = \sum \rho_i \mathbf{v}_i, \quad \rho u = \sum \rho_i u_i, \quad p = \sum p_i, \tag{1.82}$$

are mixture values for the density, momentum density, energy density, and pressure. The last of these is Dalton's law; it follows from the additivity of the normal forces on a convected solid surface.

The thermodynamic relation in (1.74), applied to P_i, reads

$$T_i\, ds_i = du_i + p_i\, d{\rho_i}^{-1}.$$

Introducing $\rho_i\, d\mathbf{r}$ into the differentials, we have

$$T_i\, d(\rho_i s_i d\mathbf{r}) = d(\rho_i u_i d\mathbf{r}) + p_i\, d(d\mathbf{r}) - g_i\, d(\rho_i d\mathbf{r}),$$

where

$$g_i \equiv h_i - T_i s_i, \qquad h_i \equiv u_i + p_i/\rho_i \tag{1.83}$$

are functions known as the *specific Gibbs* function and *specific enthalpy* respectively. Assuming that the total mass $\rho\, d\mathbf{r}$ of the fluid element is constant, and introducing the i-species concentration,

$$c_i \equiv \rho_i/\rho, \tag{1.84}$$

we have

$$T_i\, d(c_i s_i) = d(c_i u_i) + p_i\, d\rho^{-1} - g_i\, dc_i.$$

The specific entropy of the mixture is defined by

$$s = \sum c_i s_i, \tag{1.85}$$

therefore

$$ds = \sum_i \frac{1}{T_i}\{d(c_i u_i) + p_i\, d\rho^{-1} - g_i\, dc_i\}. \tag{1.86}$$

When the components are at the same temperature, this reduces to the Gibbs relation

$$T\, ds = du + p\, d\rho^{-1} - \sum g_i\, dc_i, \tag{1.87}$$

which generalizes (1.74). It is important to appreciate that entropy is a variable whose value for a given system depends on how the system is specified—the more detail that is provided, i.e. the greater the dimensionality of the thermodynamic state space, the smaller the value of s. Less detail corresponds to an increase in entropy (see Woods (1986), Chapter 4 for an elaboration of this point).

1.13 Motion of a fluid element

Suppose that a convected point $P_c(\mathbf{r}, t)$ moves with a velocity $\mathbf{v}(\mathbf{r}, t)$, then a neighbouring convected point $Q_c(\mathbf{r} + \mathbf{R}, t)$ has the fluid velocity

$$\mathbf{v}' = \mathbf{v} + \mathbf{R} \cdot \nabla \mathbf{v} + \mathrm{O}(R^2) \tag{1.88}$$

(see Fig. 1.5). The relative velocity $\mathbf{R} \cdot \nabla \mathbf{v}$ is the scalar product of the small distance $\mathbf{R}$ and the velocity gradient tensor $\nabla \mathbf{v}$, and in order to analyze this we require some acquaintance with second-order tensors and their properties. The following mathematical note gives sufficient details for our present purposes.

Mathematical note 1.13 *The decomposition of second-order tensors*

In general a second-order tensor A has a *symmetric* part A^s, an *antisymmetric* part A^a, a *trace* $\overset{\times}{A}$, a *vector* $\mathbf{A}^v$, and a *deviator* $\overset{\circ}{\mathsf{A}}$ defined by

$$\mathsf{A}^s \equiv \tfrac{1}{2}(\mathsf{A} + \tilde{\mathsf{A}}), \qquad \mathsf{A}^a \equiv \tfrac{1}{2}(\mathsf{A} - \tilde{\mathsf{A}}), \qquad \overset{\times}{A} \equiv \mathbf{1} : \mathsf{A}, \tag{1.89}$$

and

$$\mathbf{A}^v \equiv -\tfrac{1}{2}\mathbf{1} \times \mathbf{1} : \mathsf{A}, \qquad \overset{\circ}{\mathsf{A}} \equiv \mathsf{A}^s - \tfrac{1}{3}\mathbf{1}\,\overset{\times}{A}, \tag{1.90}$$

where $\mathbf{1}$ is the unit tensor, i.e. $\mathbf{1} \cdot \mathsf{A} = \mathsf{A}$ and $\mathsf{A} \cdot \mathbf{1} = \mathsf{A}$, and the tilde denotes the transposed tensor. Since $\mathbf{1} : \mathbf{1} = 3$ and $\mathbf{1} : \mathsf{A} = \mathbf{1} : \tilde{\mathsf{A}}$, it follows that the deviator of A has zero trace.

With double products like $\mathbf{ab} : \mathsf{A}$ we shall adopt the convention that

$$\mathbf{ab} : \mathsf{A} = \mathbf{b} \cdot \mathsf{A} \cdot \mathbf{a} = \mathsf{A} : \mathbf{ab},$$

e.g. if $\mathsf{A} = K\,\mathbf{ij}$, $\mathbf{ab} : \mathsf{A} = K(\mathbf{a} \cdot \mathbf{j})(\mathbf{b} \cdot \mathbf{i})$. Hence

$$\mathbf{1} : \mathbf{ab} = \mathbf{b} \cdot \mathbf{1} \cdot \mathbf{a} = \mathbf{b} \cdot \mathbf{a}, \qquad \mathbf{1} \times \mathbf{1} : \mathbf{ab} = \mathbf{b} \cdot \mathbf{1} \times \mathbf{1} \cdot \mathbf{a} = \mathbf{b} \times \mathbf{a},$$

$$\mathbf{r}\cdot\mathbf{1}\times\mathbf{1}\times\mathbf{1}:\mathbf{ab} = -\mathbf{r}\times(\mathbf{a}\times\mathbf{b}) = \mathbf{r}\cdot(\mathbf{ab}-\widetilde{\mathbf{ab}}) = 2\mathbf{r}\cdot(\mathbf{ab})^a,$$

or

$$(\mathbf{ab})^a = -\mathbf{1}\times(\mathbf{ab}).$$

Since A can always be expressed as the sum of three dyads, that is $\mathsf{A} = \mathbf{ab}+\mathbf{cd}+\mathbf{ef}$, it follows that

$$\overset{\times}{A} = \mathbf{a}\cdot\mathbf{b}+\mathbf{c}\cdot\mathbf{d}+\mathbf{e}\cdot\mathbf{f}, \quad 2\mathbf{A}^v = \mathbf{a}\times\mathbf{b}+\mathbf{c}\times\mathbf{d}+\mathbf{e}\times\mathbf{f},$$

$$\mathsf{A}^a = -\mathbf{1}\times\mathbf{A}^v = -\mathbf{A}^v\times\mathbf{1} = -\mathbf{1}\times\mathbf{1}\cdot\mathbf{A}^v,$$

and

$$\mathsf{A} = \overset{\circ}{\mathsf{A}} - \mathbf{A}^v\times\mathbf{1} + \tfrac{1}{3}\overset{\times}{A}\,\mathbf{1}.$$

In particular

$$\overset{\times}{\nabla\mathbf{v}} = \nabla\cdot\mathbf{v}, \quad (\nabla\mathbf{v})^v = \tfrac{1}{2}\nabla\times\mathbf{v} = \boldsymbol{\Omega},$$

and

$$\nabla\mathbf{v} = \overset{\circ}{\nabla\mathbf{v}} - \boldsymbol{\Omega}\times\mathbf{1} + \tfrac{1}{3}\mathbf{1}\,\nabla\cdot\mathbf{v}. \tag{1.91}$$

It is easily verified that for any vector $\mathbf{a}$, $\mathbf{a}\times\mathbf{1} = \mathbf{1}\times\mathbf{a}$, and in particular the second right-hand term in (1.91) can be expressed as $-\mathbf{1}\times\boldsymbol{\Omega}$.

Let B denote another second-order tensor, then as $\mathsf{A}:\mathsf{B} = \tilde{\mathsf{A}}:\tilde{\mathsf{B}}$, it follows that

$$\mathsf{A}^s:\mathsf{B}^a = \mathsf{A}^s:(-\mathsf{B}^a) = 0.$$

Also

$$\mathsf{A}^a:\mathsf{B}^a = \mathbf{A}^v\times\mathbf{1}\cdot\mathbf{1}\times\mathbf{B}^v = -2\mathbf{A}^v\cdot\mathbf{B}^v,$$

and therefore expanding each tensor, we obtain

$$\mathsf{A}:\mathsf{B} = \overset{\circ}{\mathsf{A}}:\overset{\circ}{\mathsf{B}} + \mathbf{A}^v\cdot\mathbf{B}^v + \tfrac{1}{3}\overset{\times}{A}\overset{\times}{B}.$$

For example (see (1.77)), as $\boldsymbol{\pi}$ is a symmetric tensor, $\boldsymbol{\pi}^v = 0$, and

$$\boldsymbol{\pi}:\nabla\mathbf{v} = \overset{\circ}{\boldsymbol{\pi}}:\overset{\circ}{\nabla\mathbf{v}} + \tfrac{1}{3}\overset{\times}{\pi}\,\nabla\cdot\mathbf{v}. \tag{1.92}$$

Also note that

$$\overset{\circ}{\mathsf{A}}:\mathsf{B} = \overset{\circ}{\mathsf{A}}:\mathsf{B}^s = \overset{\circ}{\mathsf{A}}:\overset{\circ}{\mathsf{B}}. \tag{1.93}$$

Returning to (1.88) and using $\mathbf{R}\cdot(\boldsymbol{\Omega}\times\mathbf{1}) = \mathbf{R}\times\boldsymbol{\Omega}\cdot\mathbf{1} = \mathbf{R}\times\boldsymbol{\Omega} = -\boldsymbol{\Omega}\times\mathbf{R}$, and $\mathbf{R}\cdot\mathbf{1} = \mathbf{R}$, we find from (1.91) that

$$\mathbf{v}' = \mathbf{v} + \boldsymbol{\Omega}\times\mathbf{R} + \tfrac{1}{3}\mathbf{R}\nabla\cdot\mathbf{v} + \mathbf{R}\cdot\overset{\circ}{\nabla\mathbf{v}} + \mathrm{O}(R^2). \tag{1.94}$$

A rigid body motion about an axis $\mathbf{l}$, rotating through a small angle θ, changes a position vector $\mathbf{R}$ fixed in the body to $\mathbf{R}+\theta\mathbf{l}\times\mathbf{R}$. The velocity of the point is therefore $\boldsymbol{\Omega}\times\mathbf{R}$, where $\boldsymbol{\Omega}$ is the angular velocity $\dot{\theta}\mathbf{l}$. Hence the second right-hand term of (1.94) represents a rigid body motion of the fluid element with an angular velocity equal to half the fluid vorticity, $\nabla\times\mathbf{v}$. We term $\boldsymbol{\Omega}$ the *spin* of the fluid element. Such motion does not strain (i.e. deform) the element, and it will not induce a stress, except in materials of

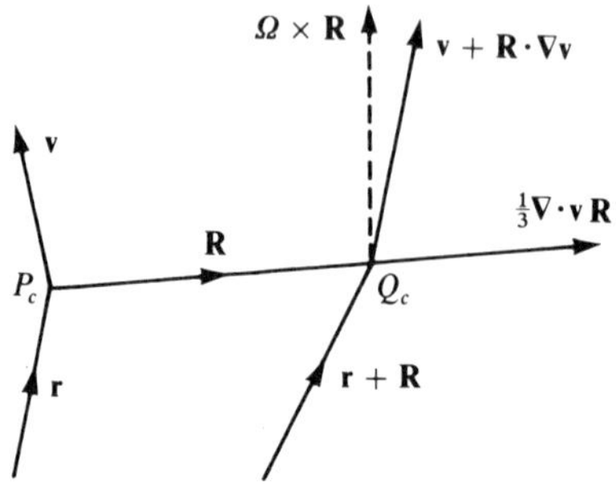

FIG. 1.5. Strain of a fluid element

unusual microstructure. The term $\boldsymbol{\Omega} \times \mathbf{R}$ can be removed from (1.94) by transforming to the convected frame P_c defined in §1.2.

Let $\hat{\mathbf{R}}$ be the unit vector along $\mathbf{R}$, then by (1.94) the 'outwards' speed of Q_c relative to P_c is $|\mathbf{R}|$ times $\frac{1}{3}\nabla \cdot \mathbf{v} + \hat{\mathbf{R}}\hat{\mathbf{R}} : \overset{\circ}{\nabla}\mathbf{v}$. If $\hat{\mathbf{R}}$ is distributed isotropically, the average of $\hat{\mathbf{R}}\hat{\mathbf{R}}$ taken over all directions radiating from P_c is $\frac{1}{3}\mathbf{1}$ (see (2.40)), and as $\mathbf{1} : \overset{\circ}{\nabla}\mathbf{v} = 0$, the average fluid speed outwards from P_c on the sphere $|\mathbf{R}| = a$ is $\frac{1}{3}a\nabla \cdot \mathbf{v}$. Thus the third right-hand term in (1.94) is due to the changing volume of the fluid element; this type of strain is called *dilatation*.

The remaining term in (1.94), representing pure straining motion without volume change, is called the *deviatoric* rate of strain. It plays a central role in kinetic theory. The symmetric part of the velocity gradient tensor, viz.

$$\mathbf{e} \equiv \tfrac{1}{2}(\nabla\mathbf{v} + \widetilde{\nabla\mathbf{v}}) = \overset{\circ}{\nabla}\mathbf{v} + \tfrac{1}{3}\mathbf{1}\,\nabla \cdot \mathbf{v} \tag{1.95}$$

is called the *rate of strain* tensor.

It is of interest (see §2.11.1) to determine the type of fluid motion for which $\mathbf{e}$ vanishes. Suppose that $\mathbf{e} = 0$ in some region, then (1.91) gives $\nabla\mathbf{v} = -\boldsymbol{\Omega} \times \mathbf{1}$. Save for one arbitrary constant, this linear, homogeneous equation for $\mathbf{v}$ has a unique solution, which by substitution is readily confirmed to be

$$\mathbf{v} = \mathbf{v}_0 + \boldsymbol{\Omega} \times \mathbf{r}, \qquad (\boldsymbol{\Omega} = \text{const.}) \tag{1.96}$$

where $\mathbf{v}_0$ is the arbitrary constant. Thus the fluid velocity is the same as the velocity of a rigid body moving with a screw motion.

Exercises 1

1.1 A fluid motion along OY has a velocity v_y dependent only on the distance x measured along OX. The random molecular motions transfer momentum between the planes $x = a, x = a + \lambda$, where λ is the mean free path. Show that this transfer occurs at the rate $\mu\, dv_y/dx$ $(\mu \approx \frac{1}{2}\rho\bar{c}\lambda)$ per unit area. Interpret this as a fluid force.

1.2 Explain why the concepts of temperature and pressure must be underpinned by particle collisions. Would you expect a gas consisting of neutrinos to have either a pressure or a pressure gradient?

1.3 Show that the root-mean-square molecular speed of a gas at standard conditions is given by $C = 1.0637 \times 10^{-10} m^{-1/2}$, where m is the mass of a molecule. Show that in oxygen, $C \approx 460\,\mathrm{ms}^{-1}$.

1.4 Show that the number density, n_0 of hydrogen molecules at STP (i.e. standard conditions) is $2.6868 \times 10^{25}\,\mathrm{m}^{-3}$ (known as *Loschmidt's* number) and that the average distance apart of these molecules is $\approx 3.34 \times 10^{-9}\,\mathrm{m}$.

1.5 In a diatomic gas some of the molecules are dissociated into separate atoms, the degree of dissociation δ being defined as the ratio of the mass of the monatomic portion to the total mass $\mathcal{M}$ of the system. With the help of Dalton's law, show that the equation of state is

$$pV = (\delta + 1)(\mathcal{M}/M_2)\mathcal{R}T$$

where M_2 is the molecular weight of the diatomic component.

1.6 Show that in a one-component gas, in the usual notation,

$$T\,ds = c_v\,dT + p\,d\rho^{-1} = c_p\,dT - \rho^{-1}\,dp$$

and obtain its energy equation in the form

$$\rho c_p\,\mathrm{D}T = -\mathrm{D}p - \boldsymbol{\pi} : \nabla\mathbf{v} + \nabla\cdot(\kappa\nabla T).$$

In what circumstances does this reduce to the diffusion equation

$$\frac{\partial T}{\partial t} = \frac{\kappa}{\rho c_p}\nabla^2 T\,?$$

1.7 Prove that the fluid acceleration $\mathbf{a} \equiv \mathrm{D}\mathbf{v}$ is related to the vorticity $\boldsymbol{\zeta}$ by

$$\mathbf{a} = \frac{\partial \mathbf{v}}{\partial t} + \nabla\tfrac{1}{2}v^2 + \boldsymbol{\zeta}\times\mathbf{v},$$

and that in an ideal, isotropic fluid with a conservative body force ($\mathbf{F} = -\nabla\phi$)

$$\rho\mathrm{D}\left(\frac{\boldsymbol{\zeta}}{\rho}\right) - \boldsymbol{\zeta}\cdot\nabla\mathbf{v} = \nabla\times\mathbf{v} = -\nabla\left(\frac{1}{\rho}\right)\times\nabla p.$$

1.8 The fluid velocity relative to a point P_c has the same direction $\hat{\mathbf{x}}$ everywhere, and varies linearly with distance in one direction $\hat{\mathbf{y}}$, perpendicular to $\hat{\mathbf{x}}$. Indicate the decomposition of this shearing motion into rotation and straining motions, and show that the deviatoric tensor has its principal axis at 45° to $\hat{\mathbf{x}}$ and $\hat{\mathbf{y}}$.

2

THE MAXWELLIAN VELOCITY DISTRIBUTION

2.1 The velocity distribution function

A concept of central importance in kinetic theory that remains to be introduced is that of the particle velocity distribution function, $f(\mathbf{r}, \mathbf{w}, t)$. This is an extension of the notion of number density to the six-dimensional space—known as *phase-space*—obtained by adjoining physical space with velocity space. We proceed as in the first paragraph of §1.2.

By a *macroscopic point* $(\mathbf{r}, \mathbf{w}, t)$ in phase-space is meant an infinitesimal 6-D volume, $d\boldsymbol{\nu}$ $(= d\mathbf{r}\, d\mathbf{w})$, centred at the point $(\mathbf{r}, \mathbf{w}, t)$, and having an extension sufficient to contain a very large number of particles of the species under consideration. The number density in this space is denoted by $f(\mathbf{r}, \mathbf{w}, t)$, so that $f\, d\boldsymbol{\nu}$ is the number of particles in $d\boldsymbol{\nu}$. We may treat f as being a continuous and differentiable function of its arguments.

From their definitions it follows that n, the number density in physical space, is related to f by

$$n(\mathbf{r}, t) = \int f(\mathbf{r}, \mathbf{w}, t)\, d\mathbf{w}, \tag{2.1}$$

where the integral extends over the whole of velocity space. In the following all indefinite integrals over velocity space will be assumed to have the maximum range. The ratio f/n is the probability that a particle chosen at random from the $n(\mathbf{r}, t)\, d\mathbf{r}$ particles in $d\mathbf{r}$ has its velocity in the range $\mathbf{w}, \mathbf{w} + d\mathbf{w}$ at time t, so the expected, or average, value of a function $\phi(\mathbf{w})$ is given by

$$\langle \phi(\mathbf{w}) \rangle = \frac{1}{n} \int \phi(\mathbf{w})\, f\, d\mathbf{w}. \tag{2.2}$$

The notation $\langle \ldots \rangle$ employed in Chapter 1 can now be replaced by integrals over velocity space. Thus for the density ρ and the fluid momentum $\rho\mathbf{v}$ we have

$$\rho = \int mf\, d\mathbf{w}, \qquad \rho\mathbf{v} = \int m\mathbf{w}f\, d\mathbf{w}, \tag{2.3}$$

where m is the particle mass. Equation (1.7) can be written

$$p = nkT = (\gamma - 1)\rho u = \int \tfrac{1}{3}mc^2\, f\, d\mathbf{w}, \tag{2.4}$$

where the perfect gas relation of (1.58) has been included; γ is defined in (1.66). The energy and momentum fluxes of (1.37) and (1.13) are

$$\boldsymbol{\xi} = \int \tfrac{1}{2}mc^2\mathbf{c}f\,d\mathbf{w}, \tag{2.5}$$

and

$$\mathbf{M} = \int m\mathbf{c}\mathbf{c}f\,d\mathbf{w}. \tag{2.6}$$

If the dimensions of the system allow *all* the flux in (2.5) and (2.6) to be collided flux, the pressure tensor $\mathbf{p}$ and heat flux vector $\mathbf{q}$ follow from

$$\mathbf{p} = \mathbf{M}, \qquad \mathbf{q} = \boldsymbol{\xi}. \tag{2.7}$$

The state of the gas may be specified more accurately by adding further details about the locations and velocities of the particles. The function $f(\mathbf{r}, \mathbf{w}, t)/n$ gives us the probability that a particular particle—say particle '1'—lies in $d\boldsymbol{\nu}_1$, and it is known as the *one-particle* distribution function. Much more detail is provided by the *two-particle* distribution function, $F(\mathbf{r}_1, \mathbf{w}_1; \mathbf{r}_2, \mathbf{w}_2, t)$, which gives the joint probability that particle '1' lies in $d\boldsymbol{\nu}_1$ and particle '2' lies in $d\boldsymbol{\nu}_2$. This process may be extended to many-particle distribution functions.

The distribution f may be enlarged in scope in another way. It strictly applies only to monatomic molecules. Molecules comprised of several atoms and hence having internal energy, require additional coordinates to be added to the arguments of f to define their internal motions (see Chapman and Cowling (1970), Chapter 11). This complication is beyond the scope of an introductory text. However, the results obtained for monatomic molecules can often be extended to polyatomic molecules by simple adjustments, e.g. the use of the equipartition of energy to obtain (1.58), which results in the appearance of a general value for γ in (2.4). In §3.8 we shall give an account of these adjustments. Meanwhile only monatomic or structureless particles will be considered.

2.2 Kinetic entropy

Our main objective at this stage is to find the formula giving the value of f when the system is in equilibrium. For this purpose we shall start by deriving a generalization of the entropy function introduced in §1.11.1.

By (1.59) and (1.64), namely $p = nkT$ and $s = c_v \ln T - R \ln \rho + \text{const.}$, the entropy density of a monatomic gas is given by

$$\rho s = -kn \ln n + \tfrac{3}{2}kn \ln T + \rho s^*, \tag{2.8}$$

where s^* is an arbitrary constant. From the additivity of entropy (see (1.85)) it follows that in a mixture, with all species having the same mass,

$$\rho s = -k \sum n_i \ln n_i + \tfrac{3}{2}k \sum n_i \ln T_i + \rho s^*. \tag{2.9}$$

We are free to prescribe the nature of the components in (2.9).

Let us suppose that the i-th component consists only of those particles of the gas that lie in the volume element $d\boldsymbol{\nu}_i$ of velocity space where the phase space density is $f_i = f(\mathbf{r}, \mathbf{w}_i, t)$. The number density of this component is $n_i = f_i\, d\mathbf{w}_i$, so (2.9) becomes

$$\rho s = -k \sum f_i \ln f_i \, d\mathbf{w}_i + R \sum m n_i (\tfrac{3}{2} \ln T_i - \ln |d\mathbf{w}_i|) + \rho s^*.$$

The spread of velocities in the i-th component is proportional to $|d\mathbf{w}_i|^{1/3}$, and hence the associated temperature is $T_i \propto |d\mathbf{w}_i|^{2/3}$. Thus the second right-hand term in ρs reduces to ρ times a constant, and may be included in ρs^*. With a particular choice for s^* (which has no significant consequences), we arrive at

$$\rho s = -k \int f(\ln f - 1)\, d\mathbf{w}. \tag{2.10}$$

Another choice for s^* can reduce the integrand to $f \ln f$; this is sometimes adopted, but we find (2.10) more convenient.

The 'kinetic' entropy defined in (2.10) contains the 'fluid' entropy given by (2.8) as special case. To distinguish the latter we shall add the subscript '0'. By (2.8) and (1.74) s_0 satisfies

$$s_0 = \tfrac{3}{2} R \ln T - R n \ln n + s^* \qquad (R = k/m), \tag{2.11}$$

and

$$T\, ds_0 = du + p\, d(1/\rho), \tag{2.12}$$

or

$$d(\rho s_0) = \frac{1}{T} d(\rho u) - \frac{g}{T} d\rho, \tag{2.13}$$

where

$$g \equiv u + p/\rho - T s_0 \tag{2.14}$$

is the specific Gibbs function (see (1.83)). By (1.58) and (2.11),

$$g = RT \left\{ \ln n - \tfrac{3}{2} \ln(2\pi k T) - R^{-1} \left(s^* - \tfrac{5}{2} R - \tfrac{3}{2} R \ln(2\pi k) \right) \right\};$$

so with a particular choice of the disposable constant s^* we obtain

$$g = \frac{kT}{m} \ln \left\{ n \left(\frac{m}{2\pi k T} \right)^{3/2} \right\}. \tag{2.15}$$

Finally, by writing (2.14) as $\rho s_0 = -T^{-1}(g - u) + kn$, and using (2.3) and (2.4) (with $\gamma = 5/3$), we find that it can be rearranged into a form with some similarity to (2.10), viz.

$$\rho s_0 = -k \int f \left\{ \frac{m}{kT} (g - \tfrac{1}{2} c^2) - 1 \right\} d\mathbf{w}. \tag{2.16}$$

We shall shortly use this relation in establishing the form of the equilibrium distribution function.

2.3 The equilibrium velocity distribution

2.3.1 *Local equilibrium*

Consider a region S of the medium bounded by a rigid, adiabatic boundary. If the gradients of the macroscopic variables $\mathbf{v}$ and T are zero throughout S, then we say that S is in an 'equilibrium' state. As will be explained in §2.11.2, pressure gradients are allowed in equilibrium provided they are balanced by appropriate force fields such as gravity. In such a global equilibrium, as shown in §1.11.3, the total entropy of S has its maximum value, which in principle can be used to determine the equilibrium distribution function f_0.

The equilibrium concept can be applied locally as follows. By (1.69), (1.70), and (1.75) at a convected macroscopic point P_c,

$$\rho \mathrm{D} s_0 + \nabla \cdot (\mathbf{q}/T) = \sigma \geq 0, \tag{2.17}$$

where

$$\sigma = -T^{-1}\, \boldsymbol{\pi} : \nabla \mathbf{v} - T^{-2}\, \mathbf{q} \cdot \nabla T. \tag{2.18}$$

Now imagine the element P_c of volume $d\mathbf{r}$ to be isolated suddenly from the rest of the fluid by a rigid, adiabatic boundary. Then, because of the occurrence of randomizing collisions within $d\mathbf{r}$, the gradients in P_c will decay to zero in a time inversely proportional to the collision frequency. During this process the inherent stability of the system requires the entropy density to increase to the maximum value allowed by the constraints acting on P_c. The key role of collisions in this process should be noted.

2.3.2 *Variation of the entropy*

Let f denote the distribution function in P_c before its isolation, and let f_0 denote its final equilibrium value. Then the decay of the gradients can change neither the number density, nor the internal energy, i.e.

$$\rho = \int mf\, d\mathbf{w} = \int mf_0\, d\mathbf{w}, \tag{2.19}$$

and

$$\rho u = \int \tfrac{1}{2} mc^2 f\, d\mathbf{w} = \int \tfrac{1}{2} mc^2 f_0\, d\mathbf{w}. \tag{2.20}$$

It follows from (2.13), (2.19), and (2.20) that variations in ρs_0 and f are related by

$$\delta(\rho s_0) = -k \int \delta f\, \frac{m}{kT}(g - \tfrac{1}{2}c^2)\, d\mathbf{w}.$$

From (2.10) we find

$$\delta(\rho s) = -k \int \delta f \ln f\, d\mathbf{w},$$

so that

$$\delta[\rho(s - s_0)] = k \int \delta f \left\{ \frac{m}{kT}(g - \tfrac{1}{2}c^2) - \ln f \right\} d\mathbf{w}. \tag{2.21}$$

Now choose

$$f = f_0 + \delta f,$$

with

$$\ln f_0 = \frac{m}{kT}(g - \tfrac{1}{2}c^2). \tag{2.22}$$

By (2.19) and (2.20) and the fact that ρ and ρu are constant,

$$\int m\delta f\, d\mathbf{w} = 0, \qquad \int \tfrac{1}{2}mc^2\delta f\, d\mathbf{w} = 0, \tag{2.23}$$

so (2.16) yields

$$\rho s_0 = -k\int f_0(\ln f_0 - 1)\, d\mathbf{w}. \tag{2.24}$$

And (2.21) reduces to

$$\delta[\rho(s - s_0)] = -k\int \frac{(\delta f)^2}{f_0}\, d\mathbf{w} \leq 0. \tag{2.25}$$

Equations (2.10), (2.24), and (2.25) show that the kinetic and fluid entropies are the same at $f = f_0$ and differ in the neighbourhood of f_0 only by a term proportional to $(f - f_0)^2$. Considered as functionals of f, ρs, and ρs_0 have the same maximum values. Also, as f_0 is independent of ∇T and $\nabla \mathbf{v}$, the choice $f = f_0$ implies that these gradients are zero, i.e. that the element is in equilibrium.

By (2.15) and (2.22), the equilibrium velocity distribution is given by

$$f_0 = n\left(\frac{m}{2\pi kT}\right)^{3/2} \exp(-\tfrac{1}{2}mc^2/kT) \qquad (\mathbf{c} \equiv \mathbf{w} - \mathbf{v}). \tag{2.26}$$

This is the most important result in kinetic theory. It was first obtained by Maxwell and published in 1860 in the Philosophical Magazine under the title *Illustrations of the Dynamical Theory of Gases*.

2.4 Maxwell's discovery of f_0

Although Maxwell's determination of the distribution function f_0 that now bears his name was flawed, it was a bold step that marked the beginning of a new era in physics. With it he introduced statistical concepts and methods for the first time in the subject, ideas that lead directly to Boltzmann's statistical mechanics and eventually to quantum theory.

2.4.1 *Maxwell's statistical treatment*

It was Clausius' invention of the mean free path concept in 1858 that triggered Maxwell's work on the distribution function. Clausius was responding to Buys-Ballot's criticism of molecular theory, namely that molecules could not have the high velocities indicated by kinetic theory, otherwise a pungent gas like ammonia would permeate a room in a tiny fraction of the several minutes actually observed. Upon reading Clausius' paper, Maxwell realised that the accepted notion of a single speed for all particles could not be correct. Collisions that impede the dispersal of particular molecules would also result in their having a spread of velocities, and the problem was to find the function describing this distribution. Maxwell had a knowledge of statistics and in particular was familiar with the 'normal' distribution in the theory of errors. His argument (with the symbols changed to our notation) was as follows:

Prop IV. To find the average number of particles whose velocities lie between given limits, after a great number of collisions among a great number of equal particles.

Let N be the whole number of particles. Let c_x, c_y, c_z be the components of the velocity of each particle in three rectangular directions, and let the number of particles for which c_x lies between c_x and $c_x + dc_x$, be $NF(c_x)\,dc_x$, where $F(c_x)$ is a function of c_x to be determined.

The number of particles for which c_y lies between $c_y + dc_y$ will be $NF(c_y)\,dc_y$; and the number for which c_z lies between c_z and $c_z + dc_z$ will be $NF(c_z)\,dc_z$, where F always stands for the same function.

Now the existence of the velocity c_x does not in any way affect that of the velocities c_y or c_z, since these are all at right angles to each other and independent, so that the number of particles whose velocity lies between c_x and $c_x + dc_x$, and also between c_y and $c_y + dc_y$, and also between c_z and $c_z + dc_z$, is

$$NF(c_x)\,F(c_y)\,F(c_z)\,dc_x\,dc_y\,dc_z.$$

If we suppose the N particles to start from the origin at the same instant, then this will be the number in the element of volume $(dc_x\,dc_y\,dc_z)$ after unit of time, and the number referred to unit volume will be

$$NF(c_x)\,F(c_y)\,F(c_z).$$

But the directions of the coordinates are perfectly arbitrary, and therefore this number must depend on the distance from the origin alone, that is

$$F(c_x)\,F(c_y)\,F(c_z) = \phi(c_x^2 + c_y^2 + c_z^2).$$

Solving this functional equation, we find

$$F(c_x) = C\,e^{Ac_x^2}, \qquad \phi(c^2) = C^3\,e^{Ac^2}.$$

Maxwell then applies physical arguments to choose the constants A and C and arrives at the conclusion that the number of particles with velocities lying in $c_x, c_x + dc_x$ is

$$dN_x = N\frac{1}{\alpha\sqrt{\pi}}\,e^{-c_x^2/\alpha^2}\,dc_x. \tag{2.27}$$

The main deficiency in the argument is Maxwell's assumption that the three resolved components of velocity could be regarded as being independent variables. This requires proof. It should emerge as a conclusion rather than be admitted as self-evident. Another shortcoming is that particle collisions appear to have played no role. But because of its statistical content, (2.27) is a landmark in nineteenth century physics.

2.4.2 *Role of collisions*

Eight years passed before Maxwell offered another derivation based on the dynamics of molecular collisions. In his comprehensive work 'On the Dynamical Theory of Gases' (*Phil. Trans.*, Vol. CLVII, 1867), referring to his assumption about the independence of the velocity components, he wrote:

As this assumption may appear precarious, I shall now determine the form of the [distribution] function in a different manner.

He considered collisions between molecules of the 'first' kind, with speeds c_1, c_1' before and after the collisions, with molecules of the 'second' kind, with corresponding speeds c_2 and c_2'. He tacitly introduced the assumption that the probability of a collision was proportional to the product of the distributions, $f(c_1)\,f(c_2)$, and showed that for equilibrium f had to satisfy

$$f(c_1)\,f(c_2) = f(c_1')\,f(c_2'). \tag{2.28}$$

On comparing the logarithm of (2.28) with the conservation of energy, namely

$$\tfrac{1}{2}m_1 c_1^2 + \tfrac{1}{2}m_2 c_2^2 = \tfrac{1}{2}m_1 {c_1'}^2 + \tfrac{1}{2}m_2 {c_2'}^2,$$

he arrived again at the distribution given in (2.27).

While this approach introduced collisions and established the equilibrium of Maxwell's distribution f_0, it left open the questions of uniqueness and stability. Five years later Boltzmann obtain an equation determining the rate of change of f—now termed a 'kinetic equation'—and with its aid was able to settle these issues (see §7.4.3).

2.5 Maximizing the kinetic entropy

The derivation of f_0 given in §2.3.2 was based on the fact that stability required the fluid entropy ρs_0 to reach a maximum at equilibrium. We then used the relation between the kinetic and fluid entropies given in (2.25) to transfer this property to ρs. Sometimes it is possible to establish the maximum principle directly for ρs without reference to the fluid equations. This requires the collision or source term appearing in the kinetic equation for f to have a particular form—one that makes it 'irreversible'. We shall defer the account of these topics until §7.4.2, and for present purposes accept that the entropy density defined in (2.10), i.e.

$$\rho s = -k \int f(\ln f - 1)\, d\mathbf{w} \tag{2.29}$$

is a maximum at $f = f_0$.

The first step in maximizing ρs is to find its variation $\delta(\rho s)$, subject to any constraints that may apply to the distribution function. In the present application there are two constants, namely the number density and the internal energy (see discussion leading to (2.19) and (2.20)):

$$n = \int f\, d\mathbf{w}, \qquad \rho u = \int \tfrac{1}{2}mc^2 f\, d\mathbf{w}. \tag{2.30}$$

We incorporate these restrictions by introducing Lagrangian multipliers α and β;

$$\delta(\rho s) = -k \int \delta\{f(\ln f - 1) + \alpha f + \beta c^2 f\}\, d\mathbf{w} = -k \int (\ln f + \alpha + \beta c^2)\, \delta f\, d\mathbf{w}.$$

Since the variation in f is now arbitrary, $\delta(\rho s)$ is zero only if the integrand vanishes, i.e. if f has the value f_0 given by

$$f_0 = \exp(-\alpha - \beta c^2). \tag{2.31}$$

The constants α and β are determined by using the constraints in (2.30).

From (2.31) the first constraint in (2.30) requires that

$$n = \exp(-\alpha) \int_{-\infty}^{\infty}\int_{-\infty}^{\infty}\int_{-\infty}^{\infty} \exp\left\{-\beta(c_x^2 + c_y^2 + c_z^2)\right\} dc_x\, dc_y\, dc_z,$$

where we have removed the displacement $d\mathbf{v}$ from $d\mathbf{w}$, i.e. set $d\mathbf{w} = d\mathbf{c}$. Now

$$\int_{-\infty}^{\infty} \exp(-\beta c^2)\, dx = \frac{\pi^{1/2}}{\beta^{1/2}}, \qquad \int_{-\infty}^{\infty} x^2 \exp(-\beta x^2)\, dx = \frac{\pi^{1/2}}{2\beta^{3/2}}, \qquad (2.32)$$

so the expression for n yields

$$\exp(-\alpha) = n\,(\beta/\pi)^{3/2}.$$

Also by (2.4) the second constraint in (2.30) becomes

$$\tfrac{3}{2}nkT = e^{-\alpha}\int_{-\infty}^{\infty}\int_{-\infty}^{\infty}\int_{-\infty}^{\infty} \tfrac{1}{2}m(c_x^2+c_y^2+c_z^2)\exp\left\{-\beta(c_x^2 + c_y^2 + c_z^2)\right\} dc_x\, dc_y\, dc_z.$$

The average value of c_x^2 is

$$\overline{c_x^2} \equiv \frac{1}{n}\int c_x^2\, f_0\, d\mathbf{c} = \left(\frac{\beta}{\pi}\right)^{3/2}\int_{-\infty}^{\infty}\int_{-\infty}^{\infty}\int_{-\infty}^{\infty} c_x^2 \exp\{-\beta(c_x^2 + c_y^2 + c_z^2)\}\, dc_x\, dc_y\, dc_z,$$

or by (2.32)

$$\overline{c_x^2} = 1/2\beta.$$

Similarly $\overline{c_y^2} = \overline{c_z^2} = (2\beta)^{-1}$. Thus by (2.4), $kT = m(2\beta)^{-1}$, whence $\beta = m/(2kT)$.

The formula in (2.31) now reads

$$f_0 = n\left(\frac{m}{2\pi kT}\right)^{3/2}\exp\left(-\frac{mc^2}{2kT}\right) \qquad (\mathbf{c} = \mathbf{w} - \mathbf{v}), \qquad (2.33)$$

in agreement with the derivation in §2.3.2. It remains to show that the stationary point is actually a maximum; this is left as an exercise.

2.6 Properties of the equilibrium distribution

2.6.1 *Averages of velocity-dependent functions*

It follows from the derivation of f_0 in §2.3.2 that it is the unique distribution towards which others will relax if the system is suddenly isolated from its environment. The relaxation rate will be shown later to be proportional to the collision frequency, τ^{-1}. Let us assume that sufficient time $\mathcal{T}$ has elapsed since the system was isolated for the equilibrium state to have been reached, i.e. that $\mathcal{T} \gg \tau$.

The probability that a particle chosen at random has a velocity in the range $\mathbf{c}, \mathbf{c} + d\mathbf{c}$, is

$$\frac{f_0}{n}\, d\mathbf{c} = \pi^{-3/2}\, \mathcal{C}^{-3} \exp(-\nu^2)\, d\mathbf{c}, \tag{2.34}$$

where $$\nu \equiv \frac{c}{\mathcal{C}}, \qquad \mathcal{C} \equiv \left(\frac{2kT}{m}\right)^{1/2}.$$

Notice that (2.34) can be split into three independent factors:

$$\frac{f_0}{n}\, d\mathbf{c} = \prod_{i=x,y,z} \frac{1}{\sqrt{\pi}\mathcal{C}} \exp(-\nu_i^2)\, dc_i, \tag{2.35}$$

where $\nu_x = c_x/\mathcal{C}$, etc., which shows that the distribution of c_x is independent of the values c_y and c_z, and similarly for c_y and c_z. This provides the proof of Maxwell's assumption (see discussion following (2.27)).

Transforming from the Cartesian coordinates $(c_x,\ c_y,\ c_z)$ to the spherical polar coordinates $(c,\ \theta,\ \phi)$, we have

$$d\mathbf{c} = c^2\, dc \sin\theta\, d\theta\, d\phi = c^2\, dc\, 4\pi\, d\Omega, \tag{2.36}$$

where $4\pi\, d\Omega$ is the element of solid angle subtended at the origin. Thus the joint probability that the relative speed ν falls in ν, $\nu + d\nu$ and the vector $\mathbf{c}$ is within the element $d\Omega$, is given by

$$\frac{f_0}{n}\, d\mathbf{c} = \frac{4}{\sqrt{\pi}} \nu^2 \exp(-\nu^2)\, d\nu\, d\Omega \qquad (0 \le \nu < \infty, \quad 0 \le \Omega \le 1). \tag{2.37}$$

The average value of a function $\Phi(\nu)$ is therefore

$$\langle \Phi \rangle = \frac{4}{\sqrt{\pi}} \int_\Omega \int_0^\infty \Phi(\nu)\, \nu^2 \exp(-\nu^2)\, d\nu\, d\Omega. \tag{2.38}$$

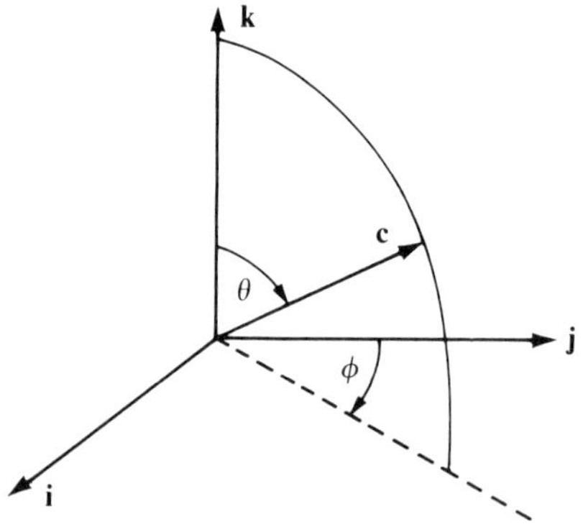

FIG. 2.1. Values of $\hat{\mathbf{c}}$ and $\hat{\mathbf{c}}\hat{\mathbf{c}}$ in cartesian coordinates.

2.6.2 *Averages of vectors and tensors*

Shortly we shall need the values of the integrals

$$\mathbf{J}_1 \equiv \int_\Omega \hat{\mathbf{c}}\, d\Omega, \qquad \mathbb{J}_2 \equiv \int_\Omega \hat{\mathbf{c}}\hat{\mathbf{c}}\, d\Omega. \tag{2.39}$$

From Fig. 2.1 we see that

$$\hat{\mathbf{c}} = \mathbf{i}\sin\theta\,\sin\phi + \mathbf{j}\sin\theta\,\cos\phi + \mathbf{k}\cos\theta$$

and therefore

$$\begin{aligned}\hat{\mathbf{c}}\hat{\mathbf{c}} &= \mathbf{ii}\sin^2\theta\sin^2\phi + \mathbf{jj}\sin^2\theta\cos^2\phi + \mathbf{kk}\cos^2\theta + (\mathbf{ij}+\mathbf{ji})\sin^2\theta\sin\phi\cos\phi \\ &\quad +(\mathbf{jk}+\mathbf{kj})\sin\theta\cos\theta\cos\phi + (\mathbf{ki}+\mathbf{ik})\sin\theta\cos\theta\sin\phi.\end{aligned}$$

From this expansion we find that

$$\mathbf{J}_1 = \frac{1}{4\pi}\int_0^{2\pi}\int_0^{\pi} \hat{\mathbf{c}}\,\sin\theta\, d\theta\, d\phi = 0,$$

and
$$\mathbb{J}_2 = \frac{1}{4\pi}\int_0^{2\pi}\int_0^{\pi} \hat{\mathbf{c}}\hat{\mathbf{c}}\,\sin\theta\, d\theta\, d\phi = \tfrac{1}{3}(\mathbf{ii}+\mathbf{jj}+\mathbf{kk}),$$

or
$$\int_\Omega \hat{\mathbf{c}}\, d\Omega = 0, \qquad \int_\Omega \hat{\mathbf{c}}\hat{\mathbf{c}}\, d\Omega = \tfrac{1}{3}\mathbf{1}, \tag{2.40}$$

where $\mathbf{1}$ is the unit tensor (see §1.5.2).

We shall also require the standard integrals

$$\frac{2}{\sqrt{\pi}}\int_0^\infty \nu^r \exp(-\nu^2)\, d\nu = \begin{cases} 1 & (r=0) \\ \frac{1}{\sqrt{\pi}}\left(\frac{1}{2}(r-1)\right)! & (r=1,3,5,\ldots) \\ \frac{1}{2}\cdot\frac{3}{2}\cdot\ldots\cdot\frac{(r-1)}{2} & (r=2,4,6,\ldots) \end{cases}. \tag{2.41}$$

2.6.3 *Pressure and distribution of speeds*

Two important applications of (2.38) are (replacing $\langle\cdots\rangle$ by a bar)

$$\overline{c} = \frac{4}{\sqrt{\pi}}\mathcal{C}\int_0^\infty \nu^3\exp(-\nu^2)\, d\nu = \frac{2}{\sqrt{\pi}}\mathcal{C} = \left(\frac{8kT}{\pi m}\right)^{1/2}, \tag{2.42}$$

and
$$\overline{\mathbf{cc}} = \frac{4}{\sqrt{\pi}}\mathcal{C}^2\int_\Omega\int_0^\infty \nu^4\exp(-\nu^2)\, d\nu\, \hat{\mathbf{c}}\hat{\mathbf{c}}\, d\Omega = \tfrac{1}{2}\mathcal{C}^2\mathbf{1}, \tag{2.43}$$

the trace of which is
$$\overline{c^2} = \tfrac{3}{2}\mathcal{C}^2 = 3kT/m = 3RT. \tag{2.44}$$

It follows from (1.15), viz. $\mathsf{p} = \rho\overline{\mathbf{cc}}$, that in equilibrium the pressure tensor is

$$\mathsf{p} = p\mathbf{1}, \qquad p = \tfrac{1}{2}\rho\mathcal{C}^2. \tag{2.45}$$

Also
$$\frac{f_0}{n} = F(\nu)\, d\nu\, d\Omega, \tag{2.46}$$

where
$$F(\nu) = \frac{4}{\sqrt{\pi}} \nu^2\, e^{-\nu^2} \qquad \left(\nu \equiv c/\mathcal{C}, \quad \mathcal{C} = (2kT/m)^{1/2}\right) \tag{2.47}$$

is the Maxwellian *speed* distribution. This function is plotted in Fig. 2.2. Its maximum occurs at $\nu = 1$, which identifies $\mathcal{C}$ as being the most probable molecular speed. Note the inequalities

$$\mathcal{C} < \bar{c} < \sqrt{\overline{c^2}}. \tag{2.48}$$

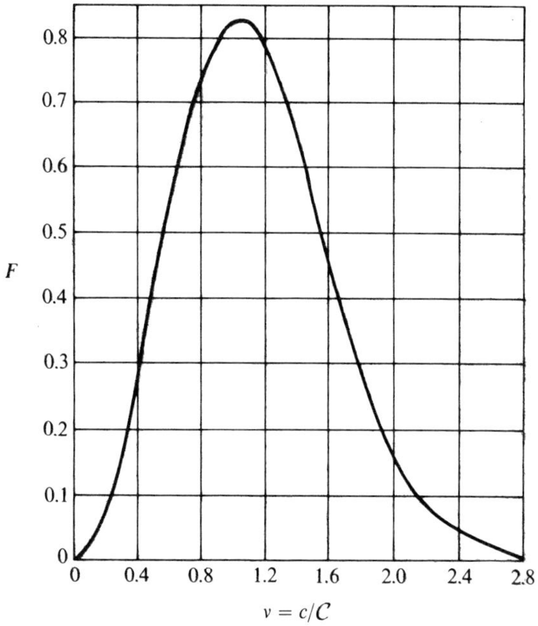

FIG. 2.2. The Maxwellian speed distribution

2.7 Dynamics of a binary collision

2.7.1 *Particle dynamics*

To obtain a general result needed in a later chapter, we shall consider an elastic collision between particles of different species, say particle P_i belonging to the i-species and particle P_j belonging to the j-species. The relative velocities will be denoted by

$$\underbrace{\mathbf{g}_{ij} \equiv \mathbf{c}_i - \mathbf{c}_j = g_{ij}\mathbf{k}_{ij},}_{\text{before collision}} \qquad \underbrace{\mathbf{g}'_{ij} \equiv \mathbf{c}'_i - \mathbf{c}'_j = g'_{ij}\mathbf{k}'_{ij},}_{\text{after collision}} \tag{2.49}$$

where $\mathbf{k}_{ij}, \mathbf{k}'_{ij}$ are unit vectors. The velocities of the centre of mass of P_i and P_j before and after the collision are

$$\mathbf{G}_{ij} = \mu_i \mathbf{c}_i + \mu_j \mathbf{c}_j, \quad \mathbf{G}'_{ij} = \mu_i \mathbf{c}'_i + \mu_j \mathbf{c}'_j, \tag{2.50}$$

where $\mu_i \equiv m_i/(m_i + m_j)$, $\mu_j \equiv m_j/(m_i + m_j)$. By conservation of momentum, the mass centre moves uniformly throughout the encounter, i.e.

$$\mathbf{G}_{ij} = \mathbf{G}'_{ij}. \tag{2.51}$$

From (2.49) and (2.50),

$$\mathbf{c}_i = \mathbf{G}_{ij} + \mu_j \mathbf{g}_{ij}, \quad \mathbf{c}_j = \mathbf{G}_{ij} - \mu_i \mathbf{g}_{ij}, \tag{2.52}$$

so that
$$\tfrac{1}{2} m_i c_i^2 + \tfrac{1}{2} m_j c_j^2 = \tfrac{1}{2}(m_i + m_j) G_{ij}^2 + \tfrac{1}{2}\left(\frac{m_i m_j}{m_i + m_j}\right) g_{ij}^2; \tag{2.53}$$

similar relations apply after the collision. Equation (2.53) expresses the theorem that the kinetic energy of the particles is equal to the kinetic energy of the total mass moving with the velocity of the centre of mass, plus the kinetic energy of the relative motion. As the first of these energies is unchanged by the collision, it follows from (2.53) and the corresponding equation following the collision, that energy is conserved throughout the encounter if

$$g_{ij} = g'_{ij}. \tag{2.54}$$

By (2.49) and (2.54),
$$\mathbf{g}_{ij} = g_{ij} \mathbf{k}_{ij}, \quad \mathbf{g}'_{ij} = g_{ij} \mathbf{k}'_{ij}, \tag{2.55}$$

hence the collision merely rotates $\mathbf{g}_{ij}$ into $\mathbf{g}'_{ij}$ without changing its magnitude. There is no constraint on the unit vector $\mathbf{k}'_{ij}$, which means that all directions of reflection or rebound, as viewed in the centre of mass frame, are equally likely (see Fig. 2.3).

2.7.2 *Liouville's law for elastic collisions*

At this point we need a result from the theory of integration.

Mathematical note 2.7 *Transformation of multiple integrals*

If the variables of integration in the multiple integral

$$\iiint \ldots F(x_1, x_2, \ldots, x_n)\, dx_1\, dx_2 \ldots dx_n$$

are changed to a set $y_1, y_2, \ldots, y_n$, then by a standard result of integration theory, the integral is transformed to

$$\iiint \ldots \mathcal{F}(y_1, y_2, \ldots, y_n)\, |J|\, dy_1\, dy_2 \ldots dy_n,$$

where
$$\mathcal{F}(y_1, y_2, \ldots, y_n) \equiv F(x_1, x_2, \ldots, x_n)$$

and J denotes the Jacobian determinant

$$J = \frac{\partial(\mathbf{x})}{\partial(\mathbf{y})} = \frac{\partial(x_1, x_2, \ldots, x_n)}{\partial(y_1, y_2, \ldots, y_n)} = \begin{vmatrix} \frac{\partial x_1}{\partial y_1}, & \frac{\partial x_2}{\partial y_1}, & \cdots & , \frac{\partial x_n}{\partial y_1} \\ \frac{\partial x_1}{\partial y_2}, & \frac{\partial x_2}{\partial y_2}, & \cdots & , \frac{\partial x_n}{\partial y_2} \\ \cdots & \cdots & \cdots & \cdots \\ \frac{\partial x_1}{\partial y_n}, & \frac{\partial x_2}{\partial y_n}, & \cdots & , \frac{\partial x_n}{\partial y_n} \end{vmatrix}.$$

If the vectors $\mathbf{x}$ and $\mathbf{y}$ are similarly partitioned, e.g. $\mathbf{x} = (\mathbf{x}', \mathbf{x}'')$, where $\mathbf{x}' = (x_1, x_2, \ldots, x_m)$, $\mathbf{x}'' = (x_{m+1}, \ldots, x_n)$, and likewise for $\mathbf{y}'$ and $\mathbf{y}''$, we may write $J = \partial(\mathbf{x}', \mathbf{x}'')/\partial(\mathbf{y}', \mathbf{y}'')$. It follows from the basic properties of determinants that

$$\frac{\partial(\mathbf{x}' + k\mathbf{x}'', \mathbf{x}'')}{\partial(\mathbf{y}', \mathbf{y}'')} = \frac{\partial(\mathbf{x}', \mathbf{x}'')}{\partial(\mathbf{y}', \mathbf{y}'')} = \frac{\partial(\mathbf{x}', \mathbf{x}'' + h\mathbf{x}')}{\partial(\mathbf{y}', \mathbf{y}'')}, \tag{2.56}$$

where k and h are constants.

Notice that we can write formally

$$dx_1\, dx_2 \cdots dx_n = |J|\, dy_1\, dy_2 \cdots dy_n, \tag{2.57}$$

provided it is understood that this relation is used only under the appropriate integral signs.

Consider the Jacobians

$$J \equiv \frac{\partial(\mathbf{G}_{ij}, \mathbf{g}_{ij})}{\partial(\mathbf{c}_j, \mathbf{c}_i)}, \qquad J' \equiv \frac{\partial(\mathbf{G}'_{ij}, \mathbf{g}'_{ij})}{\partial(\mathbf{c}'_j, \mathbf{c}'_i)}. \tag{2.58}$$

By (2.52) and (2.56),

$$J = \frac{\partial(\mathbf{c}_i - \mu_j \mathbf{g}_{ij}, \mathbf{g}_{ij})}{\partial(\mathbf{c}_j, \mathbf{c}_i)} = \frac{\partial(\mathbf{c}_i, \mathbf{c}_i - \mathbf{c}_j)}{\partial(\mathbf{c}_j, \mathbf{c}_i)} = \frac{\partial(\mathbf{c}_i, \mathbf{c}_j)}{\partial(\mathbf{c}_i, \mathbf{c}_j)} = 1.$$

Similarly $J' = 1$. By (2.57) the 6-D volume element $d\mathbf{c}_i\, d\mathbf{c}_j$ can be represented by $|J^{-1}|\, d\mathbf{g}_{ij}\, d\mathbf{G}_{ij}$, where J is defined in (2.58). Hence, as $|J| = 1$,

$$d\mathbf{c}_i\, d\mathbf{c}_j = d\mathbf{g}_{ij}\, d\mathbf{G}_{ij} = g_{ij}^2\, d\mathbf{g}_{ij}\, d\mathbf{k}_{ij}\, d\mathbf{G}_{ij}, \tag{2.59}$$

where we have expressed the volume element $d\mathbf{g}_{ij}$ in spherical polar coordinates. From (2.51), (2.54), and (2.59) it follows that

$$d\mathbf{k}'_{ij}\, d\mathbf{c}_i\, d\mathbf{c}_j = g_{ij}^2\, dg_{ij}\, d\mathbf{k}_{ij}\, d\mathbf{k}'_{ij}\, d\mathbf{G}_{ij} = d\mathbf{k}_{ij}\, d\mathbf{c}'_i\, d\mathbf{c}'_j, \tag{2.60}$$

a relationship known as *Liouville's law* for elastic collisions.

2.8 Collision frequency in a Maxwellian distribution

2.8.1 *The average relative velocity*

In order to generalize the analysis leading to (1.1), viz. $\lambda \approx 1/(\pi\sigma^2 n)$, for the mean free path in a gas with *hard* molecules of diameter σ, we now consider a molecule P_i of the i-species moving with speed $\mathbf{c}_i$ through n_j molecules per unit volume of the j-species. Denote the molecular diameters by σ_i, σ_j; then the sphere of influence for collisions between the two species has a cross-section $\pi\sigma_{ij}^2$, where $\sigma_{ij} = \frac{1}{2}(\sigma_i + \sigma_j)$. The number of collisions per second between P_i and those i-species molecules with velocity $\mathbf{c}_j$ is therefore $\pi\sigma_{ij}^2 g_{ij} n_j$, where $g_{ij} = |\mathbf{c}_i - \mathbf{c}_j|$. To obtain the mean collision frequency, we need the average value of g_{ij} taken over the velocity distributions of both species. Suppose $\langle g_{ij}\rangle$ is this average, then the unlike-molecule collision frequency per unit volume of the gas is given by the average

$$\langle \nu_{ij}\rangle = n_i n_j \langle g_{ij}\rangle \pi\sigma_{ij}^2 \qquad (\sigma_{ij} = \tfrac{1}{2}(\sigma_i + \sigma_j)). \tag{2.61}$$

We shall assume that both species have Maxwellian velocity distributions with the same temperature and fluid velocity. Hence

$$\langle g_{ij}\rangle = \iint g_{ij} \frac{f_{0i}}{n_i}\frac{f_{0j}}{n_j}\, d\mathbf{c}_i\, d\mathbf{c}_j, \tag{2.62}$$

where from (2.34),

$$\frac{f_{0i}}{n_i}\, d\mathbf{c}_i = \pi^{-3/2}\mathcal{C}_i^{-3}\exp(-c_i^2/\mathcal{C}_i^2)\, d\mathbf{c}_i \qquad \left(\mathcal{C}_i \equiv (2kT/m_i)^{1/2}\right), \tag{2.63}$$

with a similar expression applying to the j-species. Thus

$$\langle g_{ij}\rangle = \frac{1}{(\pi\mathcal{C}_i\mathcal{C}_j)^3}\iint g_{ij}\exp\left\{-\left(\frac{c_i^2}{\mathcal{C}_i^2} + \frac{c_j^2}{\mathcal{C}_j^2}\right)\right\} d\mathbf{c}_i\, d\mathbf{c}_j. \tag{2.64}$$

By (2.53) and (2.59),

$$\langle g_{ij}\rangle = \frac{1}{(\pi\mathcal{C}_i\mathcal{C}_j)^3}\int g_{ij}\exp\left(-\frac{m_{ij}}{2kT}g_{ij}^2\right) d\mathbf{g}_{ij}\int \exp\left(-\frac{m_i + m_j}{2kT}\,G_{ij}^2\right) d\mathbf{G}_{ij}, \tag{2.65}$$

where

$$m_{ij} \equiv \frac{m_i m_j}{m_i + m_j} \tag{2.66}$$

is termed the *reduced mass*.

Consider the volume integral

$$\begin{aligned} I_S(\alpha) \equiv \frac{1}{4\pi}\int x^S e^{-\alpha x^2}\, d\mathbf{x} &= \int_0^\infty x^{2+S}e^{-\alpha x^2}\, dx \\ &= \frac{1}{\alpha^{(3+S)/2}}\int_0^\infty y^{2+S}e^{-y^2}\, dy, \end{aligned} \tag{2.67}$$

where s is an integer. By (2.41),

$$I_0(\alpha) = \frac{\sqrt{\pi}}{4\alpha^{3/2}}, \qquad I_1(\alpha) = \frac{1}{2\alpha^2}.$$

With these values in the expression for $\langle g_{ij}\rangle$, we find

$$\langle g_{ij}\rangle = \left(\frac{8kT}{\pi m_{ij}}\right)^{1/2} = \left(\frac{8kT}{\pi m_i} + \frac{8kT}{\pi m_j}\right)^{1/2} = (\bar{c}_i^2 + \bar{c}_j^2)^{1/2}, \tag{2.68}$$

of which the last form follows from (2.42).

2.8.2 *Collision frequencies*

The collision frequency per unit volume is given by (2.61) and (2.68):

$$\langle \nu_{ij} \rangle = 2 n_i n_j \sigma_{ij}^2 \left(\frac{2\pi kT}{m_{ij}} \right)^{1/2}. \tag{2.69}$$

In particular if the molecules are alike and $m_i = m_j = m$, $\langle g_{ij} \rangle$ becomes

$$\langle g \rangle = \sqrt{2}\,\bar{c} \qquad (\bar{c} = (8kT/\pi m)^{1/2}), \tag{2.70}$$

and the collision frequency *per molecule* is

$$\nu = \sqrt{2}\pi n \bar{c} \sigma^2, \tag{2.71}$$

corresponding to a mean free path of

$$\lambda = \frac{1}{\sqrt{2}\pi n \sigma^2}, \tag{2.72}$$

which is not very different from the estimate in (1.1).

2.9 A molecular beam in a Maxwellian gas

2.9.1 *Average relative velocity*

Suppose that a beam of hard i-molecules moves with a constant speed through a background of j-molecules in Maxwellian equilibrium. The collision frequency between a typical beam molecule and a background molecule moving with velocity $\mathbf{c}_j$ is $n_j \pi \sigma_{ij}^2 g_{ij}$, where

$$g_{ij} = |\mathbf{c}_i - \mathbf{c}_j| = |c_i^2 + c_j^2 - 2c_i c_j \cos\theta|^{1/2} \qquad (0 \leq \theta \leq \pi), \tag{2.73}$$

θ being the angle between $\mathbf{c}_i$ and $\mathbf{c}_j$. It follows from the square of (2.73) that

$$g_{ij}\, dg_{ij} = c_i c_j \sin\theta\, d\theta \qquad (c_i \sim c_j \leq g_{ij} \leq c_i + c_j). \tag{2.74}$$

We shall adopt spherical polar coordinates, (c, θ, ϵ) in velocity space, with the fixed vector $\mathbf{c}_i$ along the axis as shown in Fig. 2.3. The point C in the figure is the centre of mass, and moves with the constant velocity $\mathbf{G}_{ij}$. As explained in §2.7.1, the effect of an elastic collision is to rotate $\mathbf{g}_{ij}$ through an angle χ, known as the *scattering angle*, without altering its magnitude. With hard molecules, all possible orientations of $\mathbf{g}_{ij}$ are equally probable.

The j-molecules are moving in all directions with equal probability, so those with velocity vectors $\mathbf{c}_j$ lying within an element $4\pi\, d\Omega = \sin\theta\, d\theta\, d\epsilon$ of solid angle (see (2.36)) are a fraction $d\Omega$ of all molecules with the speed c_j. Hence by (2.74) the average of g_{ij} over all directions of the vector $\mathbf{c}_j$ is

$$\tilde{g}_{ij} = \frac{1}{4\pi} \int_0^{2\pi} d\epsilon \int_0^{\pi} g_{ij} \sin\theta\, d\theta = \frac{1}{2c_i c_j} \int_{c_i \sim c_j}^{c_i + c_j} g_{ij}^2\, dg_{ij}.$$

Therefore

$$\tilde{g}_{ij} = \begin{cases} c_i + c_j^2/3c_i & \text{if} \quad c_i \geq c_j, \\ c_j + c_i^2/3c_j & \text{if} \quad c_i < c_j. \end{cases} \tag{2.75}$$

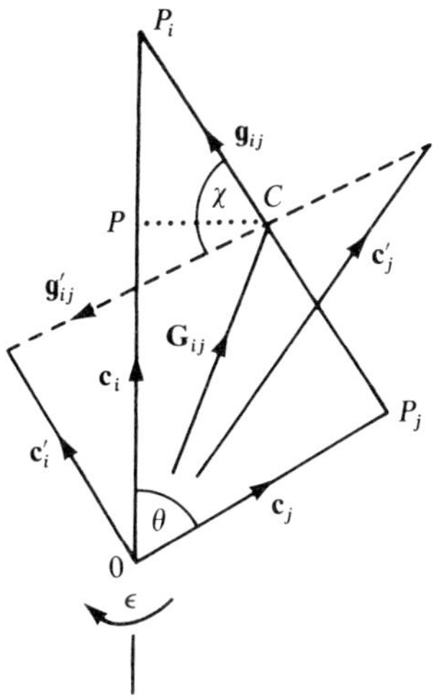

FIG. 2.3. Relative velocities of colliding particles

Next averaging $\tilde{g}_{ij}$ over a Maxwellian distribution of speeds c_j, by (2.47) and (2.75) we obtain

$$\bar{g}_{ij} = \frac{4}{\sqrt{\pi}} \int_0^\infty \tilde{g}_{ij}\, \nu^2 e^{-\nu^2}\, d\nu \qquad (\nu \equiv c_j/\mathcal{C}_j,\ \ \mathcal{C}_j \equiv (2kT/m_j)^{1/2})$$

$$= \frac{4}{\sqrt{\pi}} \int_0^x \left(c_i + \frac{\mathcal{C}_j^2}{3c_i}\nu^2 \right) \nu^2 e^{-\nu^2}\, d\nu + \frac{4}{\sqrt{\pi}} \int_x^\infty \left(\mathcal{C}_j \nu + \frac{c_i^2}{3\mathcal{C}_j \nu} \right) \nu^2 e^{-\nu^2}\, d\nu,$$

where $x = c_i/\mathcal{C}_j$. The integrands contain terms of the type $\nu^r \exp(-\nu^2)$; those with odd r can be integrated via the transformation $\nu^2 = y$, and the others can be reduced by integration by parts and expressed in terms of the error function

$$\Phi(x) \equiv \frac{2}{\sqrt{\pi}} \int_0^x e^{-y^2}\, dy \qquad (\Phi(\infty) = 1). \tag{2.76}$$

The final result is

$$\bar{g}_{ij} = \frac{c_i}{\sqrt{\pi}} \frac{E(x)}{x^2}, \tag{2.77}$$

where

$$E(x) \equiv x e^{-x^2} + \tfrac{1}{2}\sqrt{\pi}(2x^2 + 1)\Phi(x). \tag{2.78}$$

2.9.2 *Collision frequencies*

The collision frequency for a beam molecule can now be expressed as

$$\nu_i = n_j S_{ij} \bar{g}_{ij} = \frac{1}{\sqrt{\pi}} n_j S_{ij} c_i \frac{E(x_j)}{x_j^2} \qquad (S_{ij} = \pi\sigma_{ij}^2,\ \ x_j = c_i/\mathcal{C}_j), \tag{2.79}$$

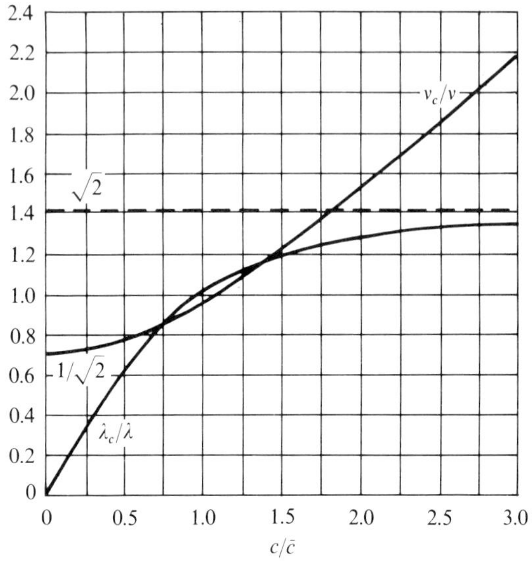

FIG. 2.4. The ratios λ_c/λ and ν_c/ν for a molecule with speed c

where S_{ij} is the mutual cross section for i- and j-molecules. If the background gas is a mixture of several kinds of molecule, (2.79) is generalized to the sum

$$\nu_i = \frac{1}{\sqrt{\pi}} \sum_j n_j S_{ij} c_i \frac{E(x_j)}{x_j^2}. \tag{2.80}$$

When the i and j-molecules are identical, (2.79) gives the value

$$\nu_c = n\pi\sigma^2\overline{g} = \sqrt{\pi}n\sigma^2 cE(x)/x^2$$

for the collision frequency experienced by a molecule moving with velocity c through the background Maxwellian distribution of similar molecules. By (2.42) and (2.71),

$$\nu_c = \frac{\nu}{2\sqrt{2}} \frac{E(x)}{x} \qquad \left(x = \frac{2}{\sqrt{\pi}} \frac{c}{\bar{c}}\right), \tag{2.81}$$

where ν is an average collision frequency. The corresponding mean free path is $\lambda_c = c/\nu_c$, so in terms of the average mean free path, $\lambda = \bar{c}/\nu$, we have

$$\lambda_c = \frac{\sqrt{2\pi}\, x^2}{E(x)} \lambda. \tag{2.82}$$

In Fig. 2.4 the ratios ν_c/ν and λ_c/λ are plotted as functions of $c/\bar{c}$.

2.10 Persistence of velocity

2.10.1 *Definition of persistence*

When a particle experiences a typical collision, its initial velocity is not always reversed, but often continues to have a component in the original direction of motion. A sequence of collisions is then required to fully destroy the initial momentum. Hence the transport of momentum through the gas is associated with a time τ_1 and corresponding distance λ_1, each larger than the collision interval τ and the mean free path λ. This phenomenon is termed the 'persistence of velocity'. An expression for the persistence in a single collision can be calculated as follows.

Referring to Fig. 2.3, we recall that following a collision between hard molecules, the velocity $\mathbf{G}_{ij}$ of their centre of mass (point C in velocity space) is unaltered, whereas the orientation of their relative velocity $\mathbf{g}'_{ij}$ is randomly distributed, i.e. all directions are equally likely. The particle P_j has velocities $\mathbf{c}_i$ and $\mathbf{c}'_i$ before and after the collision, so it is natural to define the persistence ratio ϖ_i as being the average value of $(\mathbf{c}'_i \cdot \mathbf{c}_i)/c_i^2$:

$$\varpi_i = \langle \mathbf{c}'_i \cdot \mathbf{c}_i / c_i^2 \rangle. \tag{2.83}$$

We shall obtain this average in stages.

The first average we shall find is over all directions of $\mathbf{g}'_{ij}$. Post-collision symmetry about C in Fig. 2.3 means that the average of $\mathbf{c}'_i \cdot \mathbf{c}_i / c_i$ equals the speed OP, where P is the foot of the perpendicular from C to the vector $\mathbf{c}_i$. As C divides $\mathbf{g}_{ij}$ in the ratio $m_i : m_j$, we find

$$OP = c_j \cos\theta + \frac{m_i}{m_i + m_j}(c_i - c_j \cos\theta) = \mu_i c_i + \mu_j c_j \cos\theta.$$

Next we average OP over all directions θ of the vector $\mathbf{c}_j$, subject to the constraint that a collision has occurred. The probability of a collision is proportional to the product of the relative speed g_{ij} (which is unchanged by the collision) and the element of solid angle $\sin\theta\, d\theta\, d\epsilon$ in which $\mathbf{c}_j$ falls. Normalizing this probability distribution, we obtain

$$\begin{aligned}\overline{OP} &= \int_0^\pi (\mu_i c_i + \mu_j c_j \cos\theta)\, g_{ij} \sin\theta\, d\theta \Big/ \int_0^\pi g_{ij} \sin\theta\, d\theta \\ &= (\mu_i - \mu_j) c_i + \frac{\mu_i}{2\tilde{g}_{ij}} \int_0^\pi (c_i + c_j \cos\theta) g_{ij} \sin\theta\, d\theta,\end{aligned}$$

where $\tilde{g}_{ij}$ is given by (2.75). Using (2.74) and (2.75) to eliminate θ in favour of g_{ij}, and introducing the speed ratio,

$$\kappa \equiv c_i / c_j, \tag{2.84}$$

we obtain the persistence ratio, $\varpi_{ij}(\kappa) = \overline{OP}/c_i$, or

$$\varpi_{ij}(\kappa) = \mu_i + \tfrac{1}{5}\mu_j \begin{cases} (1 - 5\kappa^2)/\{\kappa^2(3\kappa^2 + 1)\} & (\kappa > 1), \\ (\kappa^2 - 5)/(\kappa^2 + 3) & (\kappa < 1). \end{cases} \tag{2.85}$$

With particles having the same mass, we find that ϖ is related to κ as in Table 2.1.

Table 2.1 *The mean persistence*

κ	0	$\frac{1}{4}$	$\frac{1}{2}$	$\frac{2}{3}$	1	$1\frac{1}{2}$	2	4	∞
ϖ	0.333	0.339	0.354	0.368	0.400	0.441	0.473	0.492	0.500

2.10.2 *Mean persistence ratio*

In order to average the persistence of the ratio κ over the range $(0, \infty)$, we require the distribution of κ. This can be found as follows. The collision frequency per unit volume between the i-molecules in the velocity space element $d\mathbf{c}_i$ and the j-molecules in $d\mathbf{c}_j$ is (see (2.37))

$$\pi\sigma_{ij}^2 g_{ij} f_{0i}\, d\mathbf{c}_i\, f_{0j}\, d\mathbf{c}_j = 16\sigma_{ij}^2 n_i n_j \nu_i^2 \nu_j^2 g_{ij} \exp\{-(\nu_i^2 + \nu_j^2)\} d\nu_i\, d\nu_j\, d\Omega_i\, d\Omega_j,$$

where $\nu_i = c_i/\mathcal{C}_i$, $\nu_j = c_j/\mathcal{C}_j$. On integrating this expression, first over all angles between $\mathbf{c}_i$ and $\mathbf{c}_j$, then over all directions of $\mathbf{c}_i$, we get the collision frequency

$$16\sigma_{ij}^2\, n_i n_j \nu_i^2 \nu_j^2\, \tilde{g}_{ij} \exp\{-(\nu_i^2 + \nu_j^2)\}\, d\nu_i\, d\nu_j,$$

where $\tilde{g}_{ij}$ is given by (2.75). With $\kappa \equiv c_i/c_j = (m_j/m_i)^{1/2}\nu_i/\nu_j$, this frequency becomes

$$\tfrac{16}{3}\sigma_{ij}^2 n_i n_j \left(\frac{m_i}{m_j}\right)^{3/2} \mathcal{C}_j \nu_j^6\, e^{-\nu_j^2(m_i\kappa/m_j+1)} \begin{cases} \kappa(3\kappa^2+1) & (1<\kappa) \\ & d\kappa\, d\nu_j \\ \kappa^2(\kappa^2+3) & (\kappa \le 1). \end{cases}$$

Integrating this over ν_j, which ranges from zero to infinity, and dividing the result by the total collision frequency $\langle \nu_{ij} \rangle$ given in (2.69), we arrive at the distribution function for κ, viz:

$$F(\kappa)\, d\kappa = \frac{5m_i^2 m_j^2}{2(m_i+m_j)^{1/2}} \frac{1}{(m_i\kappa^2+m_j)^{7/2}} \begin{cases} \kappa(3\kappa^2+1) & (1<\kappa<\infty) \\ & d\kappa \\ \kappa^2(\kappa^2+3) & (0<\kappa \le 1). \end{cases} \tag{2.86}$$

The mean persistence ratio now follows upon using the distribution in (2.86) to find the average of (2.85). The two integrals involved are readily evaluated. The result is

$$\varpi_{ij} = \tfrac{1}{2}\mu_i + \tfrac{1}{2}\mu_i^2 \mu_j^{-1/2} \ln\{(\mu_j^{1/2}+1)/\mu_i^{1/2}\}. \tag{2.87}$$

As m_i/m_j increases from zero to infinity, ϖ_{ij} increases from zero to unity. The limiting values are physically evident: very light particles bounce off heavy particles with all directions equally likely ($\varpi_{ij} = 0$), whereas heavy particles are unaffected by collisions with light particles ($\varpi_{ij} = 1$). If the particles have the same mass, $\mu_i = \mu_j = 1/2$, and (2.87) gives

$$\varpi = \tfrac{1}{4}\{1 + \frac{1}{\sqrt{2}} \ln(1 + \sqrt{2})\} \approx 0.406. \tag{2.88}$$

It should be remembered that the above theory applies only to hard molecules. For soft molecules, i.e. those with extended force fields (see discussion in §1.1.1), one needs to define what is meant by a 'collision', and until this is done, the persistence ratio cannot be calculated. For example, if grazing encounters with only slight deflections are reckoned to be 'collisions', the persistence ratio will be just a little smaller than unity. Alternatively, if, as in plasma theory, a collision is taken to be an event that deflects a particle through 90°, the persistence ratio can be negative.

2.11 The Boltzmann distribution law

2.11.1 *Constraints on the macroscopic variables*

The equilibrium distribution (2.26),

$$f_0 = n\left(\frac{m}{2\pi kT}\right)^{3/2} \exp\{-m(\mathbf{w} - \mathbf{v})^2/2kT\}, \tag{2.89}$$

applies at a point $P(\mathbf{r})$. While its derivation did not explicitly restrict the variables n, $\mathbf{v}$, or T to being uniform in space, evidently the entropy production rate must vanish in equilibrium, otherwise the entropy density will not have attained its maximum value. (See derivation in §2.3.1)

Making use of the symmetry of the viscous stress tensor $\boldsymbol{\pi}$, and adopting the expansion (see Mathematical note 1.13)

$$\nabla\mathbf{v} = \overset{\circ}{\mathbf{e}} - \boldsymbol{\Omega} \times \mathbf{1} + \tfrac{1}{3}\mathbf{1}\,\nabla\cdot\mathbf{v} \qquad (\boldsymbol{\Omega} \equiv \tfrac{1}{2}\nabla \times \mathbf{v}), \tag{2.90}$$

we have from (2.17) and (2.18),

$$T\sigma = -\overset{\times}{\pi}\,\nabla\cdot\mathbf{v} - \overset{\circ}{\boldsymbol{\pi}} : \overset{\circ}{\mathbf{e}} - \mathbf{q}\cdot\nabla\ln T \geq 0. \tag{2.91}$$

As the 'thermodynamic forces' $\nabla\cdot\mathbf{v}$, $\overset{\circ}{\mathbf{e}}$ and ∇T tend independently to zero, so also must the associated 'fluxes' $\overset{\times}{\pi}$, $\overset{\circ}{\boldsymbol{\pi}}$ and $\mathbf{q}$, otherwise the inequality could not be satisfied for *all* values of the gradients. Thus, for small deviations from equilibrium, the inequality implies that there are linear relations between the fluxes and the gradients. It follows that, in general, equilibrium is possible only if

$$\nabla\cdot\mathbf{v} = 0, \qquad \overset{\circ}{\mathbf{e}} = 0, \qquad \nabla T = 0. \tag{2.92}$$

The qualification 'in general' is required because $\overset{\times}{\pi}$ has its principal origin in a relaxation process involving the transfer of energy between the internal structure of a molecule and its kinetic energy of translation (see Woods 1986). Thus with monatomic molecules, this flux is zero at all times, and the first constraint in (2.92) is not required.

By (1.94) and (1.95) the restrictions on the fluid velocity in (2.92) are met if the velocity is given by

$$\mathbf{v} = \mathbf{v}_0 + \boldsymbol{\Omega} \times \mathbf{r} \qquad (\boldsymbol{\Omega} \equiv \tfrac{1}{2}\nabla \times \mathbf{v} = \text{const.}), \tag{2.93}$$

where $\mathbf{v}_0$ is arbitrary constant. By (2.90) and (2.92), in steady conditions

$$\mathrm{D}\mathbf{v} = \mathbf{v} \cdot \nabla \mathbf{v} = \boldsymbol{\Omega} \times \mathbf{v} = \boldsymbol{\Omega} \times \mathbf{v}_0 + \boldsymbol{\Omega} \times (\boldsymbol{\Omega} \times \mathbf{r}) = \nabla \Psi, \tag{2.94}$$

where
$$\Psi = \boldsymbol{\Omega} \times \mathbf{v}_0 \cdot \mathbf{r} + \tfrac{1}{2}\{\boldsymbol{\Omega} \times (\boldsymbol{\Omega} \times \mathbf{r})\} \cdot \mathbf{r},$$

which is easily verified with the help of the relation $\nabla \mathbf{r} = \mathbf{1}$.

2.11.2 *Conservative body forces*

The third of (2.92) requires a uniform temperature through the gas, so that in equilibrium,

$$\nabla \cdot \mathsf{p} = \nabla \cdot (p\mathbf{1}) = kT\nabla n. \tag{2.95}$$

We shall assume that the body force $\mathbf{F}$ in the equation of fluid motion, (1.47), has a potential Φ, i.e. $\mathbf{F} = -\nabla \Phi$. Then by (2.94) and (2.95) this equation can be written

$$\nabla \Psi + RT\nabla \ln n = -\nabla \Phi,$$

whence
$$n = n_0 \exp\left\{-\frac{m}{kT}(\Phi + \Psi)\right\}, \tag{2.96}$$

where n_0 corresponds to zero potential. By (1.41) and (2.93) the density gradient must be orthogonal to $\mathbf{v} = \mathbf{v}_0 + \boldsymbol{\Omega} \times \mathbf{r}$. Although (2.96) was first derived by Maxwell, it is generally known as 'Boltzmann's distribution law'. Combining it with (2.89), we obtain

$$f_0 = n_0 \left(\frac{m}{2\pi kT}\right)^{3/2} \exp\left\{-\frac{m}{kT}\left(\tfrac{1}{2}c^2 + \Phi + \Psi\right)\right\}. \tag{2.97}$$

2.12 Effusion and transpiration of gases

2.12.1 *Effusion*

Consider the flux of molecules through an element of area dS in the OXY-plane and located at the origin O, as shown in Fig. 2.5. In spherical polar coordinates (r, θ, ϕ), the element of volume at a point P is $r^2 \sin\theta \, d\theta \, d\phi \, dr$, and the solid angle subtended at P by dS is $dS \cos\theta / 4\pi r^2$. Assuming an isotropic distribution in direction of motion of the molecules at P, there will be $(dS \cos\theta / 4\pi r^2)(f \, d\mathbf{c}) \, r^2 \sin\theta \, d\theta \, d\phi \, dr$ molecules lying in the volume element at P and with velocities in the velocity element $d\mathbf{c}$, that, in the absence of collisions, will pass through dS at some later time. Thus the total rate at which molecules pass through dS in the negative OX-direction is

$$\frac{dS}{4\pi}\left(\int cf \, d\mathbf{c}\right) \int_0^{\pi/2} \int_0^{2\pi} \sin\theta \cos\theta \, d\theta \, d\phi = \tfrac{1}{4}\bar{c} n \, dS.$$

Hence the mass flux per unit area, say $\mathcal{F}$, is $\frac{1}{4}nm\bar{c} = \frac{1}{4}\rho\bar{c}$. Thus by (2.42),

$$\mathcal{F} = \tfrac{1}{4}\rho\bar{c} = \rho\sqrt{\frac{kT}{2\pi m}} \approx 0.40\rho\sqrt{RT}. \tag{2.98}$$

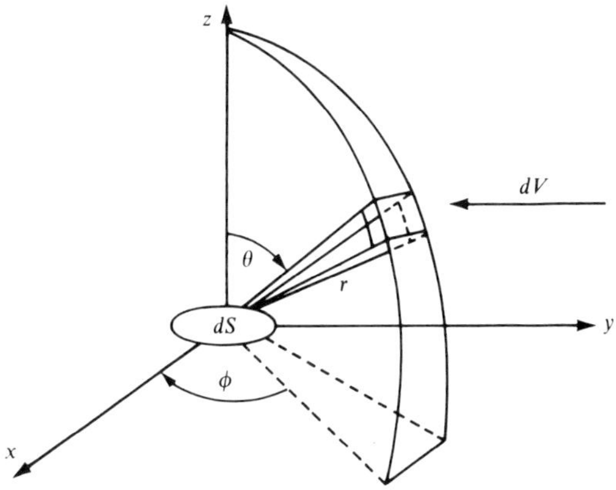

FIG. 2.5. Efflux through a small hole

We can interpret $\mathcal{F}$ as being the rate of efflux of a collisionless gas through an aperture into a vacuum. It is known to be accurate provided the mean free path exceeds the diameter of the hole by a factor of about ten.

Fluid pressure is negligible in effusion. When pressure is important, a classical result in gas dynamics is that the mass flux per unit area through a nozzle (choked flow) is

$$\mathcal{F}_c = \left\{\gamma\left(\frac{2}{\gamma+1}\right)^{(\gamma+1)/(\gamma-1)}\right\}^{1/2}\rho\sqrt{RT} \approx 0.68\rho\sqrt{RT}, \tag{2.99}$$

the approximate result being for air ($\gamma \approx 1.4$). Comparison of (2.98) and (2.99) shows that pressure considerably increases the exit velocity of the molecules. In a description at the microscopic level, the escaping molecules are hastened in their transit through the aperture by collisions with molecules coming after them. This provides a good example of the relationship between pressure and particle collisions, and shows that without collisions the pressure is zero (see discussion in §1.5.)

2.12.2 *Transpiration*

If the single aperture is replaced by the large number of orifices provided by the interstices of a plug of porous material, and if particle collisions can be ignored, equation (2.98) still applies, at least approximately, and the phenomenon is called 'transpiration'. If, as shown in Fig. 2.6, a vessel is divided into parts A and B by a porous wall, it follows that equilibrium is reached when

$$\frac{p_A}{p_B} = \left(\frac{T_A}{T_B}\right)^{1/2}. \tag{2.100}$$

If the pressure in B is initially zero, the molecules will not—on the above theory—lose energy on crossing the plug, i.e. in the final state, $T_A = T_B$.

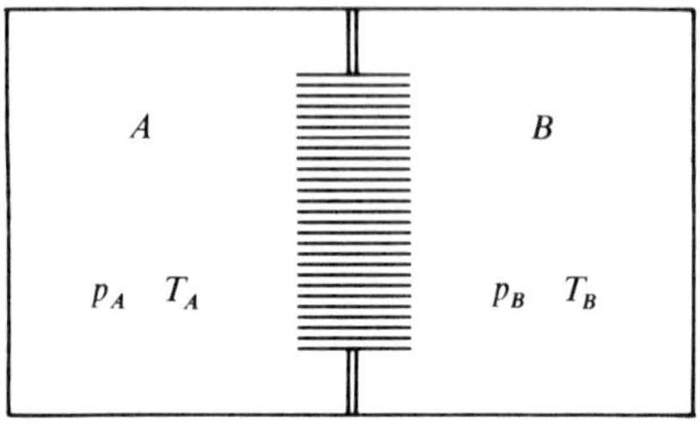

FIG. 2.6. Transpiration

However, this theory ignores (i) the cohesion of the gas, which reduces the kinetic energy of the molecules by providing an attractive force to overcome, and so makes $T_A < T_B$, and (ii) collisions that accelerate the molecules as they move into B, and makes $T_A > T_B$. The net change of temperature may be positive or negative, and is called the Joule–Thomson effect.

Exercises 2

2.1 Show that f satisfies $\int \mathbf{c} f \, d\mathbf{w} = 0$, and include this constraint in the derivation of f_0 given in §2.5.

2.2 Prove that the stationary value of the entropy density given by $f = f_0$ is a maximum.

2.3 Find the probability that a particle chosen at random has its x-component of speed in $c_x, c_x + dc_x$. What fraction of molecules of a gas have positive x-components greater than $2\mathcal{C}$? (Answer: 1.6×10^{-12}).

2.4 Show that the most probable energy of molecules having a Maxwellian velocity distribution is $\frac{1}{2}kT$.

2.5 What fraction of the molecules of a gas have speeds exceeding some given value v_0? (Answer: $n[1 - \Phi(x) + 2\pi^{-1/2} x \exp(-x^2)]$, $x = v_0/\mathcal{C}$).

2.6 Let

$$\varphi \equiv -\tau_2(\nu^2 - 5/2)\mathbf{c} \cdot \mathbf{A} - \tau_1 \boldsymbol{\nu\nu} : \overset{\circ}{\mathbf{B}},$$

(the notation being as in 2.6.1), where $\mathbf{A}$ is a constant vector and $\mathbf{B}$ is a constant tensor. Prove that

$$\int f_0 \varphi \{1, \mathbf{c}, c^2\} \, d\mathbf{c} = 0,$$

and deduce that $f_0(1 + \varphi)$ is a possible form for the distribution function f.

2.7 Show that the energy flux can be expressed as

$$\mathbf{q} = p\mathcal{C} \frac{4}{\sqrt{\pi}} \int_\Omega \int_0^\infty \varphi \nu^5 \exp(-\nu^2) \, d\nu \, \hat{\mathbf{c}} \, \mathbf{d\Omega},$$

where $\varphi = (f - f_0)/f_0$ and $\hat{\mathbf{c}}$ is unit vector parallel to $\mathbf{c}$. Use the result in exercise 2.6 and equation (1.79) to show that $\mathbf{A} = \nabla \ln T$ and $\kappa = 5kp\tau_2/2m$.

2.8 Following the method of exercise 2.7, show that $\mathbf{B} = \overset{\circ}{\nabla \mathbf{v}}$ and $\mu = p\tau_1$, where $\mathbf{B}$ is the tensor defined in exercise 2.6.

2.9 Show that in the absence of collisions, the rms speed of the molecules that escape through an aperture is $(4/3)^{1/2}$ times the rms speed of those that do not escape.

3

ELEMENTARY KINETIC THEORY

3.1 Maxwell's two transport theories

Calculating the transport of fluid momentum and energy by what Maxwell termed 'molecular agitation' has been a central problem in macroscopic physics for more than a century. And while its solution for neutral gases was largely completed by the early part of this century, it is only recently that progress has been made for magnetoplasmas.

There are two distinct methods of determining transport, both of which are due to Maxwell. The first is non-local and depends on the *displacement* of molecules between collisions; the particle agitation tends to equalize conditions—temperature, fluid velocity, etc.—at the two ends of each free path by transporting properties from one end to the other. The second is a local theory, based on a differential equation similar to (1.38), involving the velocity distribution function $f(\mathbf{r}, \mathbf{w}, t)$, and expressing the conservation of molecular properties. In Boltzmann's extension of the local theory the property is taken to be f itself. Once f is known, the required transport laws for $\mathbf{q}$ and p follow from (2.5), (2.6), and (2.7). Maxwell avoided determining f, aiming instead directly at the required velocity moments.

3.1.1 *Using the mean free path*

We shall illustrate Maxwell's first theory, published in *Phil. Mag.* **19**, 1860, in the article 'Illustrations of the Dynamical Theory of Gases', by considering viscosity. In a neutral monatomic gas, with small departures from equilibrium, the viscous stress tensor is proportional to the deviatoric rate of strain (see (1.77), (1.81), and the description following (2.91)):

$$\boldsymbol{\pi} = -2\mu \overset{\circ}{\nabla} \mathbf{v} \qquad (\mu \geq 0). \tag{3.1}$$

This relation is a generalization of Newton's formula for the stress in sheared flow. Its *form* does not depend on kinetic theory—it could be inferred as a phenomenological law from the non-negative dissipation $-\boldsymbol{\pi} : \nabla \mathbf{v}$, on assuming $\boldsymbol{\pi}$ to have zero trace (see (2.91)). It is one of the roles of kinetic theory to deduce an expression for the shear viscosity μ from the intermolecular force law for the gas in question. But in fact things usually run in the opposite direction—one uses empirical knowledge about μ to determine an appropriate force law, or with hard molecules to determine their diameters.

Maxwell's first theory used Clausius' concept of the mean free path, λ.

With its help he arrived at the formula

$$\mu = \tfrac{1}{3}\rho\bar{c}\lambda \qquad (\lambda = 1/(\sqrt{2}\pi n\sigma^2)), \tag{3.2}$$

the value of λ being that calculated in §2.8.2. Notice that μ is independent of n. Maxwell was surprised at this. He wrote

A remarkable result here presented to us in this equation..., is that if this explanation of gaseous friction be true, the coefficient of viscosity is independent of the density. Such a consequence of mathematical theory is very startling, and the only experiment I have met with on the subject does not seem to confirm it.

But some years later Maxwell verified his theory on this point by conducting his own experiments.

3.1.2 *The coefficient of viscosity*

A brief description of the derivation of (3.2) is as follows. (A full treatment will be given in §3.6.3.) Suppose that the fluid velocity is perpendicular to OZ and sheared along OZ. By (2.98) the flux of molecules with a component of peculiar velocity parallel to OZ is proportional to $n\bar{c}$. Between collisions they are displaced along OZ a distance proportional to λ, and since on average they have the same velocity perpendicular to OZ as the fluid, their peculiar velocity (Maxwell's 'agitation') transports momentum along OZ at a rate per unit area proportional to $n\bar{c}\lambda\, d(mv)/dz$. This gives rise to a friction force $\mu\, dv/dz$, so that, with the appropriate constants of proportionality, μ is given by (3.2).

Substituting $\bar{c}$ from (2.42), we can write (3.2) as

$$\mu = a\,\frac{1}{3}\,\frac{2}{\pi\sigma^2}\left(\frac{mkT}{\pi}\right)^{1/2}, \tag{3.3}$$

where a is unity. As will be shown in §6.8.2, the first step in advanced transport theory replaces $a = 1$ by $a = 15\pi/32$. The complete theory supplies a further factor of 1.0160, so that $a \approx 1.4962$, and (3.2) is replaced by

$$\mu = 0.499\rho\bar{c}\lambda. \tag{3.4}$$

Thus, while the mean-free-path method gives the correct functional dependence for μ, it appears to be unable to give the correct constant. The reason for this failure will be described in §3.3.2.

3.1.3 *Maxwell's transfer equation*

Appreciating the defects of mean-free-path methods, which became particularly obvious with energy transport, in his 1867 paper 'On the Dynamical Theory of Gases' Maxwell introduced a very different approach, with the distribution function $f(\mathbf{r}, \mathbf{w}, t)$ in the key role. His new method was based on his equation of transfer for a molecular property ϕ. This equation is (1.38), which by (1.30), (1.32) and $\rho = mn$, can be expressed as

$$\frac{\partial}{\partial t}(n\overline{\phi}) + \nabla\cdot(n\mathbf{v}\overline{\phi} + n\overline{\mathbf{c}\phi}) = n\,\frac{\delta\phi}{\delta t}, \tag{3.5}$$

using bars to denote averages and writing the source for ϕ—the right-hand term—in Maxwell's notation. In this form the source includes losses due to body force accelerations (cf. Table 1.1, p. 15).

Maxwell calculated $\delta\phi/\delta t$ for molecules for which the repulsive force between pairs a distance r apart is of the form $Kr^{-\nu}$, where K and ν are constants. Then came a stroke of great fortune. He found that the relative speed g occurred in his formula for $\delta\phi/\delta t$ in the form $g^{(\nu-5)/(\nu-1)}$ so that the choice $\nu = 5$ gave a dramatic simplification. In particular it enabled him to bypass the difficult problem of determining f, a problem that was to remain unsolved, even approximately, for more than forty years. Maxwell justified setting $\nu = 5$ by reference to some experiments on viscosity, which he believed showed that $\mu \propto T$ (see §3.2.1). But 'Maxwellian molecules', as this model is now labelled, are only rarely found to adequately represent real gases (see Table 3.1). Nevertheless, some of the relations that can be deduced from the model are surprisingly close to those applying with other force laws.

The next advance was made by Boltzmann, who in 1872 published a non-linear, integro-differential equation for the distribution function f and used it to establish the uniqueness and stability of Maxwell's equilibrium distribution. We shall describe this work in Chapter 7.

3.2 Molecules represented by point centres of force

3.2.1 *The inverse power law*

For intermolecular force laws of the type $Kr^{-\nu}$, where r is the distance between the molecules' centres and K, ν are constants, an effective molecular diameter σ_{eff} can be defined as follows. In a head-on encounter between two molecules, we can define a 'diameter' σ to be the distance of closest approach. In such a collision, the kinetic energy $\frac{1}{2}mg^2$ of the relative motion is transformed into the potential energy $K\sigma^{-\nu+1}$, i.e. $\frac{1}{2}mg^2 = K\sigma^{-\nu+1}$. For collisions between molecules in an equilibrium distribution, $\overline{g^2} \propto T$ (cf. (2.44) and exercise 3.6). Hence

$$\sigma = \left(\frac{2K}{m}\right)^{1/(\nu-1)} g^{-2/(\nu-1)}, \qquad \sigma_{eff} \propto T^{-1/(\nu-1)}. \tag{3.6}$$

The collision interval for hard-core molecules, say $\tau_0 = \lambda/\overline{c}$, follows from (2.71), which gives $n\tau_0 \propto (\overline{c}\sigma^2)^{-1}$. We may therefore introduce an effective collision interval τ for the point centres of force, which from (3.6) and $\overline{c} \propto T^{1/2}$, has a temperature dependence given by

$$n\tau \propto T^{S-1} \qquad \left(s \equiv \frac{1}{2} + \frac{2}{\nu - 1}\right). \tag{3.7}$$

Hard-core molecules are obtained in the limit $\nu = \infty$, when $s = 1/2$.

Equation (3.2) can be written $\mu \propto \rho\overline{c}^2\tau_0 \propto p\tau_0$, and it will be shown later (§3.6.2) that in the general case $\mu \propto p\tau$. This is also true of the thermal conductivity κ defined in (1.79):

$$\mu = \text{const. } p\tau, \qquad \kappa = \text{const. } c_v p\tau. \tag{3.8}$$

Much of 19th century research in the kinetic theory of gases was directed at finding accurate values for the two constants appearing in (3.8).

On combining (3.7) and (3.8), we obtain the laws

$$\mu = A_1 T^s, \qquad \kappa = A_2 T^s \qquad (A_1, A_2 \text{ constants}). \tag{3.9}$$

With Maxwellian molecules, $\nu = 5$, $s = 1$, which makes $n\tau$ a constant and μ and κ proportional to T. The relation for μ was derived in 1900 by Lord Rayleigh using dimensional analysis (see exercise 3.3).

Estimates of s, and hence of ν, can be made from measurements of μ at different temperatures. The viscosity of all gases is found to rise with temperature at a rate more rapid than the hard-core law, $\mu \propto T^{1/2}$. The power law is good in some cases, e.g. for hydrogen, helium, and neon the values quoted for s in Table 3.1 hold over the temperature range 200K to 500K, whereas with argon, nitrogen and carbon dioxide the law is satisfactory only over a more restricted range. For most gases s decreases with increasing temperature; thus as molecules collide more vigorously, penetrating further into each others' repulsive fields, they behave more like hard molecules.

3.2.2 *Other force laws*

The last observation suggests a viscosity law of the type

$$\mu = \frac{5}{16\sigma^2}\left(\frac{mkT}{\pi}\right)^{1/2} \Big/ \left(1 + \frac{S}{T}\right), \tag{3.10}$$

where S is known as Sutherland's constant, after the author of the model. The numerator follows from (3.3) with $a = 15\pi/32$, and the modification is equivalent to adding a weak attractive force between hard spheres. Although fairly successful for some gases, e.g. for neon (S= 64), it fails badly for others, such as hydrogen and helium.

Several generalizations of the power law have been proposed, a well-known one being due to Lennard-Jones. His expression for the potential energy of interaction takes into account both the softness of molecules and their mutual attraction at large distances. It is

$$V(r) = 4\epsilon\left\{(\sigma/r)^{12} - (\sigma/r)^6\right\} \qquad (\epsilon, \sigma \text{ constants}). \tag{3.11}$$

By appropriate choices of ϵ and σ, fair agreement can be achieved with the experimental values of transport properties for many gases.

The alternative to assigning values derived from experiments to ϵ and σ—and similar parameters in other force laws—is the solution of the very difficult quantum mechanical many-body problem, required to find the short range interaction potential. It is only with simple atoms like hy-

Table 3.1 *Some molecular data at STP*
(μ and s from experiment, λ from (3.4) and σ from (3.2))

Gas	μ (10^{-7}poise)	λ (10^{-6}cm)	σ (10^{-8}cm)	s	ν
hydrogen	871	11.8	2.74	0.668	12.9
helium	1943	18.6	2.18	0.657	13.7
methane	1077	5.16	4.14	0.836	7.0
acetylene	970	4.51	4.43	0.998	5.0
neon	3095	13.22	2.59	0.661	13.4
nitrogen	1734	6.28	3.75	0.738	9.4
ethane	900	3.15	5.30	0.970	5.26
oxygen	2003	6.79	3.61	0.773	8.4
argon	2196	6.66	3.64	0.811	7.5
carbon dioxide	1448	4.19	4.59	0.933	5.6
methyl bromide	1310	2.58	5.85	1.10	4.33

drogen and helium that even an approximate calculation is possible, and while the results obtained are qualitatively correct, they are quantitatively incorrect. The only case where a phenomenological input can be avoided—at least at the present time—is with interactions between charged particles, where the long-range Coulomb force ($\nu = 2$ in (3.7)) is dominant. There is some irony in the fact that in this case, where the force law is unequivocal, the ready appearance of turbulence in plasmas cancels this advantage in many situations.

3.3 The effective collision interval

3.3.1 *Relaxation times*

The question of what value to assign to the effective collision time appearing in (3.7) and (3.8) remains to be discussed. Consider viscosity. The relevant microscopic time scale is the time it takes a typical molecule P to lose an excess—or to gain a deficit—of transverse momentum, δM say, as it migrates across layers of sheared fluid. With 'soft' molecules, e.g. those with $\nu < 10$ or so (see Table 3.1), the force fields have a large range, making grazing collisions much more common than those giving a substantial change in direction. A large number of grazing collisions in sequence will be required to make an appreciable reduction in δM, and the mean-free-path concept loses its significance in this situation.

A method more appropriate to soft molecules is to identify an 'effective collision time' with the relaxation time τ_ϕ for the loss of an excess $\delta\phi$ of a property ϕ from a bunch of molecules as they move with a velocity c relative to a convected frame. This time is defined by

$$\tau_\phi \equiv \left| \frac{d}{dt} \ln \delta\phi \right|^{-1}, \tag{3.12}$$

so that
$$\frac{d(\delta\phi)}{dt} = -\frac{\delta\phi}{\tau_\phi}. \tag{3.13}$$

Alternatively, we may adopt the relaxation length, $\lambda_\phi = c\tau_\phi$,
$$\frac{d(\delta\phi)}{dx} = -\frac{\delta\phi}{\lambda_\phi} \qquad (x = ct), \tag{3.14}$$
as an effective 'mean free path' for the property ϕ. Of course (3.13) and (3.14) represent only the relaxation process $\delta\phi \to 0$; in general, complete expressions for $d(\delta\phi)/dt$ and $d(\delta\phi)/dx$ will also contain some source terms that prevent $\delta\phi$ from vanishing.

The microscopic scales τ_ϕ and λ_ϕ can either be assumed to be phenomenological, and hence inferred from experiment, or they may be deduced from a deeper theory based on assumed force laws between colliding molecules. But even with the latter seemingly more theoretical approach, in practice a phenomenological element is generally unavoidable, as remarked in the final paragraph of §3.2.2.

Let $\delta\phi_0$ denote the value of $\delta\phi$ at $t = 0$, $x = 0$, then (3.14) has the solution
$$\delta\phi = \delta\phi_0 \exp(-t/\tau_\phi) = \delta\phi_0 \exp(-x/\lambda_\phi), \tag{3.15}$$
allowing (3.13) and (3.14) to be written
$$d(\delta\phi) = -\delta\phi_0 \exp(-t/\tau_\phi)\, dt/\tau_\phi. \tag{3.16}$$
In (3.16) $\exp(-x/\lambda_\phi)$ is the probability that a given molecule transports the property ϕ through a distance x, and dx/λ_ϕ is the probability that this molecular property is lost in the interval dx. Hence $\exp(-x/\lambda_\phi)\, dx/\lambda_\phi$ is the joint probability that the property in question reaches x and is lost in dx. For example, take ϕ to be the number of molecules in a gas of hard-core molecules, moving in a given stream with velocity c. In this case λ_ϕ is the usual mean free path, λ_0 say, and the probability $P(x)\, dx$ that a given molecule reaches x and then has a collision in dx is
$$P(x)\, dx = \exp(-x/\lambda_0)\, dx/\lambda_0 \qquad (0 < x < X), \tag{3.17}$$
where X is the distance to the boundary of the gas. It is usually the case that $X \gg \lambda_0$, allowing X to be replaced by infinity. The time scale associated with λ_0 is $\tau_0 = \lambda_0/\bar{c}$.

3.3.2 *Velocity persistence*

Because of velocity persistence (see §2.10), the time scale for momentum transport by hard-core molecules is not τ_0, but a slightly longer time, $\tau_1 \approx 5\tau_0/4$. This can be established as follows.

Suppose that a fluid flow in planes normal to the axis OZ is linearly sheared in the OZ-direction, and that at O a 'typical' molecule P is one with zero peculiar velocity in these planes, i.e. is moving with the fluid. Its initial peculiar velocity $\mathbf{c}$ is parallel to OZ and on average its first collision will occur at a time τ_0 later, i.e. on the plane $z = c\tau_0$. Because of persistence ϖ in its velocity along OZ, P will experience a second collision on $z = c\tau_0 + \varpi c\tau_0 = c\tau_0(1 + \varpi)$, but since P's excess momentum δM is now being carried and dispersed by *two* molecules, the corresponding relaxation time is not $\tau_0(1 + \varpi)$ but $\tau_0(1 + \frac{1}{2}\varpi)$. In this calculation we are assuming that, on average, δM is shared equally between the two molecules. The third collision for P occurs on $z = c\tau_0(1+\varpi+\varpi^3)$, but now *four* molecules are involved in dispersing δM, giving an increment $\frac{1}{4}\varpi^2\tau_0$ to the relaxation time. Continuing this cascade process, we obtain

$$\tau_1 = \tau_0\{1 + \tfrac{1}{2}\varpi + \tfrac{1}{4}\varpi^2 + \cdots + \left(\tfrac{1}{2}\varpi\right)^n + \cdots\},$$

i.e.
$$\tau_1 = \frac{\tau_0}{1 - \frac{1}{2}\varpi} \approx 1.255\tau_0 \approx \tfrac{5}{4}\tau_0, \tag{3.18}$$

where we have adopted the average value of ϖ, namely 0.406, given by (2.88).

Maxwell's formula (3.2) can be written $\mu = \frac{1}{3}\rho\bar{c}^2\tau_0$, the form of which suggests that a more accurate expression would be*

$$\mu = \tfrac{1}{3}\rho\overline{c^2}\tau_1.$$

By (2.44), (2.45), and (3.18), this can be written $\mu = \frac{1}{3}a\rho\bar{c}\lambda_0$, where $a = 15\pi/32$, in agreement with the first approximation of advanced kinetic theory, and less than 2 per cent in error from the exact result given in (3.4). It follows that the original elementary theory is inaccurate for two reasons: it adopts the wrong average of the molecular velocities, and it neglects velocity persistence.

3.3.3 *Energy relaxation time*

The infinite cascade of collisions that dissipates the momentum excess δM also disperses most of the excess kinetic energy, $\delta E = \delta(\frac{1}{2}mc_z^2)$ directed along OZ. However, after a time τ_1, this energy will be distributed isotropically over three coordinates, according to

$$\delta E = \tfrac{1}{2}mc_x'^2 + \tfrac{1}{2}mc_y'^2 + \tfrac{1}{2}mc_z'^2 = \left(\tfrac{2}{3}\delta E\right)_{x,y} + \left(\tfrac{1}{3}\delta E\right)_z.$$

Thus only $\frac{2}{3}\delta E$ is actually dispersed out of the original beam in the time τ_1. It follows from (3.14) that

$$\frac{d(\delta E)}{dt} = -\frac{\frac{2}{3}\delta E}{\tau_1} = -\frac{\delta E}{\tau_2},$$

where τ_2 is the energy relaxation time. Hence by (3.2),

$$\tau_2 = \tfrac{3}{2}\tau_1 \approx \tfrac{15}{8}\tau_0 \qquad \left(\tau_0 = \lambda_0/\bar{c} = 1/(\sqrt{2}\pi n\sigma^2\bar{c})\right), \tag{3.19}$$

relations which prove to be quite accurate (see §3.6).

*That this is the correct method of averaging will be confirmed in §3.6.1.

The increment $\frac{1}{2}\tau_1$ to the energy relaxation time can be described as being due to 'energy persistence'. After the time τ_1 has elapsed—we refer here to the behaviour of an 'average' molecule—the beam molecules will be scattered by the ambient molecules into an isotropic distribution, having given up all their initial excess momentum. At first sight it appears that the beam's initial *directed* energy should also have been dispersed, but what in fact happens is that some of this energy continues to be transported in the original direction by the collided ambient molecules.

When the beam molecules are much lighter than the ambient molecules, and collisions between beam molecules are infrequent enough to be neglected, both velocity and energy persistence can be ignored. The result for velocity follows directly from (2.87); that for energy is a consequence of the fact that light molecules colliding with heavy molecules lose very little energy as a result (cf. equations in §1.3.1).

3.4 The peculiar velocity: no temperature gradients

At present we shall ignore molecular accelerations due to temperature gradients, dealing with this matter in the next section. We have now reached the central topic in elementary kinetic theory, namely the determination of molecular motions over distances of the order of a mean free path.

3.4.1 *Frame-indifference of the heat flux vector and the pressure tensor*

In §1.2 we defined the peculiar velocity $\mathbf{c}$ of a molecule P to be the velocity of P measured in a *convected* frame P_c, i.e. one that is carried bodily with the fluid, and whose axes rotate with the local angular velocity of the fluid. In §1.13 we showed that this spin $\boldsymbol{\Omega}$ equals half of the fluid vorticity,

$$\boldsymbol{\Omega} = \tfrac{1}{2}\nabla \times \mathbf{v}, \tag{3.20}$$

where $\mathbf{v}$ is the fluid velocity measured in the laboratory frame L. First we shall explain why it is necessary to add spin to point convection in defining the peculiar velocity.

In a laboratory frame L$'$ that rotates with an angular velocity $\boldsymbol{\omega}$ relative to L, the fluid velocity is $\mathbf{v}'$, where (see Fig. 3.1) $\mathbf{v}' = \mathbf{v} + \boldsymbol{\omega} \times \mathbf{r}$. In L$'$ the spin is

$$\boldsymbol{\Omega}' = \tfrac{1}{2}\nabla \times \mathbf{v}' = \boldsymbol{\Omega} - \tfrac{1}{2}\nabla \times (\boldsymbol{\omega} \times \mathbf{r}) = \boldsymbol{\Omega} - \tfrac{1}{2}(\boldsymbol{\omega}\nabla \cdot \mathbf{r} - \boldsymbol{\omega} \cdot \nabla\mathbf{r}),$$

and since $\nabla \cdot \mathbf{r} = 3$ and $\nabla\mathbf{r} = \mathbf{1}$, it follows that

$$\boldsymbol{\Omega}' = \boldsymbol{\Omega} - \boldsymbol{\omega}. \tag{3.21}$$

Thus the spin is 'frame-dependent', that is, its magnitude and direction depend on the choice of laboratory frame. It follows that the only way of defining $\mathbf{c}$ that leaves it and its time derivative independent of the reference frame, is to specify it in that frame in which the spin is zero; this is the frame that we have termed a 'convected frame' in §1.2.

When the distribution function f is expressed as a function of $(\mathbf{r}, \mathbf{c}, t)$,

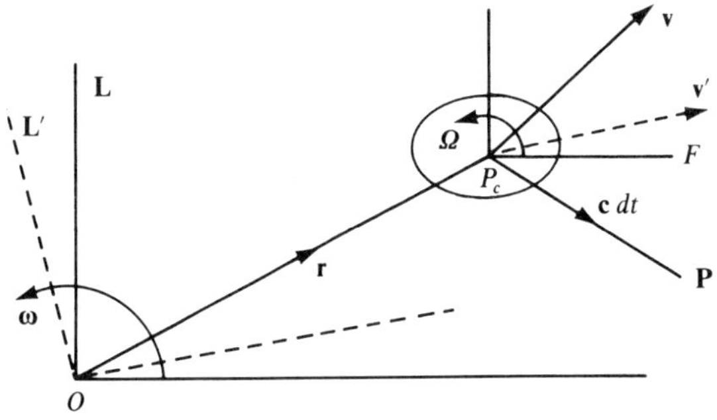

FIG. 3.1. Convected frame

the heat flux $\mathbf{q}$ and the pressure tensor p (see §2.1) follow from

$$\mathbf{q} = \int \tfrac{1}{2} m c^2 \mathbf{c}\, f\, d\mathbf{c}, \tag{3.22}$$

and

$$\mathsf{p} = \int m \mathbf{c}\mathbf{c}\, f\, d\mathbf{c}. \tag{3.23}$$

Since $\mathbf{c}$ is frame-indifferent, and f—being a scalar—is likewise, it follows from (3.22) and (3.23) that $\mathbf{q}$ and p also have this property. By the definition of diffusion given in §1.7.1, this identifies $\mathbf{q}$ and p as being purely diffusive processes.

3.4.2 *Rate of change of the peculiar velocity*

Corresponding to the peculiar velocity $\mathbf{c} = \mathbf{w} - \mathbf{v}$, we now introduce the 'agitation' acceleration, $\mathcal{F}$, defined as being the difference between the particle acceleration $\dot{\mathbf{w}}$ and its value averaged over all the particles of the same species:

$$\mathcal{F} \equiv \dot{\mathbf{w}} - \overline{\dot{\mathbf{w}}}, \qquad \overline{\mathcal{F}} = 0. \tag{3.24}$$

Long-range forces like those due to gravity, macroscopic electric potentials and so on, affect $\dot{\mathbf{w}}$ and $\overline{\dot{\mathbf{w}}}$ alike and cannot contribute to $\mathcal{F}$. This acceleration is due to the short-range, impulsive forces due to molecular collisions. The fluid acceleration $\mathrm{D}\mathbf{v}$ is equal to the average value of the particle acceleration, so that (3.24) can be written

$$\mathcal{F} = \dot{\mathbf{w}} - \mathrm{D}\mathbf{v}, \qquad \overline{\dot{\mathbf{w}}} = \mathrm{D}\mathbf{v} \qquad \left(\mathrm{D} \equiv \frac{\partial}{\partial t} + \mathbf{v}\cdot\nabla\right). \tag{3.25}$$

A particle P leaves the convected frame P_c with an initial peculiar velocity $\mathbf{c}$ relative to it. Let $d\mathbf{c}$ denote the change in P's velocity after a time dt, as seen in P_c, a frame that rotates with angular velocity $\boldsymbol{\Omega}$ relative to the laboratory frame L. Thus measured in L, the total change in $\mathbf{c}$ is $d\mathbf{c} + \boldsymbol{\Omega} \times (\mathbf{c}\, dt)$ (see Fig. 3.1). This equals $d\mathbf{w} - d\mathbf{v}$, i.e.

$$d\mathbf{c} + \boldsymbol{\Omega} \times \mathbf{c}\, dt = d\mathbf{w} - d\mathbf{v} = \dot{\mathbf{w}}\, dt - \left(\frac{\partial \mathbf{v}}{\partial t} + \mathbf{w} \cdot \nabla \mathbf{v}\right) dt$$

$$= \dot{\mathbf{w}}\, dt - (\mathrm{D}\mathbf{v} + \mathbf{c} \cdot \nabla \mathbf{v})\, dt,$$

or

$$d\mathbf{c} = (\dot{\mathbf{w}} - \mathrm{D}\mathbf{v})\, dt - \mathbf{c} \cdot (\nabla \mathbf{v} + \boldsymbol{\Omega} \times \mathbf{1}) dt.$$

By (3.25) and (1.91) this can be written

$$d\mathbf{c} = -\mathbf{c} \cdot \mathbf{e}\, dt + \mathcal{F}\, dt \qquad \left(\mathbf{e} \equiv \tfrac{1}{2}(\nabla \mathbf{v} + \widetilde{\nabla \mathbf{v}})\right), \tag{3.26}$$

whence

$$\dot{\mathbf{c}} = -\mathbf{c} \cdot \mathbf{e} + \mathcal{F}, \tag{3.27}$$

giving the rate of change of $\mathbf{c}$ in the frame P_c. Notice that the term $-\mathbf{c} \cdot \mathbf{e}$ arises from the *definition* of $\mathbf{c}$; as P moves through the sheared fluid, there are continuous changes in the velocity of the ambient fluid, so that the origin from which $\mathbf{c}$ is measured is likewise changing.

3.4.3 *The scattering force*

The agitation acceleration, or equivalently the scattering force per unit mass, can be split into a friction term opposing P's motion and a diffusion term orthogonal to it. The friction term is proportional to P's velocity relative to the mean velocity of all the other molecules. If at time $t = 0$, P starts its trajectory with velocity $\mathbf{c}$ relative to P_c, then immediately prior to $t = \tau$ say, when collisions start to moderate its motion, the speed relative to the average molecular motion is altered by dilatation to $\mathbf{c}(1 - \frac{1}{3}\tau\nabla \cdot \mathbf{v})$ (see §1.13). For example, if $\nabla \cdot \mathbf{v}$ is positive, then on average P will experience a reduced velocity relative to the average molecular motion owing to the expansion of the fluid element. It follows that the friction term has the form $-\tau^{-1}\mathbf{c}(1 - \frac{1}{3}\tau\nabla \cdot \mathbf{v})$. In general τ will depend on $|\mathbf{c}| = c$, but for the present we shall ignore this and treat τ as being constant, but having different values, depending on whether it is the relaxation of momentum or energy flux that is under consideration. The diffusion term, $\mathcal{F}_d$ say, randomizes P's lateral motion (measured from its original trajectory), and as our present interest is in the transport of momentum in the direction of the original motion, we can drop $\mathcal{F}_d$.

From (1.95), namely $\overset{\circ}{\mathbf{e}} \equiv \overset{\circ}{\nabla}\mathbf{v} = \mathbf{e} - \frac{1}{3}\mathbf{1}\, \nabla \cdot \mathbf{v}$, we can now write (3.27) as

$$\dot{\mathbf{c}} + \frac{1}{\tau}(\mathbf{1} + \tau\overset{\circ}{\mathbf{e}}) \cdot \mathbf{c} = 0, \tag{3.28}$$

which has the solution

$$\mathbf{c} = e^{-t/\tau} \exp(-t\overset{\circ}{\mathbf{e}}) \cdot \mathbf{c}_0, \tag{3.29}$$

where $\mathbf{c}_0$ is the value of $\mathbf{c}$ at $t = 0$, and the exponential operator is defined by

$$\exp(-t\overset{\circ}{\mathbf{e}}) \cdot \mathbf{c}_0 = (\mathbf{1} - t\overset{\circ}{\mathbf{e}} + \tfrac{1}{2}t^2\, \overset{\circ}{\mathbf{e}} \cdot \overset{\circ}{\mathbf{e}} - \cdots) \cdot \mathbf{c}_0.$$

As $t/\tau \to \infty$, $\mathbf{c}$ tends to zero, i.e. the particle P 'stops', but of course, this refers only to P's motion in a given direction. Our aim here is to construct a simple model, from which transport can be estimated.

3.4.4 *The Knudsen number*

We now adopt the constraint

$$\epsilon \equiv \max\{\tau_1 \|\nabla \mathbf{v}\|,\ \tau_1 |\mathrm{D} \ln T|,\ \tau_2 |\mathbf{c} \cdot \nabla \ln T|\} \ll 1, \tag{3.30}$$

where in the present case only the first argument of 'max' is involved. The neglect of terms of higher order in ϵ is known as the *principle of local action* (see §1.3). The physical significance of (3.30) is that the *microscopic* time and length scales, i.e. τ_1 (or τ_2, see (3.19)) and λ_2 ($= \tau_2 c$) are restricted to being small compared with the *macroscopic* time and length scales appearing in the arguments of 'max'. Suppose, for example, that we were applying the theory to determine the thickness of a shock wave, e.g. the distance ℓ over which the temperature increases rapidly through the shock, then for the constraint in (3.30) to be valid, we should require ℓ to be at least several mean free paths long. We shall return to this particular example in §8.7 in order to illustrate the importance of including terms of second order in ϵ at high Mach numbers.

The ratio ϵ is termed the 'Knudsen number' in recognition of Knudsen's pioneering work on a variety of gas flow problems at low pressures. These included transpiration (§2.12.2), molecular flow past surfaces, and heat transfer between plates separated by a distance less than a mean free path.

When the Knudsen number is small, (3.29) reduces to

$$\mathbf{c} = \mathbf{c}_0 \cdot (\mathbf{1} - t\overset{\circ}{\mathbf{e}})e^{-t/\tau}. \tag{3.31}$$

Before taking the theory further, we need to understand how temperature gradients can influence the value of $\mathbf{c}$.

3.5 Temperature gradients and peculiar velocities

In order to assign a value to the initial velocity $\mathbf{c}_0$ in (3.31), it is necessary to allow for the directional anisotropy in molecular speeds resulting from the presence of temperature gradients. This task has presented a serious difficulty from the beginning of the subject (circa 1859), the problem being that the obvious anisotropic expression for $\mathbf{c}_0$ does not satisfy $\langle \mathbf{c} \rangle = 0$ (see exercise 3.4). It thus introduces a spurious fluid motion relative to the assumed convected frame, and thereby confuses conduction and convection. It was largely to avoid this problem that Maxwell abandoned mean-free-path methods in his seminal work 'On the Dynamical Theory of Gases'.

3.5.1 *Changes in the ambient temperature*

It was shown in §2.6.2 that in equilibrium the peculiar velocity $\mathbf{c}$ is given by $\mathcal{C}\boldsymbol{\nu}$, where $\mathcal{C} \equiv (2kT/m)^{1/2}$, and $\boldsymbol{\nu}$ is an isotropic vector, whose magnitude has the probability density

$$F(\nu)\,d\nu = \frac{4}{\sqrt{\pi}}\,\nu^2\,\exp(-\nu^2)\,d\nu \qquad (0 \le \nu < \infty). \tag{3.32}$$

We shall now suppose that over the mean free path, $\lambda_1 = c\tau_1$, the temperature changes by a small amount δT. Consider a bunch of molecules $\mathcal{B}$ starting from a point $P_c(\mathbf{r}, t)$ and belonging to the equilibrium distribution at this point. The molecules will have relative speeds in the range $\nu, \nu + d\nu$ and have $\boldsymbol{\nu}$ lying in an element $4\pi d\Omega$ of solid angle in velocity space. When they arrive at $Q_c(\mathbf{r} + \mathbf{c}\tau_1, t + \tau_1)$, a point that we shall assume is attained immediately *prior* to their experiencing a collision, their number $\mathcal{N}$ and energy $\mathcal{E}$; viz.

$$\mathcal{N} = nF(\nu)\,d\nu\,d\Omega, \qquad \mathcal{E} = \tfrac{1}{2}mc^2\mathcal{N}, \tag{3.33}$$

will be the same as at P_c. Of course some of the individual molecules belonging to $\mathcal{B}$ will experience collisions *en route* to Q_c, but to simplify the description without violating the essential physics, we shall suppose that *all* the molecules experience their first collision after leaving P_c at a point just past Q_c.

Now $\mathcal{N}$ is the correct number in $\mathcal{B}$ for equilibrium at P_c, but at Q_c the stream $\mathcal{B}$ will have an excess (or deficit) $\delta\mathcal{N}$ of particles relative to equilibrium at Q_c. To calculate $\delta\mathcal{N}$ we need an expression for δn. From $p = nkT$, this is $\delta n/n = \delta p/p - \delta T/T$. However, the pressure gradient force is already responsible for the motions of the fluid elements at P_c and Q_c, and since $\mathbf{c}$ is measured relative to these elements, it cannot be influenced by δp. Hence in the present application we may take $\delta n/n = -\delta T/T$ (see §3.9.4). It now follows from $\nu^2 = mc^2/2kT$, (3.32), and (3.33) that, provided $|\delta T/T| \ll 1$,

$$\delta\mathcal{N} = (\nu^2 - \tfrac{5}{2})\frac{\delta T}{T}\mathcal{N}. \tag{3.34}$$

As we shall see shortly, the constraint on the size of δT is equivalent to the restriction in (3.30). At the collisions beyond Q_c, the excess $\mathcal{N}$ will be eliminated and the distribution restored to the local Maxwellian.

Relative to Q_c, $\mathcal{B}$ will have an excess (or deficit) of energy

$$\delta\mathcal{E} = \tfrac{1}{2}mc^2\delta\mathcal{N}, \tag{3.35}$$

to be dispersed by collisions beyond Q_c. The accumulated disequilibrium at Q_c can be avoided if the *initial* particle energy at P_c, say $\frac{1}{2}mc_0^2$ is appropriately modified. This requires that

$$\delta(\tfrac{1}{2}mc_0^2)\mathcal{N} = -\delta\mathcal{E} = -\tfrac{1}{2}mc^2\delta\mathcal{N}.$$

Hence by (3.34),

$$\frac{\delta c_0}{c_0} = -\tfrac{1}{2}(\nu^2 - \tfrac{5}{2})\frac{\delta T}{T}, \tag{3.36}$$

a change that amounts to shifting the disequilibrium from Q_c to P_c. In the absence of temperature gradients, $\mathbf{c}_0 = \mathcal{C}\boldsymbol{\nu}$, where the relative velocity $\boldsymbol{\nu}$

is isotropic with a magnitude distributed as in (3.32). We conclude from (3.36) that in the presence of temperature gradients, $\mathbf{c}_0$ is given by

$$\mathbf{c}_0 = \mathcal{C}\boldsymbol{\nu}(1 + \delta c_0/c_0) = \mathcal{C}\boldsymbol{\nu}\{1 - \tfrac{1}{2}(\nu^2 - \tfrac{5}{2})\delta T/T\}. \tag{3.37}$$

3.5.2 *The non-equilibrium peculiar velocity*

The rate of change of a property ϕ in a frame following the bunch $\mathcal{B}$ is

$$\left(\frac{d\phi}{dt}\right)_{\mathcal{B}} = \frac{\partial\phi}{\partial t} + \mathbf{w}\cdot\nabla\phi = \left(\frac{\partial\phi}{\partial t} + \mathbf{v}\cdot\nabla\phi\right) + \mathbf{c}\cdot\nabla\phi = \mathrm{D}\phi + \mathbf{c}\cdot\nabla\phi. \tag{3.38}$$

Thus the relative change in ϕ during the collision interval τ_1 is

$$\delta\phi/\phi = \tau_1(\mathrm{D}\ln\phi + \mathbf{c}\cdot\nabla\ln\phi),$$

and in particular

$$\delta T/T = \tau_1(\mathrm{D}\ln T + \mathbf{c}\cdot\nabla\ln T). \tag{3.39}$$

Notice from (3.30) and (3.39) that the constraint $|\delta T/T| \ll 1$, which we introduced earlier, is equivalent to $\epsilon \ll 1$.

By (3.31), (3.37), and (3.39) our final expression for the non-equilibrium peculiar velocity reads

$$\mathbf{c} = \mathcal{C}\boldsymbol{\nu}\cdot(\mathbf{1} - t\overset{\circ}{\mathbf{e}})e^{-t/\tau} - \tfrac{1}{2}\tau_1(\nu^2 - \tfrac{5}{2})\big(\mathcal{C}\boldsymbol{\nu}\mathrm{D}\ln T + \mathcal{C}^2\boldsymbol{\nu\nu}\cdot\nabla\ln T\big)e^{-t/\tau} + O(\epsilon^2), \tag{3.40}$$

where the relaxation time τ remains to be assigned.

In §2.6.2 we showed that in equilibrium,

$$\left.\begin{aligned} &\tfrac{1}{2}\langle\nu^{2r}\rangle = \tfrac{1}{2}\cdot\tfrac{3}{2}\cdot\tfrac{5}{2}\cdots\tfrac{2r+1}{2} \qquad (r = 1, 2, 3, \cdots) \\ &\langle\hat{\mathbf{c}}\rangle \quad = 0, \qquad \langle\hat{\mathbf{c}}\hat{\mathbf{c}}\rangle = \tfrac{1}{3}\mathbf{1}. \end{aligned}\right\} \tag{3.41}$$

Using these results and the fact that $\mathbf{1} : \overset{\circ}{\mathbf{e}} = 0$, we can readily verify that with $\mathbf{c}$ given by (3.40), $\langle\mathbf{c}\rangle$ is zero, as required by its definition.

3.6 Pressure and viscosity

3.6.1 *The pressure tensor*

The pressure tensor is given by (1.12):

$$\mathsf{p} = \rho\langle\mathbf{cc}\rangle^*, \tag{3.42}$$

where * denotes the collided flux. By (3.40),

$$\rho\mathbf{cc} = \rho\mathcal{C}^2\boldsymbol{\nu\nu}\cdot\left(\mathbf{1} - 2t\overset{\circ}{\nabla\mathbf{v}}\right)e^{-t/\tau_1} + Q(\boldsymbol{\nu}) + O(\epsilon^2), \tag{3.43}$$

where $Q(\boldsymbol{\nu})$ is odd in the isotropic vector $\boldsymbol{\nu}$ and we have set $\tau = 2\tau_1$. This equation gives the momentum flux that persists for a time t. It has its maximum value at $t = 0$, after which it is steadily attenuated by collisions at a rate determined by the momentum relaxation time τ_1.

The exponential term in (3.43) can be interpreted as being the probability that—for the purposes of momentum transport—a given molecule P remains uncollided for a time t. And since dt/τ_1 is the probability that P experiences a collision in the time interval dt,

$$P(t)\,dt = e^{-t/\tau_1}\,dt/\tau_1 \qquad (0 < t < t^*) \tag{3.44}$$

is the joint probability that P survives until time t and is then scattered out of the stream (cf. (3.17)). The range in (3.44) extends to the time t^* it takes P to reach the boundary of the system. In order to obtain the collided flux from (3.43), it follows from (3.44) that we divide it by τ_1 and integrate over the full range of t. Doing this and also averaging over the molecular speeds, we find

$$\rho\langle\mathbf{cc}\rangle^* = \rho\mathcal{C}^2\langle\boldsymbol{\nu\nu}\rangle\cdot\Big\{(1-e^{-t^*/\tau_1})\mathbf{1}-2\tau_1\left[1-e^{-t^*/\tau_1}(1+t^*/\tau_1)\right]\overset{\circ}{\nabla\mathbf{v}}\Big\}+O(\epsilon^2), \tag{3.45}$$

since the average of $Q(\boldsymbol{\nu})$ is zero. (See Mathematical note 5.6.)

By (3.41), $\rho\mathcal{C}^2\langle\boldsymbol{\nu\nu}\rangle = \frac{1}{2}\rho\mathcal{C}^2\mathbf{1} = nkT\,\mathbf{1}$, and it follows from (3.44) and (3.45) that

$$\mathsf{p} = p\mathbf{1} - 2\mu\,\overset{\circ}{\nabla\mathbf{v}} + O(\epsilon^2), \tag{3.46}$$

where

$$p = nkT(1 - e^{-t^*/\tau_1}), \tag{3.47}$$

and

$$\mu = nkT\tau_1\big(1 - [1 + t^*/\tau_1]e^{-t^*/\tau_1}\big). \tag{3.48}$$

3.6.2 *The perfect gas law and collisions*

In the usual case of a collisional gas, $\tau_1 \ll t^*$ and we may set the ratio t^*/τ_1 equal to infinity and reduce the above relations to their familiar forms

$$p = nkT, \qquad \mu = p\tau_1. \tag{3.49}$$

In his 1860 paper on kinetic theory, Maxwell obtained an equation equivalent to the differential form of (3.47), which he then integrated over the infinite range to obtain the usual form of the perfect gas law in (3.49). He was not concerned with the case of a 'collisionless' gas ($t^*/\tau_1 \approx 0$), for which p and μ are zero, but he clearly appreciated the essential connection between pressure and molecular collisions. It is important to realise that the perfect gas law is contingent on the gas being *collisional*, a requirement that is too frequently overlooked.*

*For example see the so-called barometric derivation of Maxwell's distribution (Present 1958, p. 87), in which collisions are neglected, while pressure is retained in a key role; also see almost any text on plasma physics, where sections devoted to 'collisionless' plasmas adopt without question, not only the pressures $p_\parallel$ and $p_\perp$ defined in §1.6.2, but also the non-vanishing of their gradients. These problems are a consequence of identifying pressure as being momentum flux and thus ignoring the role of Newton's second law of motion. In most circumstances this lacuna has no consequences because collisions are present in any case. But there are important exceptions, see §§5.8.2 and 7.11.2.

3.6.3 *Viscosity*

The expression for μ given in (3.49) is due to Maxwell (1867), but he found it by a different route, after he had abandoned his elementary free-path model. For the fifth-power force law (see §3.2.1), he showed that μ/p was the relaxation time for fluid stress, where the value of μ was derived from his equation of transfer, (3.5). For hard-core molecules an expression for τ_1 is known, so it can be used to determine μ.†

By (3.19) and (3.49) we find

$$\mu = \tfrac{1}{3}\rho\overline{c^2}\tau_1 = \frac{5}{16\sigma^2}\left(\frac{mkT}{\pi}\right)^{1/2}, \tag{3.50}$$

which, as remarked in §3.1.2, is within 1.6 per cent of the exact value.

The viscous stress tensor defined in (1.22), viz. $\boldsymbol{\pi} = \mathbf{p} - p\mathbf{1}$, is given by

$$\boldsymbol{\pi} = -2\mu\overset{\circ}{\nabla}\mathbf{v}\,. \tag{3.51}$$

This relation gives the flux of fluid momentum across a sheared fluid flow, and it may be written as a diffusion equation in the standard form (see (1.33)),

$$J_\phi = -\rho\kappa_\phi\nabla\phi, \tag{3.52}$$

where by (3.51),

$$\kappa_{\mathbf{v}} = 2\mu/\rho = \mathcal{C}^2\tau_1 = \lambda_1^2/\tau_1 \qquad (\lambda_1 = \mathcal{C}\tau_1) \tag{3.53}$$

(cf. (1.34)). The ratio μ/ρ is known as the 'kinematic viscosity'; 'momentum diffusivity' would be a suitable alternative.

3.7 Thermal conductivity

3.7.1 *The heat flux vector*

By (2.5) the energy flux vector in a monatomic gas is

$$\boldsymbol{\xi} = \tfrac{1}{2}\rho\langle\mathbf{c}c^2\rangle, \tag{3.54}$$

where the right-hand side is the uncollided flux. The heat flux vector is obtained by using the collided flux,

$$\mathbf{q} = \tfrac{1}{2}\rho\langle\mathbf{c}c^2\rangle^*, \tag{3.55}$$

and represents the heating rate by molecular collisions.

When $\mathbf{c}c^2$ is evaluated from (3.40), it is found that the terms involving the fluid velocity occur as odd functions of $\boldsymbol{\nu}$, and hence vanish on averaging. Therefore they can be omitted. The even terms in $\boldsymbol{\nu}$ are

$$\mathbf{c}c^2 = -\tfrac{3}{2}\tau_1\mathcal{C}^4\nu^4(\nu^2 - \tfrac{5}{2})\hat{\mathbf{c}}\hat{\mathbf{c}}\cdot\nabla\ln T\, e^{-t/\tau_2} + O(\epsilon^2), \tag{3.56}$$

where $\mathcal{C}^2 = 2kT/m$, we have used (3.19) and set $\tau = 3\tau_2$ to obtain the correct energy relaxation time. Averaging this with the help of (3.41) we find

† In practice the experimental values of μ are used in (3.51) to find σ (see Table 3.1, p. 53).

$$\mathcal{E} = -\tfrac{5}{2}RnkT\tau_2\nabla T\, e^{-t/\tau_2}. \tag{3.57}$$

To find the collided flux, we follow a method similar to that used in 3.6.1 for the pressure tensor. Thus we divide by τ_2 and integrate over the range $0 < t < t^*$, where t^* is the time it takes the bunch of molecules in question to reach the boundary. By (3.55) this gives

$$\mathbf{q} = -\kappa\nabla T, \tag{3.58}$$

where the coefficient of thermal conductivity is

$$\kappa = \tfrac{5}{2}RnkT\,\tau_2\big(1 - e^{-t^*/\tau_2}\big). \tag{3.59}$$

3.7.2 *Thermal diffusivity*

Notice that in a collisionless gas ($t^*/\tau_2 \approx 0$), $\mathbf{q}$ is zero whereas the energy flux is not. In the usual case, $t^*/\tau_2 \gg 1$, it follows from (3.49) and (3.59) that

$$\kappa = \tfrac{5}{2}c_v\mu \qquad (\mu = p\tau_1\ c_v = \tfrac{3}{2}R). \tag{3.60}$$

This classical relation between conductivity and viscosity is another of Maxwell's achievements, although an arithmetical error—discovered by Boltzmann—gave him 5/3 in place of 5/2. He deduced the relation from his advanced transport theory for the special case of Maxwellian molecules (see §3.2.1).

Heat flux is the diffusion of internal energy c_vT. The relation in (3.58) may be expressed as a differential equation of the form given in (3.52):

$$\mathbf{q} = -\rho\chi\nabla(c_vT), \tag{3.61}$$

where by (3.19), the thermal diffusivity is given by

$$\chi = \frac{\kappa}{\rho c_v} = \tfrac{5}{4}(2RT)\tau_1 = \tfrac{5}{6}\mathcal{C}^2\tau_2 = \tfrac{5}{6}\lambda_2/\tau_2 \qquad (\lambda_2 \equiv \mathcal{C}^2\tau_2). \tag{3.62}$$

3.7.3 *The Eucken ratio*

There is an close relationship between the viscosity and the thermal conductivity of a gas. Eucken (1913) proposed a formula for the ratio*

$$\mathrm{f} \equiv \frac{\kappa}{\mu c_v} \tag{3.63}$$

for polyatomic gases that we shall discuss in §3.8.1. For monatomic gases the advanced kinetic theory developed at the beginning of the 20th century by Enskog and Chapman gives values of f surprisingly close to the value of 2.5 appearing in (3.60), regardless of the molecular force law. Over the important range $5 < \nu < 13$ for the power ν of the force law (see Table 3.1, p. 53), the ratio varies from 2.5 at $\nu = 5$ to 2.511. Hard-core molecules ($\nu = \infty$) give f = 2.522. The cause of this consistency can be traced to

*In gas dynamics the Prandtl number, $Pr = c_p\mu/\kappa = \gamma/\mathrm{f}$, is generally used to characterize the relative importance of momentum diffusivity to thermal diffusivity.

two factors in our mean-free-path model, neither of which depends on the force law. As explained at the end of §3.3, $\tau_2/\tau_1 = 3/2$, which is the first factor. The second is the ratio $\overline{c^4}/\overline{c^2} = 5/3$, which appears because from (3.55) and (3.42) κ and μ depend on $\overline{c^4}$ and $\overline{c^2}$ respectively.

The simplest form of elementary kinetic theory ignores the distinction between τ_2 and τ_1 and substitutes molecular velocities by $\bar{c}$, but adopts $\frac{1}{2}\overline{c^2}$ for energy. The outcome is to replace each of the two factors just mentioned by unity, giving f = 1. Many unsuccessful attempts were made last century—before the advent of the Chapman–Enskog theory—to increase f by various improvements in the mean-free-path treatment of hard-core molecules. Meyer's method of averaging the mean free path with a weighting factor, described by Jeans (1921, p. 299), gave f still much too small, at ~ 1.4. At the time experiments on a range of polyatomic gases gave values of f scattered about ~ 1.7, so the discrepancy was not thought excessive. And Maxwell's erroneous 5/3 appeared better than satisfactory, until Boltzmann's intervention. However, once it was realised that the molecules in polyatomic gases could transfer their internal energy during collisions, Maxwell's (corrected) f = 2.5 was restricted to monatomic gases. And this value was soon confirmed experimentally for the rare gases argon, helium and neon.

3.8 Heat conductivity in polyatomic gases

The nature of the internal structure of polyatomic gases remained a pressing but unresolved problem for several decades after the 1870s. The bright lines in molecular spectra were evidently related to internal structure, and implied some kind of resonance. Similarly, the specific heats data could be explained only in terms of internal energy distributed over a finite number of modes. These phenomena resisted explanation within the prevailing paradigm, although many models were proposed. These included Boltzmann's dumbbell model, clusters of 'billiard balls', and linked vortex atoms. But it was not until the development of quantum theory that the restriction of internal vibrations to a finite number of modes was understood. Together with the concept of the equipartition of energy between modes—already employed by Maxwell and Boltzmann—the new theory gave equation (1.58), which may be written

$$u = c_v T = c_v' T + c_v'' T \qquad \left(c_v' = \tfrac{3}{2}R, \quad c_v'' = \tfrac{1}{2}FR\right), \tag{3.64}$$

where by (1.66),

$$F = \frac{5-3\gamma}{\gamma-1}, \tag{3.65}$$

and is the number of internal degrees of freedom.

3.8.1 *Eucken's model*

By (1.37),

$$\mathbf{q} = \rho\langle \mathbf{c}\tfrac{1}{2}c^2\rangle + \rho\langle \mathbf{c}\varepsilon\rangle, \tag{3.66}$$

where ε is the internal molecular energy, which in equilibrium has the value

$$\varepsilon_0 = c_v''T = \frac{5-3\gamma}{2(\gamma-1)}RT. \tag{3.67}$$

In the case of non-equilibrium, we shall adopt the following model. Assume that at the point $P_c(\mathbf{r}, t)$, ε has the value it would have in equilibrium at the point $Q_c(\mathbf{r} - \mathbf{c}\tau_i, t - \tau_i)$, where τ_i is the time required for the internal energy to reach equilibrium with the translatory modes. We shall discuss its relation to τ_1 shortly. As $\varepsilon \propto T$, it follows that (cf. (3.39))

$$\varepsilon = c_v''T(1 - \tau_i \mathrm{D} \ln T - \tau_i \mathbf{c} \cdot \nabla \ln T) + O(\epsilon^2), \tag{3.68}$$

where T is the temperature at $P_c(\mathbf{r}, t)$. Let $\mathbf{q}_i \equiv \rho\langle \mathbf{c}\varepsilon\rangle$, then since $\langle \mathbf{c}\rangle = 0$, we find from (3.68) that

$$\mathbf{q}_i = -\rho c_v''T\tau_i\langle \mathbf{cc}\rangle \cdot \nabla T,$$

or by (2.43)

$$\mathbf{q}_i = -c_v''p\tau_i\nabla T. \tag{3.69}$$

The specific heat appearing in (3.63) is denoted by c_v' in (3.64). Hence the total heat flux, i.e. the sum of (3.58) and (3.69), can be written

$$\mathbf{q} = -\kappa\nabla T, \tag{3.70}$$

where

$$\kappa = \tfrac{5}{2}c_v'p\tau_1 + c_v''p\tau_i. \tag{3.71}$$

The case $\tau_i = \tau_1$ of this formula is due to Eucken (1913). He adopted the translatory component of κ by analogy with its expression for monatomic gases and assumed that the molecular internal energy ε would be transported similarly to momentum.

3.8.2 *The Eucken ratio for various gases*

Writing τ_i in the form $\tau_1(1+\zeta)$, and using (3.63), (3.64), and (3.65), we obtain

$$\mathrm{f} \equiv \frac{\kappa}{\mu c_v} = \tfrac{1}{4}(9\gamma - 5) + \tfrac{1}{2}\zeta(5 - 3\gamma), \tag{3.72}$$

a slight generalization of Eucken's original expression. More elaborate generalizations have been developed that distinguish between different forms of internal energy, each with its own relaxation time (see Chapman and Cowling 1970, Chapter 13).

In Table 3.2 equation (3.72) is compared with experiment for three types of gas. For monatomic gases the agreement is good. And gases with negligible dipole moments find fair agreement with Eucken's assumption that $\tau_i = \tau_1$ $(\zeta = 0)$. For these gases ζ has the average 0.10, giving the choice $\tau_i = 1.1\tau_1$ some merit; otherwise gases like deuterium and carbon dioxide need individual consideration. Gases with large dipole moments have $\tau_i \approx \frac{1}{2}\tau_1$. This reduction in τ_i has been attributed to interactions between dipole moments, which are able to accelerate the energy relaxation.

Table 3.1 *The Eucken ratio at 0 °C*

	Gas	f(exp)	$\frac{1}{4}(9\gamma - 5)$	ζ
Monatomic gases	helium	2.45	2.50	0
	neon	2.52	2.50	0
	argon	2.48	2.50	0
	xenon	2.58	2.50	0
Gases with negligible dipole moments	hydrogen	2.02	1.92	0.26
	deuterium	2.07	1.90	0.43
	methane	1.77	1.695	0.14
	acetylene	1.54	1.54	0
	carbon monoxide	1.91	1.90	0.03
	nitrogen	1.96	1.91	0.13
	ethylene	1.48	1.56	−0.13
	air	1.96	1.90	0.15
	oxygen	1.94	1.90	0.10
	carbon dioxide	1.64	1.68	−0.07
	nitrous oxide	1.73	1.68	0.09
	chlorine	1.825	1.80	0.05
Gases with large dipole moments	ammonia	1.455	1.715	−0.50
	hyd. sulphide	1.49	1.77	−0.58
	hyd. chloride	1.655	1.90	−0.61
	sulphur dioxide	1.50	1.64	−0.24

3.9 Pressure gradients and the peculiar velocity

3.9.1 *The pressure gradient paradox*

The force acting on unit mass of gas due to spatial changes in the pressure tensor $\mathbf{p}$ is

$$\mathbf{P} = -\frac{1}{\rho}\nabla \cdot \mathbf{p} = -\frac{1}{\rho}\nabla p - \frac{1}{\rho}\nabla \cdot \boldsymbol{\pi}. \tag{3.73}$$

As described in §1.5.1, this force is due to an imbalance of particle collisions acting on a volume element $d\mathbf{r}$ of the gas. For example, from $p = nkT$ and (3.73) we see that a temperature gradient not balanced by an opposing density gradient will produce a force $\mathbf{P}$. At a molecular level of description, this is explained as follows. Suppose that $d\mathbf{r}$ lies between elements $d\mathbf{r}_1$ and $d\mathbf{r}_2$ and that $d\mathbf{r}_1$ is hotter than $d\mathbf{r}_2$, then the molecules within $d\mathbf{r}$ will experience more energetic collisions with the molecules arriving from $d\mathbf{r}_1$ than with those from $d\mathbf{r}_2$ and the net effect will be to drive $d\mathbf{r}$ down the temperature gradient. Similarly, a number density gradient in the same direction produces more frequent collisions in $d\mathbf{r}$ with molecules from $d\mathbf{r}_1$ than with those from $d\mathbf{r}_2$ and $d\mathbf{r}$ is driven down the density gradient.

A paradox emerges if we try to relate pressure gradient forces to binary collisions between molecules within $d\mathbf{r}$ alone, i.e. we ignore the boundary conditions imposed from $d\mathbf{r}_1$ and $d\mathbf{r}_2$. As shown in §2.7.1, the velocity of the centre of mass of two colliding molecules, P_1 and P_2, is unchanged by their interaction, which means that collisions *within* $d\mathbf{r}$ cannot alter its momentum. We need some method of assigning to each molecule within $d\mathbf{r}$ a force proportional to $\mathbf{P}$ to represent the effect of *external* collisions on the volume element $d\mathbf{r}$.

Individual molecules can experience the effect of the force $\mathbf{P}$ only impulsively, at the instant of colliding with other molecules. In a neutral gas where the collisional forces have a very small range—of the order of molecular diameters—the molecules move freely between collisions. Consider what happens to a test molecule P within $d\mathbf{r}$. If it collides with another member of $d\mathbf{r}$, it suffers a large impulsive force having a random direction that we term a 'scattering' impulse. Next, suppose that it collides first with a molecule P_1 from $d\mathbf{r}_1$ and immediately after with P_2 from $d\mathbf{r}_2$ (the order is not important) and we sum the effects of these encounters. As we shall establish, the net effect on P will be to give it a scattering impulse plus a small non-random impulse parallel to $\mathbf{P}$. And because the volume elements can be arbitrarily chosen, all collisions experienced by molecules like P can be analyzed into random and non-random components.

3.9.2 *Distribution of the excess momentum*

The origin of $\mathbf{P}$ can be understood as follows. The whole interaction may be considered to be a 'collision' between P_1 and P_2, with P in an intermediary role. If $d\mathbf{r}_1$ is hotter than $d\mathbf{r}_2$, an average molecule P_1 will be moving towards $d\mathbf{r}$ with a speed greater than that of P_2 approaching from the other side. The centre of mass of the P_1 P_2 'collision' will have a momentum down the temperature gradient in excess of the average of the particles within $d\mathbf{r}$. Some of this excess momentum is passed on to P. More generally, the difference in momentum between particles entering $d\mathbf{r}$ from opposite sides is transported into $d\mathbf{r}$ by collisions. We could describe these as 'surface' collisions although of course they occur within $d\mathbf{r}$. So far as momentum transport is concerned, they might as well occur on a fictitious boundary enclosing $d\mathbf{r}$. A similar argument may be advanced with density gradients.

Based on the above description, our method of dealing with pressure gradients is to distribute the imbalance of the impulsive forces due to boundary collisions on $d\mathbf{r}$ over all the molecules within $d\mathbf{r}$ by giving each one a small directed impulse parallel to $\mathbf{P}$. Thus the collisional forces on a typical molecule will consist of a large scattering impulse, random in direction, so that its vectorial average over a large number of collisions is negligible, plus an impulse parallel to $\mathbf{P}$. The scattering impulse follows from two-particle dynamics in which momentum is conserved, and the directed impulse represents the external surface forces. We shall show shortly that the directed impulse is equal to $\tau\mathbf{P}$ per unit mass, where τ is the average time between successive collisions.

That the directed impulse is essential follows from the fact that without it the average position of the molecules in $d\mathbf{r}$ would be unchanged, whereas this same collection of molecules viewed as a fluid element would be accelerated down the pressure gradient, giving an obvious contradiction.

3.9.3 *Magnitude of the directed impulse*

Now suppose that in the laboratory frame L, the molecule P is subject to a body force $\mathbf{F}$ per unit mass and a sequence of impulses $\hat{\mathbf{I}}_1, \hat{\mathbf{I}}_2, \ldots \hat{\mathbf{I}}_n$ per unit mass, applied at times $t_1, t_2, \ldots t_n$, where $0 < t_1 < t_2 < \ldots < t_n < t$. Then at time t its velocity $\mathbf{w}$ is

$$\mathbf{w} = \mathbf{w}_0 + t\mathbf{F} + \sum_{i=1}^{n} \hat{\mathbf{I}}_i,$$

$\mathbf{w}_0$ being its initial value. Divide $\hat{\mathbf{I}}_i$ into a directed component $\hat{\mathbf{P}}$ independent of i, and a random component $\hat{\mathbf{S}}_i$, and introduce the average time $\tau = t_n/n$ between impulses:

$$\mathbf{w} = \mathbf{w}_0 + t\mathbf{F} + \frac{t}{\tau}\hat{\mathbf{P}} - \left(\frac{t - t_n}{\tau}\right)\hat{\mathbf{P}} + \sum_{i=1}^{n} \hat{\mathbf{S}}_i. \tag{3.74}$$

At large times, i.e. when $t \gg \tau$, $(t - t_n)/\tau$ is O(1) and the time derivative of (3.74) is approximately

$$\dot{\mathbf{w}} = \mathbf{F} + \frac{1}{\tau}\hat{\mathbf{P}} + \sum_{i=1}^{n} \left(\frac{d}{dt}\hat{\mathbf{S}}_i\right). \tag{3.75}$$

The last term is the sum of large, randomly-oriented impulsive forces. (In §3.4.2 we termed it the 'agitation' acceleration.) Its average over all the molecules in $d\mathbf{r}$ is zero. Hence the average of (3.75) is

$$\overline{\dot{\mathbf{w}}} = \mathrm{D}\mathbf{v} = \mathbf{F} + \frac{1}{\tau}\hat{\mathbf{P}}, \tag{3.76}$$

$\overline{\dot{\mathbf{w}}}$ being identical with the fluid acceleration $\mathrm{D}\mathbf{v}$ (see (3.25)). Comparing (3.76) with the equation of fluid motion in §1.9.2, we deduce that

$$\hat{\mathbf{P}} = \tau\mathbf{P}. \tag{3.77}$$

3.9.4 *Effect of pressure gradients on molecular velocities*

Since $\mathbf{c} - \mathbf{v}_0 = (\mathbf{w} - \mathbf{v}) - (\mathbf{w}_0 - \mathbf{v}_0) = \mathbf{w} - \mathbf{w}_0 - t\mathrm{D}\mathbf{v}$, by (3.74) and (3.77) we can write (3.74) as

$$\mathbf{c} = \mathbf{c}_0 - (t - \tfrac{1}{2}\tau)\mathbf{P} \qquad (0 < t < \tau), \tag{3.78}$$

where $\mathbf{c}_0$ now includes the impulses prior to $t = 0$ and is the value of $\mathbf{c}$ in the middle of the collision interval. In the particular case that $\mathbf{c}_0$ and $\mathbf{P}$ are parallel, the temporal variation of $\mathbf{c}$ is as shown in Fig. 3.2(a). Following an impulse at $t = 0$, $\mathbf{c}$ has an initial magnitude $c_0 + \frac{1}{2}\tau P$. Then as P moves through a mean free path, the reference frame P_c is accelerated with the fluid, leaving P to fall behind, until at $t = \tau$ another impulse enables it to overtake P_c, and so on. In the absence of random scattering, this leap-frogging would maintain an average value for c of c_0. The process evidently

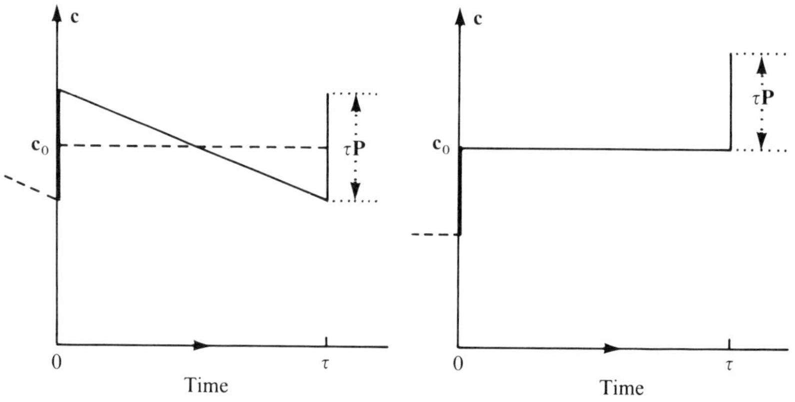

FIG. 3.2. The effect of pressure gradients on molecular velocities: (a) convected frame, (b) free flight frame.

makes no direct contribution to the transport of energy or momentum.

In Fig. 3.2(b) we show the corresponding variation of w with time, where w is the molecular speed measured in a frame moving only with the acceleration $\mathbf{F}$, i.e. with the acceleration of P during its *free flight*.

We find it useful to adopt the following terminology. By a 'collision' we shall mean a synthesis of a local binary collision in which momentum is conserved, and an impulse representing the multiplicity of 'surface' collisions responsible for the pressure gradient. The first component we will call a 'scattering' collision, implying a randomly oriented interaction, and the second component, which sends the molecules involved in a privileged direction, a 'streaming' collision.

From the above description we deduce the following important distinction between an element $d\boldsymbol{\nu}_S$ of 6-D phase space that has the fluid acceleration $\mathbf{F} + \mathbf{P}$, and an element $d\boldsymbol{\nu}$ that has only the free-flight acceleration $\mathbf{F}$. Because of the leap-frogging of P and P_c, molecules are not removed from $d\boldsymbol{\nu}_S$ by the force $\mathbf{P}$. Of course—depending on the size of $d\boldsymbol{\nu}_S$—they may move in and out of it, but their permanent departure from $d\boldsymbol{\nu}_S$ is the result of *scattering* collisions only. On the other hand, because the element $d\boldsymbol{\nu}$ fails to keep up with the average molecular motion, there are streaming losses from it in addition to the normal scattering losses. The importance of this fact for kinetic theory will be made clear in §5.8.2 (see exercise 3.9).

Exercises 3

3.1 Momentum is being transferred across a surface S in a gas by molecules with a mean free path λ. Show that on average this transfer is completed at a perpendicular distance $\frac{1}{3}\lambda$ from S. What constraint does this imply for the size of a *fluid* element?

3.2 Find a *physical* explanation of why the coefficient of viscosity is independent of density. Does the viscosity tend to infinity as the mean free path increases indefinitely? [Answer: no].

3.3 Assuming that the effective molecular diameter σ_{eff} depends on $m, \bar{\mathbf{c}}$, and the parameters K and ν in the force law $Kr^{-\nu}$, apply dimensional analysis to establish that $\sigma_{eff} \propto \bar{\mathbf{c}}^{-2/(\nu-1)}$ and deduce that the coefficient of viscosity is proportional to $T^{\frac{1}{2}+2/(\nu-1)}$.

3.4 Show that if $\mathcal{C} \equiv (2kT/m)^{1/2}$, where T is the temperature at the point $P(\mathbf{r}, t)$, the values of $\mathcal{C}$ at P and $P_0 = P(\mathbf{r} - \mathbf{c}\tau_2, t - \tau_2)$ are related by

$$\mathcal{C}_0 = \mathcal{C}(1 - \tfrac{1}{2}\tau_2 \mathrm{D} \ln T - \tfrac{1}{2}\tau_2 \mathbf{c} \cdot \nabla T) + O(\epsilon^2).$$

Why would you expect molecules travelling from P_0 to P to belong to a distribution with T_0 as its equilibrium temperature? What makes the assumption that $\mathbf{c}_0$ in (3.31) should equal $\mathcal{C}_0\boldsymbol{\nu}$ plausible, but incorrect? [Hint: $\langle \mathbf{c}_0 \rangle = -\frac{1}{4}\tau_2 \mathcal{C}^2 \nabla \ln T$.]

3.5 What constraints are required to enable the relations $p = \frac{1}{3}\rho\langle c^2 \rangle$ and $T = p/kn$ to serve as satisfactory *definitions* of p and T?

3.6 Show that—in the notation of §2.8.1—

$$\langle g^a \rangle = \frac{2}{\sqrt{\pi}} \left(\frac{4kT}{m}\right)^{a/2} \Gamma(\tfrac{3}{2} + \tfrac{1}{2}a),$$

where Γ is the gamma function and g the relative velocity of two molecules. Deduce that the average value of the collision frequency in a gas with a collisional force law $Kr^{-\nu}$ is

$$\tau_0^{-1} = 2\sqrt{\pi} n \left(\frac{m}{2K}\right)^{-2/(\nu-1)} \left(\frac{4kT}{m}\right)^{\frac{1}{2}-2/(\nu-1)} \Gamma(2 - 2/(\nu - 1)),$$

and that the viscosity is approximately $p\tau_1$ where $\tau_1 = \frac{5}{4}\tau_0$.

3.7 Show that in hydrostatic equilibrium the earth's atmosphere has a pressure gradient given by $\nabla p = \rho \mathbf{g}$, where $\mathbf{g}$ is the acceleration of gravity. Assuming a uniform temperature, show that the number density of a component of the atmosphere at a height z is given by $n(z) = n(0)e^{-mgz/kT}$. Interpret this result in terms of particle trajectories and collisions.

3.8 Suppose that in the previous exercise at a given instant, say at $t = 0$ a demon abolishes all collisions, save those with the earth's surface. Describe the subsequent motion of the molecules on the assumption that they rebound (i) elastically and (ii) diffusely from the surface.

3.9 A phase space element $d\boldsymbol{\nu} = d\mathbf{r}\, d\mathbf{w}$ is moving with the speed $\mathbf{w}$ and the acceleration $\mathbf{F}$ of a bunch of molecules during their free flight between collisions. Now allow collisions with other molecules to steadily remove the $f\, d\mathbf{w}$ molecules within $d\boldsymbol{\nu}$. Explain why the loss rate from $d\boldsymbol{\nu}$ cannot be fully represented by a term of the type $-f/\tau$ unless the pressure gradient is zero. (See §5.8.2.)

4

PARTICLE DIFFUSION

4.1 Introduction

In §1.7.1, diffusion was defined as being a transport process due to the purely random component of molecular motion, frame indifference being identified as an essential property of the phenomenon. In the preceding chapter we studied the diffusion of momentum and energy. It remains to consider the diffusion of particle mass, or equivalently, the diffusion of particle numbers. In a gas comprising a single species, unless particles can be differentiated, the diffusion of particle number density is a meaningless concept. The property of being a particular type of particle is universal, so its migration cannot be distinguished. However, if we could label some of the particles, for example by using radioactive isotopes of some of the original species, and could trace the relative motions of such isotopes, we would have a good approximation to what is termed 'self-diffusion'.

There are four possible driving forces able to generate a flow of one gas relative to another in a gas mixture. These diffusion processes are: (i) diffusion due to non-uniformity of composition, (ii) pressure diffusion, (iii) thermal diffusion, and (iv) forced diffusion resulting from differential acceleration due to body forces.

The first of these processes is the most familiar. For example, in a binary mixture at temperature T and total pressure $p = p_1 + p_2$, a gradient in the pressure ratio $x_1 = p_1/p$ will tend to make species '1' diffuse along the gradient $-\nabla x_1$, i.e. the molecules of a species will tend to migrate to where they are relatively scarce. Pressure diffusion occurs when there is a mass difference between the species, e.g. the pressure gradient in the earth's atmosphere causes heavy molecules to sink and light ones to rise (see exercise 4.1). The same phenomenon occurs in rotating gases, with the heavy molecules diffusing outwards, away from the axis of rotation. Thermal diffusion is a more subtle phenomenon. It arises because of the variation of the collision interval with molecular speed; it therefore depends on the precise form of the molecular interaction law. In most neutral gases thermal diffusion causes the lighter particles to migrate to the hotter regions.

The clearest and most important example of forced diffusion occurs in a plasma, where the electrical and magnetic forces depend on the *sign* of the charge carried by the microscopic particles. Thus the electrons and ions are accelerated in opposite directions by these forces, the resulting diffusion giving rise to an electric current.

4.2 The general diffusion equation

4.2.1 *Relative velocity for two molecular species*

Suppose that the gas comprises two molecular species, with properties distinguished by the subscripts 'i' and 'j'. The fluids will be labelled P_i and P_j. By (1.42) the equation of motion for P_i is

$$\rho_i \frac{\partial \mathbf{v}_i}{\partial t} + \rho_i \mathbf{v}_i \cdot \nabla \mathbf{v}_i = -\nabla \cdot \mathsf{p}_i + \rho_i \mathbf{F}_i + \mathbf{R}_{ij}, \tag{4.1}$$

the additional term $\mathbf{R}_{ij}$ being the friction per unit volume acting on P_i due to the relative motion of P_j. A similar equation holds for P_j, and by Newton's third law,

$$\mathbf{R}_{ij} = -\mathbf{R}_{ji}. \tag{4.2}$$

Let

$$\rho = \rho_i + \rho_j, \qquad \mathbf{v} = (\rho_i \mathbf{v}_i + \rho_j \mathbf{v}_j)/\rho, \tag{4.3}$$

be the density and fluid velocity of the composite fluid. We shall assume that the diffusion velocities $(\mathbf{v}_i - \mathbf{v})$ and $(\mathbf{v}_j - \mathbf{v})$ are small and omit quadratic terms in these quantities. In this case (4.1) and the corresponding equation for P_j yield

$$\mathrm{D}(\mathbf{v}_i - \mathbf{v}) + (\mathbf{v}_i - \mathbf{v}_j) \cdot \nabla \mathbf{v} = (\mathbf{F}_i - \frac{1}{\rho_i}\nabla \cdot \mathsf{p}_i) - (\mathbf{F}_j - \frac{1}{\rho_j}\nabla \cdot \mathsf{p}_j) + \frac{\rho}{\rho_i \rho_j}\mathbf{R}_{ij}, \tag{4.4}$$

where $\mathrm{D} = \partial/\partial t + \mathbf{v} \cdot \nabla$.

When $\nabla \mathbf{v}$ is zero and the right-hand bracketed terms are also zero, the friction force rapidly establishes the equilibrium $\mathbf{v}_i = \mathbf{v}_j$. Let this relaxation take a time τ_{ij}, i.e. (cf. (3.13)) $\mathrm{D}(\mathbf{v}_i - \mathbf{v}_j) = -(\mathbf{v}_i - \mathbf{v}_j)/\tau_{ij}$, then by (4.4),

$$\mathbf{R}_{ij} = \frac{\rho_i \rho_j}{\rho} \frac{\mathbf{v}_j - \mathbf{v}_i}{\tau_{ij}}. \tag{4.5}$$

Assuming this relation to hold generally, we may write (4.4) in the form

$$(\mathbf{v}_i - \mathbf{v}_j) \cdot (\mathbf{1} + \tau_{ij} \nabla \mathbf{v}) + \tau_{ij} \mathrm{D}(\mathbf{v}_i - \mathbf{v}_j)$$
$$= \tau_{ij} \Big\{ (\mathbf{F}_i - \frac{1}{\rho_i}\nabla \cdot \mathsf{p}_i) - (\mathbf{F}_j - \frac{1}{\rho_j}\nabla \cdot \mathsf{p}_j) \Big\}.$$

On adopting restrictions similar to those in (3.30), we can reduce the left hand side of this relation to its dominant term, viz. $(\mathbf{v}_i - \mathbf{v}_j)$, and arrive at the general equation of diffusion,

$$\mathbf{v}_i - \mathbf{v}_j = \tau_{ij} \left\{ (\mathbf{F}_i - \frac{1}{\rho_i}\nabla \cdot \mathsf{p}_i) - (\mathbf{F}_j - \frac{1}{\rho_j}\nabla \cdot \mathsf{p}_j) \right\}. \tag{4.6}$$

4.2.2 *Diffusion coefficients*

The diffusion equation is also written in the form

$$\mathbf{v}_i - \mathbf{v}_j = -\frac{n^2}{n_i n_j} D_{ij} \, \mathbf{d}_{ij}, \tag{4.7}$$

where $n = n_i + n_j$, D_{ij} is known as the 'coefficient of mutual diffusion':

$$\mathbf{d}_{ij} \equiv \frac{\rho_i \rho_j}{p\rho}\Big\{(\mathbf{F}_j - \frac{1}{\rho_j}\nabla p_j) - (\mathbf{F}_i - \frac{1}{\rho_i}\nabla p_i)\Big\}, \tag{4.8}$$

and we have omitted the usually negligible viscosity term. Comparison of (4.6) and (4.7) gives

$$D_{ij} = \rho\tau_{ij}kT/nm_i m_j. \tag{4.9}$$

When the two species are identical, $m_i = m_j = m$ and (4.9) reduces to the coefficient of self-diffusion:

$$D_S = \tau_d kT/m, \tag{4.10}$$

where τ_d is the diffusion time.

An estimate of the relationship between τ_d and the mean free time for hard-core molecules, i.e. $\tau_0 = \lambda/\bar{c}$, can be found as follows. In §3.3.2 it was shown that, owing to velocity persistence, the microscopic time scale for momentum transport was increased from τ_0 to $\tau_1 = \tau_0/(1 - \frac{1}{2}\varpi)$, where ϖ is the persistence ratio. At each collision the excess momentum, say δM, was divided between two molecules, each carrying $\frac{1}{2}\delta M$ on to their next collisions. This dispersal produced the factor $\frac{1}{2}$ in the formula for τ_1. However, with self-diffusion the property of being an identifiable molecule is not transferred by collisions—what matters here is the total forward displacement of the labelled molecule. By an argument similar to that given in §3.3.2, this gives a diffusion time $\tau_d = \tau_0/(1 - \varpi)$. Hence

$$\tau_d = \tau_0/(1-\varpi) = \tau_1(1 - \tfrac{1}{2}\varpi)/(1-\varpi) \approx 1.342\tau_1, \tag{4.11}$$

where the numerical value corresponds to ϖ having its average value of 0.406.

From (3.49) and (4.11),

$$D_S = 1.342\mu/\rho, \tag{4.12}$$

which result is in good agreement with experiment for most common gases near room temperature. Chapman and Cowling (1970) quote values of D_S obtained from isotope diffusion experiments for twelve gases. The values of $\rho D_S/\mu$ range between 1.29 and 1.42, with an average of 1.344. Ten of the gases fall in the range 1.32 to 1.37.

4.3 Mutual diffusion of hard-core molecules

An accurate expression for the mutual diffusivity D_{ij} can be deduced for hard-core molecules by a slight modification of the theory given in §§2.7 and 2.8. The first step is to generalize the theory in §2.7.1 by replacing the peculiar velocities $\mathbf{c}_i$ and $\mathbf{c}_j$ by $\mathbf{w}_i$ and $\mathbf{w}_j$, i.e. to refer all velocities to the laboratory frame instead of the fluid frame. The reader can readily confirm that this amounts to no more than a change in notation in equation (2.49) *et seq*, since the special role of the variables $\mathbf{c}_i$ and $\mathbf{c}_j$ in the Maxwellian velocity distributions is not required until (2.63) is reached.

4.3.1 *Friction force between diffusing species*

We start with (2.49), now written as $\mathbf{g}_{ij} = \mathbf{w}_i - \mathbf{w}_j$, $\mathbf{g}'_{ij} = \mathbf{w}'_i - \mathbf{w}'_j$. As remarked at the end of §2.7.1, in a coordinate system moving with the centre of mass of colliding particles P_i and P_j, all directions of reflection or rebound are equally probable. It follows from this that the average velocity of P_i after colliding with P_j is equal to the velocity of their centre of mass. Hence the average momentum lost by P_i in a collision with P_j is

$$m_i(\mathbf{w}_i - \mathbf{G}_{ij}) = m_{ij}(\mathbf{w}_i - \mathbf{w}_j) = m_{ij}\mathbf{g}_{ij}, \tag{4.13}$$

where we have used $\mathbf{G}_{ij} = \mu_i \mathbf{w}_i + \mu_j \mathbf{w}_j$ and $m_i\mu_j = m_{ij}$ (see (2.50) and (2.66)).

As noted in the first paragraph of §2.8, there are

$$\pi\sigma_{ij}^2 g_{ij} n_j \qquad \left(\sigma_{ij} \equiv \tfrac{1}{2}(\sigma_i + \sigma_j)\right),$$

collisions per second between P_i and the j-species molecules. So by (4.13) momentum is transferred from the j-species to the i-species at the rate

$$-\pi\sigma_{ij}^2 g_{ij}\mathbf{g}_{ij}\, m_{ij} n_i\, n_j$$

per unit volume. The drag or friction $\mathbf{R}_{ij}$ appearing in (4.1) is the average of this rate, i.e.

$$\mathbf{R}_{ij} = -\pi\sigma_{ij}^2 n_i n_j m_{ij}\langle g_{ij}\mathbf{g}_{ij}\rangle. \tag{4.14}$$

4.3.2 *Average value of* $g_{ij}\mathbf{g}_{ij}$

Now (cf. (2.62) *et seq.*)

$$\langle g_{ij}\mathbf{g}_{ij}\rangle = \iint g_{ij}\mathbf{g}_{ij}\frac{f_{0i}}{n_i}\frac{f_{0j}}{n_j}\,d\mathbf{w}_i\,d\mathbf{w}_j \tag{4.15}$$

$$= \frac{1}{(\pi\mathcal{C}_i\mathcal{C}_j)^3}\iint g_{ij}\mathbf{g}_{ij}\exp\left\{-\left[\frac{(\mathbf{w}_i - \mathbf{v}_i)^2}{\mathcal{C}_i^2} + \frac{(\mathbf{w}_j - \mathbf{v}_j)^2}{\mathcal{C}_j^2}\right]\right\}d\mathbf{w}_i\,d\mathbf{w}_j,$$

where in the present case the fluid velocities $\mathbf{v}_i$ and $\mathbf{v}_j$ are different. Provided a laboratory frame can be chosen such that $(\frac{1}{2}m_i v_i^2 + \frac{1}{2}m_j v_j^2) \ll \frac{1}{2}kT$, the diffusion velocities will be small enough to permit the approximation

$$(\mathbf{w}_i - \mathbf{v}_i)^2/\mathcal{C}_i^2 + (\mathbf{w}_j - \mathbf{v}_j)^2/\mathcal{C}_j^2 \approx w_i^2/\mathcal{C}_i^2 + w_j^2/\mathcal{C}_j^2 - 2\left(\mathbf{w}_i\cdot\mathbf{v}_i/\mathcal{C}_i^2 + \mathbf{w}_j\cdot\mathbf{v}_j/\mathcal{C}_j^2\right).$$

By (2.52) (with $\mathbf{c}_i$ and $\mathbf{c}_j$ replaced by $\mathbf{w}_i$ and $\mathbf{w}_j$) we find that

$$\frac{\mathbf{w}_i\cdot\mathbf{v}_i}{\mathcal{C}_i^2} + \frac{\mathbf{w}_j\cdot\mathbf{v}_j}{\mathcal{C}_j^2} = \left(\frac{\mathbf{v}_i}{\mathcal{C}_i^2} + \frac{\mathbf{v}_j}{\mathcal{C}_j^2}\right)\cdot\mathbf{G}_{ij} + \frac{m_{ij}}{2kT}(\mathbf{v}_i - \mathbf{v}_j)\cdot\mathbf{g}_{ij}.$$

Using these expression in (4.15) we obtain

$$\begin{aligned}\langle g_{ij}\mathbf{g}_{ij}\rangle \quad = \quad & \frac{1}{(\pi\mathcal{C}_i\mathcal{C}_j)^3}\iint g_{ij}\mathbf{g}_{ij}\exp\left\{-\left(\frac{w_i^2}{\mathcal{C}_i^2} + \frac{w_j^2}{\mathcal{C}_j^2}\right)\right\} \\ & \times\exp\left\{\mathbf{A}_{ij}\cdot\mathbf{G}_{ij} + \frac{m_{ij}}{kT}(\mathbf{v}_i - \mathbf{v}_j)\cdot\mathbf{g}_{ij}\right\}d\mathbf{w}_i\,d\mathbf{w}_j,\end{aligned}$$

where $\mathbf{A}_{ij} = 2(\mathbf{v}_i/\mathcal{C}_i^2 + \mathbf{v}_j/\mathcal{C}_j^2)$. By the equations corresponding to (2.53) and (2.59), this integral becomes

$$\begin{aligned}\langle g_{ij}\mathbf{g}_{ij}\rangle &= \frac{1}{(\pi\mathcal{C}_i\mathcal{C}_j)^3}\int g_{ij}\mathbf{g}_{ij}\exp\left\{-\frac{m_{ij}}{2kT}g_{ij}^2\right\}d\mathbf{g}_{ij}\int\Big\{1+\mathbf{A}_{ij}\cdot\mathbf{G}_{ij}\\ &\quad+\frac{m_{ij}}{kT}(\mathbf{v}_i-\mathbf{v}_j)\cdot\mathbf{g}_{ij}\Big\}\exp\left\{-\frac{m_i+m_j}{2kT}G_{ij}^2\right\}d\mathbf{G}_{ij}. \quad (4.16)\end{aligned}$$

Both $\mathbf{g}_{ij}$ and $\mathbf{G}_{ij}$ are isotropic vectors. Therefore by an application of (2.40),

$$\begin{aligned}\langle g_{ij}\mathbf{g}_{ij}\rangle &= \frac{m_{ij}}{kT}\frac{1}{(\pi\mathcal{C}_i\mathcal{C}_j)^3}(\mathbf{v}_i-\mathbf{v}_j)\cdot\int g_{ij}\mathbf{g}_{ij}\mathbf{g}_{ij}\exp\left\{-\frac{m_{ij}}{2kT}g_{ij}^2\right\}d\mathbf{g}_{ij}\\ &\qquad\times\int\exp\left\{-\frac{m_i+m_j}{2kT}G_{ij}^2\right\}d\mathbf{G}_{ij}\\ &= \frac{m_{ij}}{3kT}\frac{1}{(\pi\mathcal{C}_i\mathcal{C}_j)^3}(\mathbf{v}_i-\mathbf{v}_j)\int g_{ij}^3\exp\left\{-\frac{m_{ij}}{2kT}g_{ij}^2\right\}d\mathbf{g}_{ij}\\ &\qquad\times\int\exp\left\{-\frac{m_i+m_j}{2kT}G_{ij}^2\right\}d\mathbf{G}_{ij}.\end{aligned}$$

We now require the integrals $I_0(\alpha)$ and $I_3(\alpha)$ defined in 2.8.1: we find that $I_0(\alpha) = \sqrt{\pi}/4\alpha^{3/2}$ and $I_3(\alpha) = 1/\alpha^3$. By $\mathcal{C}_i^2 = 2kT/m_i$, $\mathcal{C}_j^2 = 2kT/m_j$ and these integrals, we arrive at

$$\langle g_{ij}\mathbf{g}_{ij}\rangle = \frac{8}{3}\left(\frac{2kT}{\pi m_{ij}}\right)^{1/2}(\mathbf{v}_i-\mathbf{v}_j). \quad (4.17)$$

4.3.3 *Mutual Diffusivity*

Substituting (4.17) into (4.14) we have

$$\mathbf{R}_{ij} = \tfrac{8}{3}\pi\sigma_{ij}^2 n_i n_j(2m_{ij}kT/\pi)^{1/2}(\mathbf{v}_i-\mathbf{v}_j). \quad (4.18)$$

Then (4.5) and (4.9) yield

$$\tau_{ij} = \frac{3}{8}\frac{m_i m_j}{\rho\pi\sigma_{ij}^2}\left(\frac{\pi}{2m_{ij}kT}\right)^{1/2}, \quad (4.19)$$

and

$$D_{ij} = \frac{3}{8}\frac{1}{\pi\sigma_{ij}^2 n}\left(\frac{\pi kT}{2m_{ij}}\right)^{1/2}, \quad (4.20)$$

where $\sigma_{ij} \equiv \frac{1}{2}(\sigma_i + \sigma_j)$ and σ_i and σ_j are the diameters of the molecules. This equation can be used to determine the molecular diameters from the mutual diffusion coefficients of a triad of gases.

In the case of self-diffusion, $m_i = m_j = 2m_{ij} = m$, when by (3.19), equation (4.19) reduces to

$$\tau_d = \frac{3m}{8\pi\sigma^2 n}\left(\frac{\pi}{kT}\right)^{1/2} = \tfrac{3}{2}\tau_0 \approx \tfrac{6}{5}\tau_1. \quad (4.21)$$

Then from (3.49), (3.50), and (4.10),

$$D_S \approx 1.2\,\frac{\mu}{\rho} = \frac{3}{8\pi\sigma^2 n}\left(\frac{kT}{\pi m}\right)^{1/2}, \tag{4.22}$$

which is to be compared with our approximate mean-free-path result in (4.12). It is curious that the approximate formula is in better agreement with observations than (4.22). The reason for this is not understood. It has been suggested that the collision cross-section between unlike isotopes is slightly smaller than for like ones.

4.4 Thermal diffusion

4.4.1 *The physics of thermal diffusion*

The general diffusion equation in (4.6) or (4.7) is incomplete. The missing term depends on the temperature gradient, and it arises because in general the relaxation time τ_{ij} depends on the temperature of the diffusing gases. To illustrate the phenomenon, let us suppose that τ_{ij} decreases with increasing temperature, this being the usual case with uncharged molecules. We shall also assume that $m_i \ll m_j$, which by (2.68) allows us to neglect the thermal agitation of the j-species compared with that of the i-species. Then molecules of the i-species arriving at a convected point P_c from higher temperatures, say from the 'right' of P_c, will have a shorter collision time with the j-species molecules at P_c—treated as being stationary—than those arriving at P_c from the left. Hence, on average, the former experience a larger friction force than the latter, and this imbalance results in a net force directed to the right, acting on the i-species. This thermal force tends to drive the lighter molecules up the temperature gradient to the hotter region, and a balance will be achieved when the increased numbers in this region are sufficient to set up an equal flux in the opposite direction due to the number density gradient.

Maxwellian molecules have collision times independent of the temperature (see exercise 3.6) and so are not subject to thermal diffusion. And as it is generally assumed in elementary kinetic theory that local variations in the collision time can be neglected, the phenomenon is not revealed by the usual mean-free-path arguments. Its existence was first discovered by Enskog in 1911, and more fully explained by Chapman in 1916, each using their own development of advanced kinetic theory. The phenomenon was confirmed experimentally by Chapman and Dootson in 1916. In 1938 Clusius and Dickel applied it in their thermal diffusion column, invented for the purpose of separating isotopes. For a comprehensive review of the subject, the account given by Jones and Furry (1946) remains instructive.

When generalized to include thermal diffusion, (4.7) is usually expressed in one of two forms:

$$\mathbf{v}_i - \mathbf{v}_j = -\frac{n^2}{n_i n_j} D_{ij}(\mathbf{d}_{ij} + \kappa_T \nabla \ln T), \tag{4.23}$$

or

$$\mathbf{v}_i - \mathbf{v}_j = -\frac{n^2}{n_i n_j} D_{ij}\mathbf{d}_{ij} - D_{ij}\alpha_T \nabla \ln T, \tag{4.24}$$

where

$$\kappa_T = x_i x_j \alpha_T \qquad (x_i \equiv n_i/n,\ x_j \equiv n_j/n) \tag{4.25}$$

is termed the 'thermal diffusion ratio'. The second form employs the thermal diffusion 'factor', α_T, which has the merit of being much less dependent on composition than κ_T.

4.4.2 *The thermal diffusion factor*

To deduce an expression for α_T, we start by generalizing (4.14) to

$$\mathbf{R}_{ij} = -n_i n_j m_{ij}\langle \pi\tilde{\sigma}_{ij}^2 g_{ij}\mathbf{g}_{ij}\rangle, \tag{4.26}$$

where $\tilde{\sigma}_{ij}$ differs from σ_{ij} in being a slowly varying function of temperature (cf. (3.6)). With a corresponding variable collision time $\tilde{\tau}_{ij}$, equation (4.5) is generalized to

$$\mathbf{R}_{ij} = -\frac{\rho_i\rho_j}{\rho}\left\langle \frac{\mathbf{g}_{ij}}{\tilde{\tau}_{ij}}\right\rangle, \tag{4.27}$$

whence from (4.26),

$$\pi\tilde{\sigma}_{ij}^2 g_{ij} = \frac{m_i + m_j}{\rho}\frac{1}{\tilde{\tau}_{ij}}. \tag{4.28}$$

Next we adopt a model for the force acting between colliding molecules. If, as in §3.2.1, the molecules are assumed to be power law centres of force, we can write

$$\pi\tilde{\sigma}_{ij}^2\, g_{ij} \propto T^{1-s} \qquad \left(s = \frac{1}{2} + \frac{2}{\nu - 1}\right), \tag{4.29}$$

where ν is the index of the force law. To find the appropriate temperature for (4.29), we retrace the motions of the two colliding molecules, say P_i and P_j, back one energy collision interval $\tilde{\tau}_{ij}^e$. Let T_0 denote the ambient temperature at the point of collision, G say, then since $-\tilde{\tau}_{ij}^e(\mathbf{w}_i - \mathbf{G}_{ij})$ is the displacement of P_i relative to G at a collision time earlier, the effective temperature of P_i is the path average

$$\tfrac{1}{2}\{T_0 + (T_0 - \tilde{\tau}_{ij}^e[\mathbf{w}_i - \mathbf{G}_{ij}]\cdot(\nabla T)_0)\}.$$

Allowing for P_j's motion similarly, we arrive at the effective temperature:

$$T = T_0 - \tfrac{1}{2}\tilde{\tau}_{ij}^e(\mu_j - \mu_i)\mathbf{g}_{ij}\cdot(\nabla T)_0, \tag{4.30}$$

since by (2.52) (in which $\mathbf{c}_i$ and $\mathbf{c}_j$ are now replaced by $\mathbf{w}_i$ and $\mathbf{w}_j$),

$$\mathbf{w}_i - \mathbf{G}_{ij} + \mathbf{w}_j - \mathbf{G}_{ij} = (\mu_j - \mu_i)\mathbf{g}_{ij}.$$

We can now combine (4.29) and (4.30) to obtain

$$\begin{aligned}\pi\tilde{\sigma}_{ij}^2\,\mathbf{g}_{ij} &= [\pi\sigma_{ij}^2\,\mathbf{g}_{ij}]_0\{1-\tfrac{1}{2}\tau_{ij}^e(\mu_j-\mu_i)(1-s)\mathbf{g}_{ij}\cdot(\nabla\ln T)_0\}\\ &= \frac{m_i+m_j}{\rho}\left(\frac{1}{\tau_{ij}}\right)\{1-\tfrac{1}{2}\tau_{ij}^e(\mu_j-\mu_i)(1-s)\mathbf{g}_{ij}\cdot\nabla T\},\end{aligned} \tag{4.31}$$

by (4.28) and the suppression of the subscript '0' and the tilde denoting 'variable'. Let $\mathbf{R}_{ij}^{th}$ denote that part of $\mathbf{R}_{ij}$ due to temperature gradients, then by (4.27) and (4.31),

$$\mathbf{R}_{ij}^{th} = -\frac{\rho_i\rho_j}{2\rho}(1-s)\xi_{ij}\,\Delta m\langle\mathbf{g}_{ij}\mathbf{g}_{ij}\rangle\cdot\nabla\ln T, \tag{4.32}$$

where $\qquad \xi_{ij}\equiv\tau_{ij}^e/\tau_{ij}, \qquad \Delta m\equiv\mu_i-\mu_j=(m_i-m_j)/(m_i+m_j).$

To calculate $\langle\mathbf{g}_{ij}\mathbf{g}_{ij}\rangle$ we follow the method used for $\langle g_{ij}\mathbf{g}_{ij}\rangle$ in (4.15) *et seq.* Because of isotropy, the integral over $g_{ij}\mathbf{g}_{ij}$ in (4.16) vanishes, but the corresponding integral over $\mathbf{g}_{ij}\mathbf{g}_{ij}$ does not. And because of the constraint

$$(\tfrac{1}{2}m_iv_i^2+\tfrac{1}{2}m_jv_j^2)\ll\tfrac{1}{2}kT$$

introduced following (4.15), in the present application we may omit the terms containing $\mathbf{v}_i$ and $\mathbf{v}_j$. The outcome is

$$\langle\mathbf{g}_{ij}\mathbf{g}_{ij}\rangle = \frac{2kT}{m_{ij}}\mathbf{1}. \tag{4.33}$$

Thus from (4.28), (4.32) and (4.33),

$$\mathbf{R}_{ij}^{th} = -\frac{\rho_i\rho_j}{\rho}(1-s)\xi_{ij}\Delta m\,\frac{kT}{m_{ij}}\,\nabla\ln T. \tag{4.34}$$

With this value substituted in (4.4), we find from (4.8) and (4.9) that

$$(\mathbf{v}_i-\mathbf{v}_j)\cdot(\mathbf{1}+\tau_{ij}\nabla\mathbf{v})+\tau_{ij}\mathrm{D}(\mathbf{v}_i-\mathbf{v}_j) = -\frac{n^2}{n_in_j}D_{ij}\mathbf{d}_{ij}-D_{ij}\alpha_T\nabla\ln T, \tag{4.35}$$

where $\qquad \alpha_T=\xi_{ij}(1-s)\,\dfrac{m_i-m_j}{x_im_i+x_jm_j} \qquad \left(x_i=\dfrac{n_i}{n},\quad x_j=\dfrac{n_j}{n}\right).$ (4.36)

We remind the reader that s is the power index in the viscosity law (see §3.2.1), $\mu\propto T^S$, and that

$$1-s=\frac{\nu-5}{2(\nu-1)}, \tag{4.37}$$

where ν is the molecular force law index. The terms containing τ_{ij} in (4.35) are actually $\mathrm{O}(\tau_{ij}^2)$ (see (4.9) and (4.24), and in a theory correct to $\mathrm{O}(\tau_{ij})$ may be omitted.

4.5 Limiting cases of thermal diffusion

Of the various transport processes, thermal diffusion is the most subtle and complicated. Although our simple mean-free-path approach makes the mechanism clear, it does not give accurate expressions for κ_T and α_T. Equation (4.36) is satisfactory, especially in the limiting cases to be considered shortly. Advanced kinetic theory yields a surprisingly intricate dependence of α_T on molecular diameters, force laws, concentrations and masses. The experimental support for the theory is incomplete and sometimes weak. For details the reader is referred to the text by Chapman and Cowling (1970).

4.5.1 *Lorentzian gas*

A binary gas mixture in which the particles P_j of one component are much lighter than the particles of the other, and with a concentration sufficiently low for the mutual encounters of P_j particles to be ignored, is known as a *Lorentzian* gas.

We shall choose $m_i \gg m_j$ and consider the behaviour of the lighter gas. Because the light molecules are isotropically scattered in collisions with the heavy molecules, both their velocity and energy persistences are zero (see §§3.3.2 and 3.3.3), and so the mean free paths for all transported properties are the same. Hence $\xi_{ij} = 1$, and (4.36) becomes

$$\alpha_T = (1 - s)\frac{m_i - m_j}{x_i m_i + x_j m_j}. \tag{4.38}$$

For a Lorentzian gas we take the limits $m_j/m_i \to 0$ and $x_j \to 0$ (i.e. $x_i \to 1$). Then by (4.37) and (4.38),

$$\alpha_T = 1 - s = \frac{\nu - 5}{2(\nu - 1)}, \tag{4.39}$$

which agrees with the exact value obtained by Enskog in 1912.

4.5.2 *Isotopic gas*

With isotope mixtures $m_i \approx m_j$ and so far as collisions are concerned, the gas behaves like a single species. In (4.31), τ_{ij} is the relaxation time for momentum transport and τ^e_{ij} is the relaxation time for energy transport. By (3.19) in a single species their ratio is 3/2. We shall therefore take $\xi_{ij} = 1.5$. Hence by (4.36) and (4.37), for isotope mixtures,

$$\alpha_T = 0.75\,\frac{\nu - 5}{\nu - 1}\,\frac{m_i - m_j}{x_i m_i + x_j m_j} \approx 0.75\,\frac{\nu - 5}{\nu - 1}\,\frac{m_i - m_j}{m_i + m_j}, \tag{4.40}$$

where the second form applies if neither x_i nor x_j are very small.

From the exact theory, Jones and Furry (1946) obtained

$$\alpha_T = 0.89\,C(\nu)\,\frac{\nu - 5}{\nu - 1}\,\frac{m_i - m_j}{m_i + m_j}, \tag{4.41}$$

Table 4.1 *The isotopic thermal diffusion factor α_0 at 25 °C*

Gas	s	α_0 (exp)	α_0 (th) eqn (4.40)	α_0 (th) eqn (4.41)
hydrogen	0.668	0.48	0.50	0.55
helium	0.657	0.43	0.51	0.55
methane	0.836	0.22	0.25	0.26
ammonia	1.10	0	−0.2	−0.1
neon	0.661	0.46	0.51	0.56
nitrogen	0.738	0.37	0.39	0.42
oxygen	0.773	0.35	0.34	0.36
argon	0.811	0.24	0.28	0.30

where $C(\nu)$ is a slowly varying function that increases monotonically from 0.874 at $\nu = 9$ to 0.906 at $\nu = 15$ and finally to 1.000 at $\nu = \infty$. Thus compared with the exact theory, our expression (4.41) under-estimates α_T by between 4 per cent at $\nu = 9$ to 19 per cent at $\nu = \infty$. But it is closer to the experimental values than this theory, at least for the gases in Table 4.1, where α_0 is the coefficient in

$$\alpha_T = \alpha_0 \frac{m_i - m_j}{m_i + m_j}. \tag{4.42}$$

4.6 The diffusion heat flux

A phenomenon thermodynamically reciprocal to thermal diffusion is the energy flux in a gas mixture resulting from the difference in the force acting on unit mass of each species, namely

$$\mathbf{F}_{ij} \equiv (\mathbf{F}_i - \frac{1}{\rho_i}\nabla p_i) - (\mathbf{F}_j - \frac{1}{\rho_j}\nabla p_j). \tag{4.43}$$

Let this heat flux be denoted by $\mathbf{q}_D$; then we shall show that

$$\mathbf{q}_D = \tau_{ij}\kappa_T p \mathbf{F}_{ij}, \tag{4.44}$$

where κ_T is the thermal diffusion ratio defined in (4.25) and τ_{ij} is the relaxation time for momentum transfer.

4.6.1 *Rate of energy transfer between species*

We shall start with the expression for the rate at which energy flux is transferred from the j-species to the i-species per unit volume:

$$\mathbf{Q}_{ij} = -n_i n_j m_{ij} \langle \pi\tilde{\sigma}_{ij}^2 g_{ij} (\tfrac{1}{2} g_{ij}^2 \mathbf{g}_{ij}) \rangle, \tag{4.45}$$

which follows on replacing $\mathbf{g}_{ij}$ in (4.26) by $\frac{1}{2}g_{ij}^2\mathbf{g}_{ij}$. We next adopt an expansion for $\pi\tilde{\sigma}_{ij}^2 g_{ij}$ similar to that yielding (4.31), save that instead of starting from (4.29), we use the more general form (see (3.6)):

$$\pi\tilde{\sigma}_{ij}^2 g_{ij} \propto \left(\overline{g_{ij}^2}\right)^{1-s}. \tag{4.46}$$

Even in the absence of a temperature gradient, variations in $\overline{g_{ij}^2}$ can occur via the time dependence of g_{ij}^2. From (3.25) and (4.1) applied to each species, we obtain

$$\begin{aligned}\dot{\mathbf{g}}_{ij} = \dot{\mathbf{w}}_i - \dot{\mathbf{w}}_j &= (\mathbf{F}_i - \frac{1}{\rho_i}\nabla p_i + \mathcal{F}_i) - (\mathbf{F}_j - \frac{1}{\rho_j}\nabla p_j + \mathcal{F}_j) \\ &= \mathbf{F}_{ij} + (\mathcal{F}_i - \mathcal{F}_j), \end{aligned} \tag{4.47}$$

where in the present application we can omit the small viscous and friction terms.

The time scale τ_g for changes in $\overline{g_{ij}^2}$ is the same as for changes in the temperature. From (4.31) this is $\tau_g = \tau_{ij}^e(\mu_j - \mu_i)$. In the time τ_g, $\mathbf{g}_{ij}$ increases from $(\mathbf{g}_{ij}^{(0)} - \tau_g \dot{\mathbf{g}}_{ij})$ say, to $\mathbf{g}_{ij}^{(0)}$, with an average of $\mathbf{g}_{ij}^{(0)} - \frac{1}{2}\tau_g\dot{\mathbf{g}}_{ij}$. Similarly the average value of g_{ij}^2 is

$$\left[g_{ij}^2\right]_{av} = \left[g_{ij}^2\right]_0 - \tau_g \tfrac{1}{2}\dot{\mathbf{g}}_{ij}^2 = \left[g_{ij}^2\right]_0 - \tau_g \mathbf{g}_{ij}\cdot(\mathbf{F}_{ij} + \mathcal{F}_i - \mathcal{F}_j).$$

With this value in (4.46), in place of (4.31) we get

$$\pi\tilde{\sigma}_{ij}^2 g_{ij} = \frac{m_i + m_j}{\rho}\frac{1}{\tau_{ij}}\left\{1 - \tau_{ij}^e(\mu_j - \mu_i)(1-s)\mathbf{g}_{ij}\cdot(\mathbf{F}_{ij} + \mathcal{F}_i - \mathcal{F}_j)/g_{ij}^2\right\}. \tag{4.48}$$

Substituting into (4.45), we find

$$\mathbf{Q}_{ij} = -\frac{\rho_i\rho_j}{2\rho}\left\langle \frac{g_{ij}^2\mathbf{g}_{ij}}{\tau_{ij}}\right\rangle + \frac{\rho_i\rho_j}{2\rho}\xi_{ij}(\mu_j - \mu_i)(1-s)\langle \mathbf{g}_{ij}\mathbf{g}_{ij}\cdot(\mathbf{F}_{ij} + \mathcal{F}_i - \mathcal{F}_j)\rangle, \tag{4.49}$$

where $\xi_{ij} \equiv \tau_{ij}^e/\tau_{ij}$. On adopting the method used to evaluate $\langle g_{ij}\mathbf{g}_{ij}\rangle$ in (4.15) *et seq.*, we find that the first right-hand term yields a contribution to $\mathbf{Q}_{ij}$ proportional to $(\mathbf{v}_j - \mathbf{v}_i)$, which changes sign when i and j are interchanged. It therefore gives rise to a balancing interchange of energy flux that leaves the heat flux for the mixture unaltered. As the random collisional forces $\mathcal{F}_i$ and $\mathcal{F}_j$ appearing in the second term in (4.49) are isotropically distributed (see (3.24)), the average $\langle \mathbf{g}_{ij}\mathbf{g}_{ij}\cdot(\mathcal{F}_i - \mathcal{F}_j)\rangle$ is negligible. Thus by (4.33) we obtain

$$\mathbf{Q}_{ij}^0 = \frac{\rho_i\rho_j}{\rho}(1-s)\xi_{ij}(\mu_j - \mu_i)\frac{kT}{m_{ij}}\mathbf{F}_{ij},$$

for the average rate at which energy flux is transferred between the species due to the force difference $\mathbf{F}_{ij}$. This term is unaltered when i and j are interchanged. It follows from (4.25) and (4.36) that energy flux is absorbed by the gas *mixture* at the rate

$$\tfrac{1}{2}(\mathbf{Q}_{ij} + \mathbf{Q}_{ji}) = x_i x_j \alpha_T p \mathbf{F}_{ij} = \kappa_T p \mathbf{F}_{ij}. \tag{4.50}$$

4.6.2 *The diffusion heat flux*

The processes described in (4.35) and (4.49) have the same time scale τ_{ij}. To convert the heat-flux absorption rate into the diffusion heat flux $\mathbf{q}_D$, we multiply by τ_{ij}, giving

$$\mathbf{q}_D = \kappa_T \tau_{ij} p \mathbf{F}_{ij}, \tag{4.51}$$

in agreement with (4.44). Another form for the diffusion heat flux follows from (4.6):

$$\mathbf{q}_D = p\kappa_T(\mathbf{v}_i - \mathbf{v}_j). \tag{4.52}$$

Let $\hat{\mathbf{x}}$ denote a unit vector in such a direction that $\hat{\mathbf{x}} \cdot (\mathbf{v}_i - \mathbf{v}_j) > 0$. Then by (4.52), $\hat{\mathbf{x}} \cdot \mathbf{q}_D > 0$, and the diffusion heat flux will tend to produce an incremental temperature gradient such that $\hat{\mathbf{x}} \cdot \nabla T > 0$. It follows from (4.23) that this acts to reduce $\hat{\mathbf{x}} \cdot (\mathbf{v}_i - \mathbf{v}_j)$, thus preserving thermal stability.

The appearance of the same coefficient κ_T in (4.23) and (4.51) is a special case of a general reciprocal theorem in the thermodynamics of irreversible processes (see Woods 1986).

Exercises 4

4.1 Show that the fluid velocity of component '1' relative to component '2' in a gas mixture can be expressed as

$$\mathbf{v}_1 - \mathbf{v}_2 = \frac{\rho p}{\rho_1 \rho_2} \tau_{12} \nabla \left(\frac{n_1}{n}\right) - \tau_{12} \left(\frac{m}{m_1} - \frac{m}{m_2}\right) \frac{\nabla p}{\rho} \qquad (m = \rho/n),$$

and provide a physical explanation of each of these terms.

4.2 With the help of (2.45) and the expansion (cf. (3.39))

$$n(\mathbf{r} \pm \tau_\phi \mathbf{c}, t \pm \tau_\phi) \approx n(\mathbf{r}, t) \pm \tau_\phi (Dn + \mathbf{c} \cdot \nabla n),$$

where τ_ϕ is the relaxation time for a molecular property ϕ, show that the flux of ϕ across a surface element at $(\mathbf{r}, t)$ is

$$\mathbf{J}_\phi = -D_\phi \nabla(n\phi)$$

where $D_\phi = \frac{2}{3} C^2 \tau_\phi = \frac{4}{3} p \tau_\phi / \rho$ is the diffusivity. Why is τ_1 (see §3.3.3) the appropriate value for τ_ϕ when $\phi = 1$? Deduce that the above approach yields a value $\frac{4}{3}\mu/\rho$ for the coefficient of self-diffusion.

4.3 A mixture of hard-core molecules has two components. Its temperature and total pressure are constant and there are no body forces. Show that the relative fluid velocity of the components is

$$\mathbf{v}_2 - \mathbf{v}_1 = -\frac{3}{8\pi} \frac{1}{n_1 n_2 \sigma_{12}^2} \left(\frac{\pi k T}{2 m_{12}}\right)^{1/2} \nabla n_2.$$

4.4 In a magnetoplasma (see §9.2.1) the Lorentz force acting on the ions $(r = i)$ and the electrons $(r = e)$ is $\mathbf{F}_r = (Q_r/m_r)(\mathbf{E} + \mathbf{v}_r \times \mathbf{B})$, where $\mathbf{E}$, $\mathbf{B}$ are the electric and magnetic fields and Q_r is the particle charge. Given that the plasma is electrically neutral $(Q_i n_i + Q_e n_e = 0)$, that $m_i \gg m_e$ and that the electric current is the flux of charge, $\mathbf{j} = Q_i n_i \mathbf{v}_i + Q_e n_e \mathbf{v}_e$, show that

$$\mathbf{j} = \sigma(\mathbf{E} + \mathbf{v}_e \times \mathbf{B} + \frac{1}{e n_e} \nabla p_e) \qquad (\sigma = e^2 n_e \tau_{ei}/m_e, \quad e = -Q_e)$$

is a good approximation.

4.5 A plasma may be approximated as a Lorentzian gas, with the Coloumb force law $\nu = 2$. Show that thermal diffusion adds a term $\mathbf{j}_d$ to the electric current, where

$$\mathbf{j}_d = \sigma \frac{3}{2}\frac{k}{e} \nabla T,$$

in which $-e$ is the electron charge, the ionization number, $Z = Q_i/e$, is unity, and σ is defined in the previous exercise.

4.6 Show that if the i-th and j-th species have different temperatures, (4.16) is modified as follows:

$$\frac{m_{ij}}{2kT} g_{ij}^2 \quad \longrightarrow \quad \frac{m_{ij}}{2k}\left(\frac{1}{m_i T_i} + \frac{1}{m_j T_j}\right) g_{ij}^2,$$

$$\frac{m_i + m_j}{2kT} G_{ij}^2 \quad \longrightarrow \quad \frac{1}{2k}\left(\frac{m_i}{T_i} + \frac{m_j}{T_j}\right) G_{ij}^2,$$

$$\frac{m_{ij}}{kT}(\mathbf{v}_i - \mathbf{v}_j) \quad \longrightarrow \quad \frac{m_{ij}}{2k}\left(\frac{1}{T_i} + \frac{1}{T_j}\right)(\mathbf{v}_i - \mathbf{v}_j),$$

and that the coefficient of $(\mathbf{v}_i - \mathbf{v}_j)$ in (4.17) is replaced by

$$\frac{2m_{ij}}{3k\sqrt{\pi}}\left(\frac{1}{T_i} + \frac{1}{T_j}\right)\left\{2k\left(\frac{T_i}{m_i} + \frac{T_j}{m_j}\right)\right\}^{-3/2}\left\{\frac{m_{ij}^2}{2k}\left(\frac{1}{m_i T_i} + \frac{1}{m_j T_j}\right)\right\}^{-3}.$$

4.7 Consider the case when the i-th species is relatively massive ($m_i \gg m_j$), and $\mathcal{C}_i \ll \mathcal{C}_j$. By setting $\mathbf{v}_i$ equal to zero, show that in this case the third modification of exercise 4.6 is replaced by

$$\frac{m_{ij}}{kT}(\mathbf{v}_i - \mathbf{v}_j) \rightarrow \frac{m_j}{kT_j}(-\mathbf{v}_i)$$

and deduce that (4.20) becomes

$$D_{ij} = \frac{3}{8\pi\sigma_{ij}^2 n}\left(\frac{\pi k T_j}{2m_j}\right)^{1/2}.$$

4.8 Show that for the case described in exercise 4.7, (4.30) is replaced by

$$T_j \approx T_{0j} + \tfrac{1}{2}\tilde{\tau}_{ij}^e \mu_i \mathbf{g}_{ij} \cdot \nabla T_j, \qquad T_i \approx T_{0i},$$

and that the thermal diffusion is almost entirely due to the lighter particles.

5

INTERMEDIATE KINETIC THEORY

5.1 Introduction

In the elementary theory, the peculiar velocity $\mathbf{c}$ is treated as being a dependent variable. It depends on an initial value $\mathbf{c}_0$ chosen from the equilibrium distribution f_0, on the lapsed time since the last collision and on local temperature and fluid velocity gradients (see (3.40)). The procedure makes no use of the *actual* distribution function f, since f_0 is entirely sufficient for averaging purposes. Details of the collisional process are suppressed by adopting relaxation times τ_1 and τ_2 for momentum and energy transfer. These times could be regarded as being a phenomenological input to the theory. However, if their dependence on an assumed molecular force law is required, then advanced kinetic theory is unavoidable. We shall present two forms of this theory in the following chapters.

Before introducing the advanced theory, we shall explain a phenomenological theory intermediate between the elementary and advanced theories. Like the elementary theory, this depends on the relaxation times τ_1 and τ_2, but it is based on a differential equation for the evolution of the velocity distribution function $f(\mathbf{r}, \mathbf{c}, t)$, with $\mathbf{c}$ now in the role of an *independent* variable; it therefore more closely resembles the advanced than the elementary theory. It is in fact an approximation to the advanced theory initiated by Boltzmann and developed by Enskog. Its merit is simplicity and physical clarity, which makes it especially valuable in circumstances when the validity of the more complex advanced theory is uncertain.

5.1.1 *The kinetic equation*

Both the intermediate and advanced theories take as their starting point a relation for the rate of change of $f(\mathbf{r}, \mathbf{c}, t)$ (or $f(\mathbf{r}, \mathbf{w}, t)$) with time, known as the *kinetic equation*. The left-hand side of this equation is the rate of change of $f(\mathbf{r}, \mathbf{c}, t)$ following a volume element $d\boldsymbol{\nu}_a = d\mathbf{r}\, d\mathbf{c}$ of phase space moving with velocity $\mathbf{c}$ and acceleration $\mathbf{a}$ relative to the convected or fluid frame and the right-hand side is the rate of change of f caused by collisions. It will be shown in 5.2.2 that conservation of particle numbers requires these rates to balance, i.e.

$$\mathrm{D}f + \mathbf{c} \cdot \nabla f + \mathbf{a} \cdot \frac{\partial f}{\partial \mathbf{c}} = \left(\frac{df}{dt}\right)_{col}, \tag{5.1}$$

where D is the convective derivative in physical space:

$$\mathrm{D} \equiv \frac{\partial}{\partial t} + \mathbf{v} \cdot \nabla. \tag{5.2}$$

The left-hand side of (5.1) is termed the 'streaming derivative' and the right-hand side is called the 'collision operator'. Advanced kinetic theory aims to model $(df/dt)_{col}$ by considering the collisional process in detail, while the intermediate theory is content to ignore the detail, adopting in its place a relaxation time τ for the distribution f to decay to its equilibrium value f_0. This model, due to Bhatnager, Gross and Krook (1954), yields the very simple collision operator

$$\left(\frac{df}{dt}\right)_{col} = -\frac{1}{\tau}(f - f_0). \tag{5.3}$$

Then from (5.1), in the absence of gradients and accelerations,

$$\frac{\partial f}{\partial t} = \frac{1}{\tau}(f_0 - f),$$

so that $(f - f_0)$ decays like the exponential $e^{-t/\tau}$.

5.1.2 *Disadvantages of the relaxation model*

One disadvantage of the model is that a single relaxation time is inadequate to represent the collisional process with any accuracy. The value of τ in (5.3) depends on whether it is the momentum or energy term in f that is under consideration. The physical reasons for the distinction between the relaxation times τ_1 for momentum and τ_2 for energy have been presented in §3.3. Following a collision, a given particle P will, on average, retain a component of velocity in the direction of its original motion. Further collisions are required to 'halt' P in the sense of stopping its original motion in a given direction. This persistence of motion depends on the particle speed c, and as its momentum and energy are different functions of c, it follows that, on average, particles will transport their momentum and energy through different stopping distances and take different times to do so. The problem can be by-passed by adopting distinct values for τ depending on the diffusion process being studied.

Another, more fundamental difficulty with the BGK model—that we shall return to in §5.8—is that it is restricted to small values of the Knudsen number ϵ (see (3.30)), although this constraint is not made evident in the usual derivation. Provided gradients are 'small', diffusion is usually well described by terms first order in ϵ. But with larger gradients, such as occur in shock waves, ultrasonic sound waves, and especially in magnetoplasmas with strong magnetic fields, terms of order ϵ^2 are important.

5.2 General form of the kinetic equation

5.2.1 *The streaming derivative*

The differential of a scalar function $\psi(\mathbf{r}, \mathbf{c}, t)$ is

$$d\psi = \left(\frac{\partial \psi}{\partial t}\right)_{\mathbf{r},\mathbf{c}} dt + \left(\frac{\partial \psi}{\partial \mathbf{r}}\right)_{\mathbf{c},\, t} \cdot d\mathbf{r} + \left(\frac{\partial \psi}{\partial \mathbf{c}}\right)_{\mathbf{r},\, t} \cdot d\mathbf{c}, \tag{5.4}$$

from which it follows—in a simplified notation—that the rate of change of ψ following an element $d\boldsymbol{\nu}_a = d\mathbf{r}\, d\mathbf{c}$ of phase space moving with velocity $\mathbf{c}$ and acceleration $\mathbf{a}$, is given by*

$$\mathbb{D}_a \psi = \mathrm{D}\psi + \mathbf{c} \cdot \nabla\psi + \mathbf{a} \cdot \frac{\partial \psi}{\partial \mathbf{c}}, \tag{5.5}$$

where D is the convective derivative in physical space defined in (5.2). The streaming derivative of a *vector* function $\boldsymbol{\psi}_1(\mathbf{r}, \mathbf{c}, t)$ has a similar form, save that we need to extend the description of the motion of $d\boldsymbol{\nu}_a$ to include its *spinning* about its centre with angular velocity $\boldsymbol{\Omega} = \frac{1}{2}\nabla \times \mathbf{v}$. The reason for this was given in §3.4.1, and further relevant remarks will be found in §5.7. At this point we require some expressions for time derivatives in rotating frames.

Mathematical note 5.2 *Time derivatives in rotating reference frames*

Let $\mathbf{i}$, $\mathbf{j}$, $\mathbf{k}$ be an orthogonal triad of unit vectors rigidly attached to a frame L^* that is rotating with an angular velocity $\boldsymbol{\Omega}$ relative to a frame L. Let D and D^* denote the time derivatives in L and L^*. Then in an infinitesimal time interval dt, viewed from L, the unit vector $\mathbf{i}$ is displaced through a distance $\boldsymbol{\Omega} \times \mathbf{i}\, dt$. It follows that $\mathrm{D}\mathbf{i} = \boldsymbol{\Omega} \times \mathbf{i}$, and similarly for the other unit vectors.

For a scalar function ψ_0 it is evident that

$$\mathrm{D}^*\psi_0 = \mathrm{D}\psi_0. \tag{5.6}$$

Now consider the vector $\boldsymbol{\psi}_1 = \psi_x \mathbf{i} + \psi_y \mathbf{j} + \psi_z \mathbf{k}$. Its rate of change in L is

$$\mathrm{D}\boldsymbol{\psi}_1 = (\mathbf{i}\mathrm{D}\psi_x + \boldsymbol{\Omega} \times \mathbf{i}\psi_x) + (\mathbf{j}\mathrm{D}\psi_y + \boldsymbol{\Omega} \times \mathbf{j}\psi_y) + (\mathbf{k}\mathrm{D}\psi_z + \boldsymbol{\Omega} \times \mathbf{k}\psi_z),$$

i.e.

$$\mathrm{D}\boldsymbol{\psi}_1 = \mathrm{D}^*\boldsymbol{\psi}_1 + \boldsymbol{\Omega} \times \boldsymbol{\psi}_1,$$

where D^* is the time derivative with fixed values of $\mathbf{i}$, $\mathbf{j}$, $\mathbf{k}$, i.e. in the frame L^*. Hence the rate of change of the vector in the rotating frame is given by

$$\mathrm{D}^*\boldsymbol{\psi}_1 = \mathrm{D}\boldsymbol{\psi}_1 - \boldsymbol{\Omega} \times \boldsymbol{\psi}_1. \tag{5.7}$$

To deduce the time derivative of a second-order tensor $\boldsymbol{\psi}_2$ we proceed as follows. First consider the dyad $\mathbf{ab}$. Differentiating each vector separately,

*The barred D notation is adopted to signal the joining of physical and velocity space and the subscript 'a' to denote the acceleration relative to the fluid frame.

$$\begin{aligned}\mathrm{D}^*(\mathbf{ab}) = (\mathrm{D}^*\mathbf{a})\mathbf{b} + \mathbf{a}\mathrm{D}^*\mathbf{b} &= (\mathrm{D}\mathbf{a})\mathbf{b} - \boldsymbol{\Omega} \times \mathbf{ab} + \mathbf{a}\mathrm{D}\mathbf{b} - \mathbf{a}\boldsymbol{\Omega} \times \mathbf{b} \\ &= \mathrm{D}(\mathbf{ab}) - \boldsymbol{\Omega} \times \mathbf{ab} + \mathbf{ab} \times \boldsymbol{\Omega},\end{aligned}$$

and as the tensor $\boldsymbol{\psi}_2$ can be decomposed into a sum of nine independent dyads (cf. (1.19)), we deduce that

$$\mathrm{D}^*\boldsymbol{\psi}_2 = \mathrm{D}\boldsymbol{\psi}_2 - \boldsymbol{\Omega} \times \boldsymbol{\psi}_2 + \boldsymbol{\psi}_2 \times \boldsymbol{\Omega}. \tag{5.8}$$

In the following we shall define the streaming derivative more generally than in (5.5):

$$\mathbb{D}_a \equiv \mathrm{D}^* + \mathbf{c} \cdot \nabla + \mathbf{a} \cdot \frac{\partial}{\partial \mathbf{c}}, \tag{5.9}$$

where D^* is the operator occurring in (5.6), (5.7), or (5.8), depending on the nature of the operand.

In (5.9) we are free to select both the velocity **c** and the acceleration **a** to suit our purpose. It should be remembered that these vectors describe the motion of a *phase-space element*, and do not necessarily relate to particular molecules. Once we have chosen **c**, then molecules that happen to have the same velocity will remain in $d\boldsymbol{\nu}_a$ until removed either by a mismatch of their acceleration with **a**, or by being scattered out of $d\boldsymbol{\nu}_a$ by collisions.

5.2.2 *Conservation of particle numbers*

The number of particles in an element $d\boldsymbol{\nu}_a$ is $f\, d\boldsymbol{\nu}_a$, and in a streaming frame moving with velocity **c** and acceleration **a** relative to the fluid (or convected) frame P_c, its rate of change is $\mathbb{D}_a(f\, d\boldsymbol{\nu}_a)$. Now suppose that **a** is chosen so that the losses and gains experienced by the element $d\boldsymbol{\nu}_a$ are due only to collisions, and let $(df/dt)_{col}\, d\boldsymbol{\nu}_a$ denote this rate. Then by conservation of particle numbers,

$$\mathbb{D}_a(f\, d\boldsymbol{\nu}_a) = (df/dt)_{col}\, d\boldsymbol{\nu}_a. \tag{5.10}$$

In §1.10.1 we proved a famous relation due to Euler relating the rate of change of a volume element $d\mathbf{r}$ in physical space to the divergence of the velocity vector:

$$\mathrm{D}(d\mathbf{r}) = \nabla \cdot \mathbf{v}\, d\mathbf{r}. \tag{5.11}$$

In phase space the 6D 'velocity' is $(\mathbf{v} + \mathbf{c}, \mathbf{a})$, and its divergence is $(\partial/\partial\mathbf{r}, \partial/\partial\mathbf{c})$, so by a generalization of the argument given in §1.10.1, we obtain*

$$\mathbb{D}_a(d\boldsymbol{\nu}_a) = (\frac{\partial}{\partial \mathbf{r}} \cdot \mathbf{v} + \frac{\partial}{\partial \mathbf{c}} \cdot \mathbf{a})\, d\boldsymbol{\nu}_a, \tag{5.12}$$

the term $(\partial/\partial\mathbf{r}) \cdot \mathbf{c}$ vanishing because **r** and **c** are independent variables. Thus $\mathbb{D}_a(d\boldsymbol{\nu}_a)$ is zero provided the chosen acceleration satisfies the

*Remember that $d\boldsymbol{\nu}_a$ is a scalar.

constraint*

$$\nabla \cdot \mathbf{v} + \frac{\partial}{\partial \mathbf{c}} \cdot \mathbf{a} = 0. \tag{5.13}$$

Whereupon (5.10) reduces to the general form of the kinetic equation,

$$\mathbb{D}_a f = \mathbb{C}_a(f) \qquad \big(\mathbb{C}_a(f) \equiv (df/dt)_{col}\big), \tag{5.14}$$

in which we have introduced a more convenient notation for the collision operator.

We can divide the acceleration $\mathbf{a}$ of $d\boldsymbol{\nu}_a$ into a component $\mathbf{a}_0$ say, that is independent of $\mathbf{c}$ and a term arising from fluid shear. In the absence of collisions it follows from (3.27) that the particle acceleration due to fluid shear is†

$$\dot{\mathbf{c}} = -\mathbf{c} \cdot \mathbf{e} \qquad \Big(\mathbf{e} \equiv \tfrac{1}{2}(\nabla \mathbf{v} + \widetilde{\nabla \mathbf{v}})\Big), \tag{5.15}$$

so if $d\boldsymbol{\nu}_a$ also has this acceleration, there will be no streaming losses due to a mismatch of accelerations. With $\mathbf{a} = \dot{\mathbf{c}} + \mathbf{a}_0$,

$$\frac{\partial}{\partial \mathbf{c}} \cdot \mathbf{a} = -\frac{\partial}{\partial \mathbf{c}} \cdot (\mathbf{c} \cdot \mathbf{e}) = -\frac{\partial \mathbf{c}}{\partial \mathbf{c}} : \mathbf{e} = -\mathbf{1} : \mathbf{e} = -\nabla \cdot \mathbf{v},$$

so (5.13) is satisfied.

5.2.3 *Choices for the acceleration of the phase-space element*

There are two cases of importance in the theory. The first is when $\mathbf{a}_0 = 0$, i.e. when the phase space element has velocity $\mathbf{c}$ and acceleration $\dot{\mathbf{c}}$. We shall distinguish this choice for the acceleration by adding the subscript 's'. Thus

$$\mathbb{D}_s \equiv \mathrm{D}^* + \mathbf{c} \cdot \nabla + \dot{\mathbf{c}} \cdot \frac{\partial}{\partial \mathbf{c}} \qquad (\dot{\mathbf{c}} = -\mathbf{c} \cdot \mathbf{e}) \tag{5.16}$$

is the streaming derivative that follows $d\boldsymbol{\nu}_s$. In this case the acceleration is that experienced by a molecule, referred to a frame that is *swept* along with the fluid. This frame has the merit that, relative to $d\boldsymbol{\nu}_s$, all the collective motions of the particles vanish, leaving purely random particle activity. In particular, because $d\boldsymbol{\nu}_s$ has the same *average* acceleration as the molecules contained in it, there are no streaming losses from this element of phase space.(see §3.9.4). To distinguish it from the convective derivation D in real space, we shall term it the 'swept' (streaming) derivative.

For the second case we take $\mathbf{a}_0 = \mathbf{F} - \mathrm{D}\mathbf{v}$, where $\mathbf{F}$ is the body force per unit mass and $\mathrm{D}\mathbf{v}$ is the fluid acceleration. In the laboratory frame $\mathbf{F}$ is the acceleration of the particles in *free flight*, so $\mathbf{a}_0$ is their average acceleration *between* collisions as measured in the convected frame. From

*In the following we shall replace $\partial/\partial\mathbf{r}$ by its more familiar form, ∇.

†A formal derivation of this relation is given in Mathematical note §7.5.

the equation of fluid motion*,

$$\mathrm{D}\mathbf{v} = \mathbf{P} + \mathbf{F} \qquad \left(\mathbf{P} \equiv -\frac{1}{\rho}\nabla \cdot \mathsf{p}\right),$$

it follows that $\mathbf{a}_0 = -\mathbf{P}$. For this case we shall omit subscripts and write for the 'free' streaming derivative of a phase space element $d\boldsymbol{\nu}$,

$$\mathbb{D} = \mathrm{D}^* + \mathbf{c} \cdot \nabla + \mathbf{a} \cdot \frac{\partial}{\partial \mathbf{c}}, \tag{5.17}$$

where

$$\mathbf{a} = -\mathbf{P} + \dot{\mathbf{c}} = -\mathbf{P} - \mathbf{c} \cdot \mathbf{e}. \tag{5.18}$$

Corresponding to these choices of the acceleration of the phase space element, the kinetic equation can be written in either of two equivalent forms:

$$\mathbb{D}f = \mathbb{C}(f), \tag{5.19}$$

or

$$\mathbb{D}_S f = \mathbb{C}_S(f), \qquad \mathbb{C}_S = \mathbb{C} + \mathbf{P} \cdot \frac{\partial}{\partial \mathbf{c}}. \tag{5.20}$$

To gain a clear physical picture of the collision process, we shall find it useful to have both forms of the kinetic equation.

Equation (5.19), but without the contribution to $\mathbf{a}$ from the fluid spin, i.e. with an acceleration $-\mathbf{P} - \mathbf{c} \cdot \nabla \mathbf{v}$, is the usual form adopted for the streaming derivative. Unfortunately—as we shall explain in §8.5—this leads to errors when the theory is pressed to second order in the Knudsen number (Woods 1983a).

There are several detailed models for the collision operator $\mathbb{C}$, notably those associated with the names Boltzmann (§7.2.3) and Fokker-Planck (§10.1.2). The simple, phenomenological model due to Bhatnager, Gross and Krook—the BGK model—is explained in §5.4.

5.3 Constraints on the collision operator

5.3.1 *The moment equation*

Let $\varphi(\mathbf{c})$ denote a dynamical function, then by (2.1) and (2.2) its average is $\overline{\varphi}$, where

$$n\overline{\varphi} = \int \varphi f \, d\mathbf{c}, \qquad n = \int f \, d\mathbf{c}. \tag{5.21}$$

Multiplying (5.19) by φ and using the independence of $\mathbf{r}$, $\mathbf{c}$ and t, we get

$$\mathrm{D}^*(\varphi f) + \nabla \cdot (\mathbf{c}\varphi f) + \frac{\partial}{\partial \mathbf{c}} \cdot (\mathbf{a}\varphi f) - f \frac{\partial}{\partial \mathbf{c}} \cdot (\mathbf{a}\varphi) = \varphi \mathbb{C}(f).$$

*Since we are formulating the theory in a frame convected with the fluid, the introduction of the equation of fluid motion at this stage is a necessary step; in any case the widely held view that it is possible to *reduce* fluid dynamics to kinetic theory is incorrect (see §5.3.2). It should be remembered that the kinetic equation is no more than a balance equation for particle *numbers*, whereas the equation of fluid motion is a particular form of Newton's second law of motion. The latter cannot be deduced from the former without additional assumptions.

The next step is to integrate this equation over velocity space. By the divergence theorem the integral of the third term becomes a surface integral over the infinite sphere $|\mathbf{c}| = R \to \infty$ that vanishes provided that as $R \to \infty$, $|\varphi \mathbf{a} f| \to 0$ rapidly enough. With collisions $f \sim f_0 \sim \exp(-\beta R^2)$, so the surface integral will vanish, leaving

$$\mathrm{D}^*(n\overline{\varphi}) + \nabla \cdot (n\overline{\mathbf{c}\varphi}) - n\overline{\frac{\partial}{\partial \mathbf{c}} \cdot \mathbf{a}\varphi} = \int \varphi \mathbb{C}(f)\, d\mathbf{c}. \tag{5.22}$$

We shall omit the argument from $\mathbb{C}(f)$ in the following.

With $\mathbf{a} = -\mathbf{P} - \mathbf{c} \cdot \mathbf{e}$, we obtain the moment equation

$$\mathrm{D}^*(n\overline{\varphi}) + \nabla \cdot (n\overline{\mathbf{c}\varphi}) + n\mathbf{e} : \overline{\frac{\partial}{\partial \mathbf{c}}(\mathbf{c}\varphi)} + n\mathbf{P} \cdot \overline{\frac{\partial \varphi}{\partial \mathbf{c}}} = \int \varphi \mathbb{C}\, d\mathbf{c}. \tag{5.23}$$

Now let $\varphi = m$, the particle mass. As $\overline{\mathbf{c}} = 0$, and

$$n\mathbf{e} : \overline{\frac{\partial}{\partial \mathbf{c}} \mathbf{c} m} = \rho \mathbf{e} : \mathbf{1} = \rho \nabla \cdot \mathbf{v},$$

(5.23) becomes

$$\mathrm{D}\rho + \rho \nabla \cdot \mathbf{v} = \int m \mathbb{C}\, d\mathbf{c}. \tag{5.24}$$

With $\varphi = m\mathbf{c}$, by (2.6) and

$$n\mathbf{e} : \overline{\frac{\partial}{\partial \mathbf{c}} m\mathbf{c}\mathbf{c}} = \rho \mathbf{e} : \left(\mathbf{1}\overline{\mathbf{c}} + \overline{\mathbf{c}}\mathbf{1}\right) = 0,$$

we get

$$\nabla \cdot \mathsf{M} + \rho \mathbf{P} = \int m\mathbf{c} \mathbb{C}\, d\mathbf{c}. \tag{5.25}$$

Finally, with $\varphi = \frac{1}{2} m c^2$, by (2.5) and

$$n\mathbf{e} : \overline{\frac{\partial}{\partial \mathbf{c}}(\tfrac{1}{2} m c^2 \mathbf{c})} = \tfrac{1}{2}\rho \mathbf{e} : \{2\overline{\mathbf{c}\mathbf{c}} + \overline{c^2}\mathbf{1}\} = \mathbf{e} : \mathsf{M} + \rho u \mathbf{1} : \mathbf{e},$$

(5.23) gives

$$\mathrm{D}(\rho u) + \nabla \cdot \boldsymbol{\xi} + \rho u \nabla \cdot \mathbf{v} + \mathbf{e} : \mathsf{M} = \int \tfrac{1}{2} m c^2 \mathbb{C}\, d\mathbf{c}. \tag{5.26}$$

If the flux is collided flux, $\mathsf{M} = \mathsf{p}$, $\boldsymbol{\xi} = \mathbf{q}$, (see §2.1) and in this case it follows from the conservation laws of §1.9.2 that (5.24), (5.25), and (5.26) impose the following constraints on the collision operator $\mathbb{C}$:

$$\int \{m,\ m\mathbf{c},\ \tfrac{1}{2} m c^2\} \mathbb{C}\, d\mathbf{c} = \{0,\ 0,\ 0\}. \tag{5.27}$$

Higher moments obtained by the choices $\varphi = \mathbf{c}\mathbf{c}$ and $\varphi = \mathbf{c}c^2$ lead to expressions for p and $\mathbf{q}$; these are pursued in §§6.6 and 6.7.

5.3.2 *Moments in a collisionless gas*

It is also possible to deduce the constraints in (5.27) directly from certain forms of the collision operator. In particular, if $\mathbb{C}$ has the form adopted by Boltzmann, these equations follow directly from the conservation of particle mass, momentum and energy during collisions (see §7.3.2). We have used the macroscopic forms of the conservation laws to obtain the constraints. Had we taken the former approach, (5.24), (5,25), and (5.26) would have reduced to

$$\mathrm{D}\rho + \rho\nabla\cdot\mathbf{v} = 0, \qquad \rho\mathrm{D}\mathbf{v} + \nabla\cdot\mathbf{M} - \rho\mathbf{F} = 0,$$

and

$$\mathrm{D}(\rho u) + \nabla\cdot\boldsymbol{\xi} + \rho u\nabla\cdot\mathbf{v} + \mathbf{e}:\mathbf{M} = 0,$$

in the second of which we have used the equation of fluid motion to remove $\mathbf{P}$. This restores the moments to those that would be obtained from the kinetic equation, expressed in the laboratory frame.

Notice that while the conservation of fluid mass has been obtained from the kinetic equation by taking its zeroth velocity moment, it is not true that its first and second velocity moments are the same as the equations for the conservation of fluid momentum and energy. It remains either to introduce particle collisions or to assume that these are absent. In the former case, (2.7), viz. $\mathbf{p} = \mathbf{M}$, $\mathbf{q} = \boldsymbol{\xi}$, apply, and the usual fluid equations result. However in the second case, by Newton's second law the momentum flux across a small element of physical space cannot alter (there being no collisions) and it follows that $\nabla\cdot\mathbf{M} = 0$, reducing the equation of 'fluid' motion to $\mathrm{D}\mathbf{v} = \mathbf{F}$. Strictly there is no fluid, and the particles are accelerated individually by the body force without interations.

5.4 The BGK kinetic equation

5.4.1 *Mean-free-path derivation*

We shall follow a method similar to that used in §3.5, where we found deviations from equilibrium in the number and energy of a group of molecules in moving a mean free path from an initial equilibrium state.

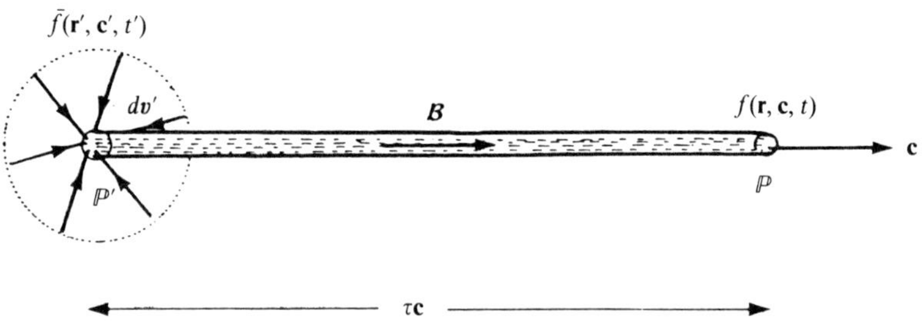

FIG. 5.1. Molecules being inversely scattered into a stream $\mathcal{B}$

Consider those molecules P that have collisions within a mean free path or so of an element $d\boldsymbol{\nu}'$ that is moving through phase space with velocity $\mathbf{c}'$, as illustrated in Fig. 5.1. Some will converge on $d\boldsymbol{\nu}'$ and there experience second collisions that direct their paths into a stream $\mathcal{B}$ with initial velocity $\mathbf{c}'$. If the point* $\mathbb{P}' = (\mathbf{r}', \mathbf{c}', t')$ is at $\mathbf{r}' = \mathbf{r} - \tau\mathbf{c}$, $\mathbf{c}' = \mathbf{c} - \tau\mathbf{a}_m$, $t - \tau$, where $\mathbf{a}_m$ is the free flight acceleration of P within $\mathcal{B}$, then after a collision-free flight lasting τ seconds (on average), the molecules will arrive at $\mathbb{P}(\mathbf{r}, \mathbf{c}, t)$ and contribute to the number density at this point. Initially the molecules converge on $\mathbb{P}'$ from all directions and with a range of speeds, so their effective phase space density at $\mathbb{P}'$ will be some average, $\bar{f}(\mathbf{r}', \mathbf{c}', t')$ say, to be determined later by considering the details of the 'inverse' collisions collecting them into the stream $\mathcal{B}$. As no molecules are lost from $\mathcal{B}$ until after $\mathbb{P}$ is passed, on the assumption that all the molecules arriving at $\mathbb{P}$ were collided into $\mathcal{B}$ one collision time earlier, we may write

$$f(\mathbf{r}, \mathbf{c}, t) = \bar{f}(\mathbf{r} - \tau\mathbf{c},\ \mathbf{c} - \tau\mathbf{a}_m,\ t - \tau). \tag{5.28}$$

Of course in reality such synchronous behaviour is highly improbable, but as a means of finding the kinetic equation the model is satisfactory since it does not violate the essential physics—we can think of $\mathcal{B}$ as being an ensemble of molecules representing the behaviour of a single 'typical' molecule. A more general model leading to the same outcome will be described shortly.

We now introduce the assumption that during P's transit from $\mathbb{P}'$ to $\mathbb{P}$, they have the *same* acceleration as the phase space element, i.e. by (5.18)

$$\mathbf{a}_m = \mathbf{a} = -\mathbf{P} - \mathbf{c} \cdot \mathbf{e}. \tag{5.29}$$

It will be shown in §5.9.3, that this is an approximation, and a more accurate relation will be derived, but it is sufficient for our present purposes.

The simplest assumption that we can make about $\bar{f}$ is that it is the equilibrium distribution f_0. We shall accept this *pro tempore*, and justify it as a satisfactory approximation at the end of §5.5.2. By *the* equilibrium distribution at P_c is meant the distribution f_0 that has in common with f both its number and energy density, i.e.

$$\int \{m,\ m\mathbf{c},\ \tfrac{1}{2}mc^2\}(f - f_0)\, d\mathbf{c} = 0, \tag{5.30}$$

where for completeness we have added the momentum constraint that follows from the definition of $\mathbf{c}$ (cf. (5.27)). But (5.30) is really a *definition*, not a limiting constraint, for if it is not satisfied, we are free to alter f_0 by adding to it an expression of the form $(\alpha + \boldsymbol{\beta} \cdot \mathbf{c} + \gamma c^2) f_0$, where α, $\boldsymbol{\beta}$ and γ are functions of $\mathbf{r}$ and t alone, chosen so that $(f - f_0[1 + \alpha + \boldsymbol{\beta} \cdot \mathbf{c} + \gamma c^2])$ *does* satisfy (5.30). This modification merely adjusts the basic fluid variables, ρ, $\mathbf{v}$ and T to their 'correct' values (see exercise 5.1).

*Note that $P_c(\mathbf{r}, t)$ is a convected, spinning point in physical space, and $\mathbb{P}(\mathbf{r}, \mathbf{c}, t)$ is a point in phase space that moves with velocity $\mathbf{c}$ relative to P_c

5.4.2 *The kinetic equation and its limitations*

It follows from (5.17) and (5.18) that the appropriate streaming derivative for the bunch $\mathcal{B}$ is

$$\mathbb{D} \equiv \mathrm{D}^* + \mathbf{c} \cdot \nabla - (\mathbf{P} + \mathbf{c} \cdot \mathbf{e}) \cdot \frac{\partial}{\partial \mathbf{c}}. \tag{5.31}$$

Returning to (5.28), with $\bar{f}$ replaced by f_0, we adopt a Taylor expansion about $\mathbb{P}(\mathbf{r}, \mathbf{c}, t)$ to obtain

$$f = f_0 - \tau \mathbb{D} f_0 + \mathrm{O}(\tau^2). \tag{5.32}$$

Notice that (cf. (3.30))

$$|\tau \mathbb{D} \ln f_0| = \mathrm{O}(\epsilon), \qquad |f - f_0| = \mathrm{O}(\epsilon), \tag{5.33}$$

so that we may write (5.32) as $f_0 = (1 + \tau \mathbb{D}) f + \mathrm{O}(\epsilon^2)$, or

$$\mathbb{D} f = \frac{1}{\tau}(f_0 - f) + \mathrm{O}(\epsilon^2). \tag{5.34}$$

Without the error term, this is known as the BGK kinetic equation (cf. (5.3)).

A general model—avoiding synchronous collisions—replaces (5.28) by

$$f(\mathbf{r},\, \mathbf{c},\, t) = \int_{-\infty}^{t} e^{-(t-t')/\tau} f_0\Big(\mathbf{r} - (t-t')\mathbf{c},\, \mathbf{c} - (t-t')\mathbf{a},\, t'\Big) \frac{dt'}{\tau}, \tag{5.35}$$

the factor $\exp\{-(t-t')/\tau\}\, dt'/\tau$ being the probability that a molecule P joins the bunch $\mathcal{B}$ in $t', t' + dt'$ and experiences a free-flight for a time $t - t'$ (see (3.44)). Equation (5.34) follows on taking the Taylor expansion of the integrand. A more accurate treatment will be given in §5.11.

It is tacitly assumed in the usual presentation of the BGK theory that equation (5.34) can be replaced by

$$\mathbb{D} f = \mathbb{C}_0(f), \qquad \Big(\mathbb{C}_0(f) \equiv \frac{1}{\tau}(f_0 - f)\Big), \tag{5.36}$$

i.e. that there is *no* error term in the equation. Also it is assumed that the fluid shear acceleration term in $\mathbb{D}$ is $-\mathbf{c} \cdot \nabla \mathbf{v}$ and not the frame-indifferent form $-\mathbf{c} \cdot \mathbf{e}$ appearing in (5.31). Neither of these assumptions affects the application of the theory to determining first-order (in ϵ) diffusivities. The importance of frame indifference will be explained in **5.7**, and in **5.9** it will be made clear why the BGK theory fails with $\mathrm{O}(\epsilon^2)$ terms.

5.5 The non-equilibrium distribution function

5.5.1 *Streaming derivative of the equilibrium distribution*

The distribution function f can be deduced from (5.32) and the equilibrium distribution (2.33), viz.

$$f_0 = \pi^{-3/2} n \mathcal{C}^{-3} \exp(-c^2/\mathcal{C}^2) \qquad (\mathcal{C} = (2kT/m)^{1/2}). \tag{5.37}$$

Using $p = nkT$ to eliminate n we get

$$\ln f_0 = \ln p - \tfrac{5}{2}\ln T - c^2/\mathcal{C}^2 + \text{const.} \tag{5.38}$$

Hence
$$\mathrm{D}\ln f_0 = \mathrm{D}\ln(pT^{-5/2}) + c^2\mathcal{C}^{-2}\mathrm{D}\ln T. \tag{5.39}$$

For a monatomic gas it follows from (1.65), with $\gamma = 5/3$, that

$$s = c_v \ln(pT^{-5/2}) + \text{const.} \tag{5.40}$$

Now the heat flux $\mathbf{q}$ and the viscous stress tensor $\boldsymbol{\pi}$ are first order in the Knudsen number ϵ, hence from (1.69),

$$\mathrm{D}s = \mathrm{O}(\epsilon). \tag{5.41}$$

Then with continuity in the form $\mathrm{D}\ln\rho = \mathrm{D}\ln n = -\nabla\cdot\mathbf{v}$ (see (1.46)), we find from the above equations and $p = nkT$ that

$$\mathrm{D}\ln n = -\nabla\cdot\mathbf{v}, \quad \mathrm{D}\ln T = -\tfrac{2}{3}\nabla\cdot\mathbf{v} + \mathrm{O}(\epsilon), \quad \mathrm{D}\ln p = -\tfrac{5}{3}\nabla\cdot\mathbf{v} + \mathrm{O}(\epsilon), \tag{5.42}$$

enabling us to reduce (5.39) to

$$\mathrm{D}\ln f_0 = -\frac{2c^2}{3\mathcal{C}^2}\nabla\cdot\mathbf{v} + \mathrm{O}(\epsilon). \tag{5.43}$$

To find $\mathbf{c}\cdot\nabla\ln f_0$ we replace D in (5.39) by $\mathbf{c}\cdot\nabla$. Thus

$$\mathbf{c}\cdot\nabla\ln f_0 = (\nu^2 - \tfrac{5}{2})\mathbf{c}\cdot\nabla\ln T + \mathbf{c}\cdot\nabla\ln p \qquad (\nu \equiv c/\mathcal{C}). \tag{5.44}$$

From (5.38),

$$\begin{aligned}\mathbf{a}\cdot\frac{\partial \ln f_0}{\partial \mathbf{c}} &= -(\mathbf{P} + \mathbf{c}\cdot\mathbf{e})\cdot\left\{-2\mathbf{c}/\mathcal{C}^2\right\} = 2\mathbf{P}\cdot\mathbf{c}/\mathcal{C}^2 + 2\mathbf{c}\cdot\mathbf{e}\cdot\mathbf{c}/\mathcal{C}^2 \\ &= -\mathbf{c}\cdot\nabla\ln p + 2\boldsymbol{\nu}\cdot\left(\mathring{\mathbf{e}} + \tfrac{1}{3}\mathbf{1}\,\nabla\cdot\mathbf{v}\right)\cdot\boldsymbol{\nu},\end{aligned} \tag{5.45}$$

Adding (5.43), (5.44), and (5.45), and noting the definition of $\mathbb{D}$ given in (5.31), we arrive at

$$\mathbb{D}\ln f_0 = (\nu^2 - \tfrac{5}{2})\mathbf{c}\cdot\nabla\ln T + 2\boldsymbol{\nu}\cdot\mathring{\mathbf{e}}\cdot\boldsymbol{\nu} \qquad (\boldsymbol{\nu} \equiv \mathbf{c}/\mathcal{C}). \tag{5.46}$$

5.5.2 *Two relaxation times*

The collision interval τ in (5.34) has the same value for momentum and energy transport, but according to the theory in §3.3 there should be two times, τ_2 and τ_1, where $\tau_2 = \frac{3}{2}\tau_1$. Had we treated ∇T and $\mathring{\mathbf{e}}$ *separately* in obtaining the Taylor expansion, we could have introduced two times in place of τ. Accepting this modification—to be justified later—we can now write (5.32) as

$$f = f_0\{1 - \tau_2(\nu^2 - \tfrac{5}{2})\mathbf{c}\cdot\nabla\ln T - 2\tau_1\boldsymbol{\nu}\cdot\mathring{\mathbf{e}}\cdot\boldsymbol{\nu} + \mathrm{O}(\epsilon^2)\}, \tag{5.47}$$

where τ_2 and τ_1 are microscopic times whose values remain to be determined.

We return now to the question of replacing $\bar{f}$ by f_0 in (5.28). The BGK value for $\bar{f}$ is obtained by the simplest form of averaging. We take the average value of the expression given in (5.47) over all directions of the vector $\mathbf{c}$. This gives $\bar{f} = f_0 + O(\epsilon^2)$, so the error in the theory is not increased if $\bar{f}$ is replaced by f_0. In §7.2 we shall give Boltzmann's accurate treatment of inverse scattering, given the force law between the molecules.

Of course we could accept (5.47) as being complete as it stands, with τ_1 and τ_2 in the role of phenomenological collision intervals. This amounts to a simplified form of advanced kinetic theory. We shall return to this model in later chapters of this text.

5.6 Viscosity and thermal conductivity

5.6.1 *The pressure tensor*

Provided the momentum flux is collided flux, the pressure tensor is given by (2.6) and (2.7):

$$\mathsf{p} = \rho\langle\mathbf{cc}\rangle = m\int \mathbf{cc}f\,d\mathbf{c}. \tag{5.48}$$

With f given by (5.47) we obtain

$$\mathsf{p} = m\int \mathbf{cc}\,f_0(1 - 2\tau_1\boldsymbol{\nu}\cdot\overset{\circ}{\mathbf{e}}\cdot\boldsymbol{\nu})\,d\mathbf{c} + O(\epsilon^2), \tag{5.49}$$

where the omitted term is odd in the vector $\mathbf{c}$ (or $\boldsymbol{\nu}$) and therefore vanishes (see (5.57) below). At this point we need expressions for integrals involving tensors of the fourth order.

Mathematical note 5.6 *Isotropic tensors of the fourth order*

Let A be any second-order tensor, then the fourth-order unit tensor U satisfies

$$\mathsf{U}:\mathsf{A} = \mathsf{A}:\mathsf{U} = \mathsf{A}. \tag{5.50}$$

This tensor can be written as a combination of two second-order tensors, $\mathsf{U} = \mathbf{1}^\diamond\mathbf{1}$ with the convention that $^\diamond$ is interpreted as indicated in

$$\mathbf{1}^\diamond\mathbf{1}:\mathsf{A} = \mathbf{1}\cdot\mathsf{A}\cdot\mathbf{1} = \mathsf{A} = \mathsf{A}:\mathbf{1}^\diamond\mathbf{1}, \tag{5.51}$$

i.e. scalar products are formed with the 'inside' vectors of $\mathbf{1}$ in each appearance.

Suppose that $\mathbf{i},\mathbf{j},\mathbf{k}$ are the mutually orthogonal unit vectors of a Cartesian coordinate system (see §1.5.2), then $\mathbf{1} = \mathbf{ii}+\mathbf{jj}+\mathbf{kk}$. Therefore

$$\begin{aligned}\mathbf{1}^\diamond\mathbf{1}:\mathsf{A} &= (\mathbf{ii}+\mathbf{jj}+\mathbf{kk})\cdot\mathsf{A}\cdot(\mathbf{ii}+\mathbf{jj}+\mathbf{kk}) \qquad (5.52)\\ &= \left[\mathbf{iiii}+\mathbf{jjjj}+\mathbf{kkkk}+\mathbf{ijji}+\mathbf{jiij}+\mathbf{jkkj}+\mathbf{kjjk}+\mathbf{kiik}+\mathbf{ikki}\right]:\mathsf{A},\end{aligned}$$

so $\mathbf{1}^\diamond\mathbf{1}$ is equal to the tensor in square brackets.

From the relations given in Mathematical note 1.13,

$$\mathbf{1}^\diamond\mathbf{1}:\mathsf{A} + \tfrac{1}{2}\mathbf{1}\times\mathbf{1}\times\mathbf{1}:\mathsf{A} - \tfrac{1}{3}\mathbf{11}:\mathsf{A} = \mathsf{A} + \mathbf{1}\times\overset{\times}{\mathsf{A}}{}^{v} - \tfrac{1}{3}\mathbf{1}A = \overset{\circ}{\mathsf{A}},$$

i.e. the tensor

$$\mathsf{K} \equiv \mathbf{1}^\diamond\mathbf{1} + \tfrac{1}{2}\mathbf{1}\times\mathbf{1}\times\mathbf{1} - \tfrac{1}{3}\mathbf{11}, \tag{5.53}$$

has the property
$$\mathbf{K} : \mathbf{A} = \mathbf{A} : \mathbf{K} = \overset{\circ}{\mathbf{A}}. \qquad (5.54)$$

There is a general theorem that all isotropic tensors of even order can be written as a sum of products of second-order unit tensors. The most general isotropic, fourth-order tensor is therefore a linear combination of any three independent fourth-order tensors that can be constructed from $\mathbf{1}$, e.g. $\mathbf{1}^{\diamond}\mathbf{1}$, $\mathbf{1} \times \mathbf{1} \times \mathbf{1}$ and $\mathbf{11}$. By (5.53) we may adopt $\mathbf{K}$ in place of $\mathbf{1}^{\diamond}\mathbf{1}$, and hence the general isotropic tensor is

$$\mathbf{L}(\alpha, \beta, \gamma) = \alpha \mathbf{K} - \tfrac{1}{2}\beta \mathbf{1} \times \mathbf{1} \times \mathbf{1} + \tfrac{1}{3}\gamma \mathbf{11}, \qquad (5.55)$$

where α, β, γ are arbitrary scalars.

Integrals of the type

$$\mathbf{J}_n = \int \overbrace{\hat{\mathbf{c}}\hat{\mathbf{c}} \ldots \hat{\mathbf{c}}}^{n\ factors}\, d\Omega \qquad (5.56)$$

occur in kinetic theory. When n is odd, it is easily verified that the integral is zero (cf. the derivation of the first of (2.40)), i.e.

$$\mathbf{J}_{2r-1} = 0 \qquad (r = 1, 2, \ldots). \qquad (5.57)$$

The even values of n that arise are 2, 4, and 6; we shall show here that

$$\mathbf{J}_2 = \tfrac{1}{3}\mathbf{1}, \qquad \mathbf{J}_4 = \tfrac{2}{15}\mathbf{K} + \tfrac{1}{9}\mathbf{11}, \qquad (5.58)$$

and treat $\mathbf{J}_6$ in §8.3.2. The first of these integrals was evaluated earlier (see (2.40)).

The integral $\mathbf{J}_4$ is isotropic, so we can write $\mathbf{J}_4 = \mathbf{L}$. As $\mathbf{1} : \mathbf{J}_4 = \mathbf{J}_2 = \tfrac{1}{3}\mathbf{1}$ also equals $\gamma\mathbf{1}$ by (5.55), $\gamma = \tfrac{1}{3}$. By (5.56), $\mathbf{J}_4 : \mathbf{A}$ is a symmetrical second-order tensor, but this is true of $\mathbf{L} : \mathbf{A}$ only if $\beta = 0$. Let $\mathbf{k}$ be unit vector parallel to the axis $\theta = 0$ of spherical coordinates, then by $\sin\theta\, d\theta\, d\phi = 4\pi\, d\Omega$,

$$\mathbf{kkkk} \,⋮\, \mathbf{J}_4 = \int \cos^4\theta\, d\Omega = \frac{1}{4\pi}\int_0^{2\pi}\left(\int_0^{\pi} \cos^4\theta \sin\theta\, d\theta\right) d\phi = \tfrac{1}{5}.$$

By (5.54) and (5.55), this component is

$$\mathbf{kkkk} \,⋮\, \mathbf{L} = \alpha \mathbf{kk} : \overset{\circ}{\mathbf{kk}} + \tfrac{1}{3}\gamma = \alpha \mathbf{kk} : (\mathbf{kk} - \tfrac{1}{3}\mathbf{1}) + \tfrac{1}{9} = \tfrac{2}{3}\alpha + \tfrac{1}{9},$$

and therefore $\alpha = \tfrac{2}{15}$, which confirms the result for $\mathbf{J}_4$ given in (5.58).

Now return to (5.49). With the help of (2.37), (2.45), (5.54), and (5.58) we find

$$\mathbf{p} = \rho \int_\Omega \int_0^\infty \mathbf{cc}(1 - 2\tau_1 \boldsymbol{\nu} \cdot \overset{\circ}{\mathbf{e}} \cdot \boldsymbol{\nu}) \frac{4}{\sqrt{\pi}} \nu^2 e^{-\nu^2}\, d\nu\, d\Omega$$

$$= \rho \mathcal{C}^2 \frac{4}{\sqrt{\pi}} \int_0^\infty (\tfrac{1}{3}\mathbf{1} - \tfrac{4}{15}\tau_1 \nu^2\, \mathbf{K} : \overset{\circ}{\mathbf{e}}) \nu^4 e^{-\nu^2}\, d\nu,$$

that is,
$$\mathbf{p} = p\mathbf{1} - 2\mu \overset{\circ}{\mathbf{e}} \qquad (\mu = p\tau_1). \qquad (5.59)$$

Thus the coefficient of viscosity is $p\tau_1$, as already derived by elementary kinetic theory in §3.6.

5.6.2 *The heat flux vector*

The heat flux tensor is given by (2.5) and (2.7):

$$\mathbf{q} = \tfrac{1}{2}m \int c^2 \mathbf{c} f \, d\mathbf{c}. \tag{5.60}$$

Hence by (5.47),

$$\mathbf{q} = \tfrac{1}{2}m \int c^2 \mathbf{c} f_0 \{-\tau_2(\nu^2 - \tfrac{5}{2})\mathbf{c} \cdot \nabla \ln T\} \, d\mathbf{c},$$

omitting a term odd in $\boldsymbol{\nu}$. Using (2.37), (2.41), and (5.58) we arrive at

$$\mathbf{q} = -\kappa \nabla T, \tag{5.61}$$

where

$$\kappa = \tfrac{5}{2} R p \tau_2 = \tfrac{15}{4R} p \tau_1 = \tfrac{5}{2} c_v \mu, \tag{5.62}$$

since by (3.19) $\tau_2 = \frac{3}{2}\tau_1$.

We have now recovered the results obtained in §§3.6 and 3.7 by the elementary theory. The merit of the present treatment is that it can be generalized to take account of terms of higher than first order in ϵ. We shall pursue this in §5.8 *et seq.* Before that we need to introduce the concept of frame indifference.

5.7 Frame indifference

5.7.1 *Definition of frame indifference*

To illustrate the principle of frame indifference—at least as we shall require it in this book—we adopt the equations of fluid motion given in (1.46) to (1.48):

$$\left.\begin{aligned} \mathrm{D}\rho + \rho \nabla \cdot \mathbf{v} &= 0, \qquad \rho \mathrm{D}\mathbf{v} + \nabla \cdot \mathsf{p} = \rho \mathbf{F}, \\ \rho \mathrm{D}u + \mathsf{p} : \nabla \mathbf{v} &= -\nabla \cdot \mathbf{q}. \end{aligned}\right\} \tag{5.63}$$

If a constant vector is added to the velocity $\mathbf{v}$ in these equations, they are unchanged, that is they are invariant under Galilean coordinate transformations. Of course the equations of physics must be coordinate-free or 'frame indifferent' in this limited sense.

Transformations to accelerated frames offer more of a challenge. Suppose that (5.63) hold in a frame L, and that in a frame L' rotating with constant angular velocity $\boldsymbol{\omega}$ relative to L, they are replaced by

$$\left.\begin{aligned} \mathrm{D}'\rho + \rho \nabla \cdot \mathbf{v}' &= 0, \qquad \rho \mathrm{D}'\mathbf{v}' + \nabla \cdot \mathsf{p}' = \rho \mathbf{F}', \\ \rho \mathrm{D}'u + \mathsf{p}' : \nabla \mathbf{v}' &= -\nabla \cdot \mathbf{q}', \end{aligned}\right\} \tag{5.64}$$

the scalars ρ and u being unchanged. Each term in (5.64) must have the same *physical* significance as the corresponding term in (5.63), i.e. $\mathbf{F}'$ is still a volume distributed body force, $\nabla \cdot \mathsf{p}'$ remains the local or 'point' form of a surface force due to particle collisions, $\mathbf{p}' : \nabla \mathbf{v}'$ is still the rate at which work is done by the fluid stress, and $\mathbf{q}'$ continues to be the transport of energy due only to random particle motions, interpretations that were essential in the formulation of the balance equations in the first instance. This invariance in physical significance is what we term the *principle of frame indifference*.*

5.7.2 *Rotating frames*

Let ψ_0, $\boldsymbol{\psi}_1$, $\boldsymbol{\psi}_2$ denote scalar, vector, and (symmetric) tensor functions of $\mathbf{r}, t$ and take the origin O to be the axis of rotation, then by the transformation rules for time derivatives (see Mathematical note 5.3),

$$\mathrm{D}'\boldsymbol{\psi}_1 = \mathrm{D}\boldsymbol{\psi}_1 - \boldsymbol{\omega} \times \boldsymbol{\psi}_1, \quad \mathrm{D}'\boldsymbol{\psi}_2 = \mathrm{D}\boldsymbol{\psi}_2 - \boldsymbol{\omega} \times \boldsymbol{\psi}_2 + \boldsymbol{\psi}_2 \times \boldsymbol{\omega}. \tag{5.65}$$

The velocity transforms according to $\mathbf{v}' = \mathbf{v} - \boldsymbol{\omega} \times \mathbf{r}$,

whence $$\nabla \mathbf{v}' = \nabla \mathbf{v} + \nabla \mathbf{r} \times \boldsymbol{\omega} = \nabla \mathbf{v} + \mathbf{1} \times \boldsymbol{\omega}, \quad \nabla \cdot \mathbf{v}' = \nabla \cdot \mathbf{v},$$

and $$\mathbf{p}' : \nabla \mathbf{v}' = \mathbf{p}' : \nabla \mathbf{v} + \mathbf{p}' : \mathbf{1} \times \mathbf{1} \cdot \boldsymbol{\omega} = \mathbf{p}' : \nabla \mathbf{v},$$

where we have used (1.90) and the symmetry of $\mathbf{p}'$.

By the above equations we find that (5.63) transforms into (5.64) provided that
$$\mathbf{F}' = \mathbf{F} - 2\boldsymbol{\omega} \times \mathbf{v} + \boldsymbol{\omega} \times (\boldsymbol{\omega} \times \mathbf{r}) + \nabla \cdot (\mathbf{p}' - \mathbf{p}),$$

and $$(\mathbf{p}' - \mathbf{p}) : \nabla \mathbf{v} + \nabla \cdot (\mathbf{q}' - \mathbf{q}) = 0,$$

or $$(\boldsymbol{\pi}' - \boldsymbol{\pi}) : \nabla \mathbf{v} + (\mathbf{q}' - \mathbf{q}) \cdot \nabla \ln T = -T \nabla \cdot \{(\mathbf{q}' - \mathbf{q})/T\}, \tag{5.66}$$

since $p = p'$.

It is clear that $\mathbf{F}'$ can retain its physical meaning as an independent body force only if $\nabla \cdot (\mathbf{p}' - \mathbf{p}) = \nabla \cdot (\boldsymbol{\pi}' - \boldsymbol{\pi}) = 0$. And as $\nabla \mathbf{v}$ and ∇T are independent gradients, it follows from (5.66) that $\boldsymbol{\pi}' = \boldsymbol{\pi}$. We now identify $(\mathbf{q}' - \mathbf{q}) \cdot \nabla \ln T$ as being the change in the thermal dissipation (cf. (1.70) and (1.75)) due to the transformation, and $(\mathbf{q}' - \mathbf{q})/T$ as being the resulting entropy flux. The dissipation rate is an intrinsic property of the medium that cannot depend on the (arbitrary) choice of coordinate system. Hence frame indifference and (5.66) require that $\mathbf{q}' = \mathbf{q}$ and we are led to

$$\mathbf{p}' = \mathbf{p}, \qquad \mathbf{q}' = \mathbf{q}. \tag{5.67}$$

*There are other versions of the principle. When it is applied directly to constitutive equations, e.g. expressions relating $\mathbf{q}$ and $\boldsymbol{\pi}$ to gradients, it is termed 'material frame indifference', and requires the response of the *same* medium to be identical for all observers. In interpreting this principle, it is important to bear in mind that spinning a material may alter its physical properties, so that it is not the same. Thus there can be a distinction between rotating a material with the observer 'fixed' and rotating the observer with a 'fixed' material.

5.7.3 *Application to kinetic theory*

Now we apply frame indifference to the kinetic equation $\mathbb{D}f = \mathbb{C}(f)$. In L' it reads $\mathbb{D}'f' = \mathbb{C}'(f')$. By (5.31), (5.36), (5.65), and $f' = f$,

$$\mathbb{D}'f' = \mathbb{C}'(f'), \qquad \mathbb{D}'f' = \mathrm{D}f + \mathbf{c}\cdot\nabla f - (\mathbf{P}' + \mathbf{c}\cdot\mathbf{e}')\cdot\frac{\partial f}{\partial \mathbf{c}},$$

where
$$\mathbf{P}' = -\frac{1}{\rho}\nabla\cdot\mathsf{p}' = -\frac{1}{\rho}\nabla\cdot\mathsf{p} = \mathbf{P},$$

and

$$\mathbf{c}\cdot\mathbf{e}' = \tfrac{1}{2}\mathbf{c}\cdot\{\nabla\mathbf{v}' + \widetilde{\nabla\mathbf{v}'}\} = \tfrac{1}{2}\mathbf{c}\cdot\{(\nabla\mathbf{v} + \mathbf{1}\times\boldsymbol{\omega}) + (\widetilde{\nabla\mathbf{v}} - \mathbf{1}\times\boldsymbol{\omega})\} = \mathbf{c}\cdot\mathbf{e}.$$

Hence
$$\mathbb{D}'f' = \mathbb{C}'(f') \Longleftrightarrow \mathbb{D}f = \mathbb{C}(f), \tag{5.68}$$

as required by the principle.

Notice that
$$\mathbf{c}\cdot\nabla\mathbf{v}' = \mathbf{c}\cdot\nabla\mathbf{v} + \mathbf{c}\times\boldsymbol{\omega},$$

so that were the acceleration assumed to be $-(\mathbf{P} + \mathbf{c}\cdot\nabla\mathbf{v})$ (see final paragraph of §5.4.2), in place of (5.68) we would find that

$$\mathbb{D}'f' = \mathbb{D}f + \boldsymbol{\omega}\times\mathbf{c}\cdot\frac{\partial f}{\partial \mathbf{c}} = \mathbb{C}(f),$$

where $\boldsymbol{\omega}$ is an arbitrary vector. To remove this arbitrariness, it is necessary to replace $\boldsymbol{\omega}$ by $\boldsymbol{\Omega}$, and to use $\nabla\mathbf{v} + \boldsymbol{\Omega}\times\mathbf{1} = \mathbf{e}$ to restore the correct (i.e. frame-indifferent) form of the streaming operator.

Frame indifference enters kinetic theory via the definition of the peculiar velocity $\mathbf{c}$ (Woods 1983a). If $\mathbf{w}$ is the particle velocity in the laboratory (L) frame, then $\mathbf{c}$ is its velocity in a frame P_c that moves both with the fluid velocity $\mathbf{v}$, *and* rotates with the local fluid spin $\boldsymbol{\Omega}$ (see §1.13). The point is that $\mathbf{w} = \mathbf{v}(\mathbf{r}, t) + \mathbf{c}$ alone does not fully specify $\mathbf{c}$, since—as the differentials $d\mathbf{w}, d\mathbf{c}$ occur in kinetic theory—it is also necessary to specify the relative orientations of the reference frames in which $\mathbf{w}$ and $\mathbf{c}$ are measured. There is no justification for measuring $\mathbf{c}$ in a frame P having a fixed orientation relative to L, for L itself is arbitrary, and making $d\mathbf{c}$ depend on the choice of L (via P) would produce frame-dependent diffusivities. The problem does not arise until $\mathrm{O}(\epsilon^2)$ terms are encountered. In the remainder of this chapter we shall be concerned with finding these terms.

5.8 Pressure gradients and the kinetic equation

5.8.1 *General form for the kinetic equation*

A more general form for the kinetic equation than (5.34) is obtained by retaining the average $\bar{f}$, introduced in (5.28):

$$\mathbb{D}f = \frac{1}{\tau}\bar{f} - \frac{1}{\tau}f + \mathrm{O}(\epsilon^2). \tag{5.69}$$

We can deduce this equation from first principles as follows. The left-hand side is the rate of change of f in $d\boldsymbol{\nu}$. This is balanced by 'direct' scattering losses at the rate f/τ and 'inverse' scattering gains at the rate $\bar{f}/\tau$. That the direct scattering losses should be proportional to the number $f\,d\boldsymbol{\nu}$ within the element is physically obvious. And that the inverse scattering gains should be proportional to *some* average value of the distribution function in the neighbourhood of $d\boldsymbol{\nu}$ follows from the fact that the incoming molecules approach $d\boldsymbol{\nu}$ from all directions and with all speeds.

With the BGK approximation, $\bar{f} \approx f_0$, we have

$$\mathbb{D}f = \frac{1}{\tau} f_0 - \frac{1}{\tau} f + \mathrm{O}(\epsilon^2). \tag{5.70}$$

5.8.2 *The effect of pressure gradients*

The first step towards reducing the error term in (5.70) is to determine how this equation should be modified in the presence of a pressure gradient force $\mathbf{P} = -\nabla \cdot \mathbf{p}/\rho$. In §3.9 the outcome of a representative collision experienced by a molecule P was divided into *scattering* and *streaming*. Scattering involves binary collisions,* with momentum conserved, and the resulting impulses, say $\hat{\mathbf{p}}$ and $-\hat{\mathbf{p}}$ per unit mass, are large and randomly distributed in direction. Streaming collisions represent the 'surface' effects on $d\boldsymbol{\nu}$, and result from distributing the momentum imbalance of these surface interactions over all the molecules in $d\boldsymbol{\nu}$. The resulting impulse per molecule is small and equal to $\tau\mathbf{P}$ per unit mass.* In §3.9 the point was made that *both* the scattering and streaming components remove molecules from the free-flight element $d\boldsymbol{\nu}$. Yet losses and gains due to collisional streaming do not appear in (5.70).

First consider direct scattering and streaming from the element $d\boldsymbol{\nu}$. Just prior to the collisions taking place—assumed for simplicity of description to be synchronous—the distribution function is $f(\mathbf{r},\, \mathbf{c},\, t)$. The total impulse per unit mass on each particle is $(\hat{\mathbf{p}} + \tau\mathbf{P})$, where $\hat{\mathbf{p}}$ is randomly distributed in direction. As before, the scattering impulse removes particles from $d\boldsymbol{\nu}$ at the rate f/τ. The streaming impulse abruptly changes the velocity of the particles within $d\boldsymbol{\nu}$ from $\mathbf{c}$ to $\mathbf{c} + \tau\mathbf{P}$, with the result that their distribution function changes to

$$f(\mathbf{r},\, \mathbf{c} + \tau\mathbf{P},\, t) = f(\mathbf{r},\, \mathbf{c},\, t) + \tau\mathbf{P} \cdot \frac{\partial f}{\partial \mathbf{c}}(\mathbf{r},\, \mathbf{c},\, t).$$

It follows that $d\boldsymbol{\nu}$, which continues to move with the original velocity $\mathbf{c}$, abruptly loses $\tau\mathbf{P} \cdot (\partial f/\partial \mathbf{c})$ particles per unit volume of phase space. As it loses no more in this fashion until the next set of collisions τ seconds later, the rate of loss due to streaming collisions is $\mathbf{P} \cdot (\partial f/\partial \mathbf{c})$. This loss occurs simultaneously with the direct scattering loss, f/τ. Adding the

*Not an essential feature, but the probability that three or more molecules collide simultaneously in a tenuous gas is so small that we can neglect it.

*The assumption that pressure is merely momentum flux so that collisions do not need to be invoked in its definition, means that streaming collisions play no role in the standard version of kinetic theory.

losses, we see that the pressure gradient force changes $-f/\tau$ in (5.70) to $-f/\tau - \mathbf{P}\cdot(\partial f/\partial\mathbf{c})$.

Inverse scattering is similarly affected by the pressure force. Now suppose that $d\boldsymbol{\nu}$ receives particles from its environment by both scattering and streaming collisions. As with the losses, the impulse on each particle is of the form $(\hat{\mathbf{p}} + \tau\mathbf{P})$. Inverse scattering impulses bring particles into $d\boldsymbol{\nu}$ at the rate f_0/τ and the simultaneous streaming impulse changes f_0 to $f_0 + \tau\mathbf{P}\cdot(\partial f_0/\partial\mathbf{c})$, which gives a total gain rate of $f_0/\tau + \mathbf{P}\cdot(\partial f_0/\partial\mathbf{c})$ to replace f_0/τ in (5.70). Another way of describing the process is to observe that just prior to the collision, $d\boldsymbol{\nu}$ has a velocity $-\tau\mathbf{P}$ relative to the arriving particles, and when this mismatch is removed by the impulsive force, there is a sudden change in the phase space density.

5.8.3 *Modified kinetic equation*

Combining the inverse and direct collisional effects, we now replace (5.70) by

$$\mathbb{D}f = \frac{1}{\tau}\left(f_0 + \tau\mathbf{P}\cdot\frac{\partial f_0}{\partial\mathbf{c}}\right) - \frac{1}{\tau}\left(f + \tau\mathbf{P}\cdot\frac{\partial f}{\partial\mathbf{c}}\right) + \mathrm{O}(\epsilon^2),$$

or

$$\mathbb{D}f = \mathbb{C}_1(f), \tag{5.71}$$

where in place of the collision operator $\mathbb{C}_0$ of BGK theory (see (5.36)) we now have

$$\mathbb{C}_1(f) = \frac{1}{\tau}(f_0 - f) + \mathbf{P}\cdot\frac{\partial}{\partial\mathbf{c}}(f_0 - f) + \mathrm{O}(\epsilon^2). \tag{5.72}$$

To verify that the new term in (5.72) is $\mathrm{O}(\epsilon^2)$, we multiply the equation by τ to non-dimensionalize the operators, and then note that $\tau\mathbf{P} = -\tau\nabla\cdot\mathsf{p}/\rho$ is $\mathrm{O}(\epsilon)$ and that by (5.47) this is also true of $(f_0 - f)$. Notice from (5.30) that $\mathbb{C}_1$ satisfies the constraints in (5.27).

By a similar argument, equation (5.69) should be replaced by

$$\mathbb{D}f = \frac{1}{\tau}(\bar{f} - f) + \mathbf{P}\cdot\frac{\partial}{\partial\mathbf{c}}(\bar{f} - f) + \mathrm{O}(\epsilon^2). \tag{5.73}$$

There is another way of presenting (5.71) that throws light on the error term. By (5.16) and (5.20) it can be written

$$\mathbb{D}_S f = \frac{1}{\tau}f_0 + \mathbf{P}\cdot\frac{\partial f_0}{\partial\mathbf{c}} - \frac{1}{\tau}f + \mathrm{O}(\epsilon^2), \tag{5.74}$$

where

$$\mathbb{D}_S = \mathrm{D}^* + \mathbf{c}\cdot\nabla + \dot{\mathbf{c}}\cdot\frac{\partial}{\partial\mathbf{c}} \qquad (\dot{\mathbf{c}} = -\mathbf{c}\cdot\mathbf{e}). \tag{5.75}$$

The swept derivative $\mathbb{D}_S$ follows an element $d\boldsymbol{\nu}_S$ that 'keeps up' with the convected fluid, and from which streaming losses do not occur (see final paragraph of §3.9.4). For this reason the *loss* term on the right-hand side is pure scattering. It follows that *if* there is a further error term on the right of (5.74), it must be in the form of a further inverse streaming term, say $\boldsymbol{\alpha}\cdot(\partial f_0/\partial\mathbf{c})$, where $\boldsymbol{\alpha}$ is an $\mathrm{O}(\epsilon)$ acceleration.

5.9 Inverse streaming from particle acceleration

5.9.1 *Matching accelerations*

Accepting the hypothesis just stated, we add $\boldsymbol{\alpha} \cdot \partial f_0 / \partial \mathbf{c}$ to the right-hand side of (5.74) to obtain

$$\mathbb{D}_S f = \frac{1}{\tau} f_0 + (\mathbf{P} + \boldsymbol{\alpha}) \cdot \frac{\partial f_0}{\partial \mathbf{c}} - \frac{1}{\tau} f + \mathrm{O}(\epsilon^3), \tag{5.76}$$

which is equivalent to

$$\mathbb{D} f = \left\{ \mathrm{D}^* + \mathbf{c} \cdot \nabla - (\mathbf{P} + \mathbf{c} \cdot \mathbf{e}) \cdot \frac{\partial}{\partial \mathbf{c}} \right\} f = \mathbb{C}(f), \tag{5.77}$$

where (cf. (5.72))

$$\mathbb{C}(f) = \frac{1}{\tau}(f_0 - f) + \mathbf{P} \cdot \frac{\partial}{\partial \mathbf{c}}(f_0 - f) + \boldsymbol{\alpha} \cdot \frac{\partial f_0}{\partial \mathbf{c}} + \mathrm{O}(\epsilon^3). \tag{5.78}$$

We see from (5.76) that $\boldsymbol{\alpha}$ has a physical origin similar to $\mathbf{P}$. Recall from the description of inverse scattering in §5.8.2 that just prior to collision, the element $d\boldsymbol{\nu}$ has a velocity $-\tau \mathbf{P}$ relative to the arriving particles. They receive a streaming impulse $\tau \mathbf{P}$ to remove this discrepancy, which increases the phase space density from f_0 to $f_0 + \tau \mathbf{P} \cdot \partial f_0 / \partial \mathbf{c}$. Likewise if $d\boldsymbol{\nu}$ has a further velocity component $-\tau \boldsymbol{\alpha}$ relative to the arriving particles, this is also eliminated impulsively, changing the density to $f_0 + \tau(\mathbf{P} + \boldsymbol{\alpha}) \cdot \partial f_0 / \partial \mathbf{c}$. So what we now require is an expression for the acceleration $\boldsymbol{\alpha}$ of the arriving particles relative to their target element $d\boldsymbol{\nu}$.

5.9.2 *Trajectories over two mean free paths*

By considering the trajectories of molecules over one collision interval τ and taking a single inverse collision into account, in §5.4 we obtained a kinetic equation correct to first order in the Knudsen number ϵ. To increase the accuracy to $\mathrm{O}(\epsilon^2)$ we are obliged to extend the procedure back in time one further collision interval τ, which is equivalent to taking *two* mean free paths into account. Because the possible trajectories just before the final mean free path are infinite in number, to simplify the treatment it is necessary to average over these trajectories and then to use this average to obtain the effect of earlier collisions on the final (inverse) collision. The calculation of pressure gradient effects in the previous section can also be viewed in this light, except that many more collisions are involved in the collective process of fluid acceleration.

For the present we shall adopt the synchronous model of molecular behaviour used in §5.4; a completely general description will be given in §5.11. The synchronous model is illustrated in Fig. 5.2, which shows a final mean free path $P_c' P_c$ and a typical penultimate path $P_c^* P_c'$, where P_c^* is a point on a sphere $\mathcal{S}$, a mean free path in radius, with centre at $P_c'' = (\mathbf{r}'', t'') = (\mathbf{r}' - \tau \mathbf{c}', t' - \tau)$, and $P_c' = (\mathbf{r}', t') = (\mathbf{r} - \tau \mathbf{c}, t - \tau)$ is the location of the inverse collision. A typical molecule m starts from a previous collision at P_c^*, after which it is found to be on one of a family of spheres shrinking to the point P_c', where it encounters another molecule m' that also had its previous collision on $\mathcal{S}$. The points P_c'', P_c', P_c are convected with the local fluid velocity, and the centre, say $\overline{\mathsf{m}}$, of the family of spheres moves from P_c'' to P_c', starting with a velocity $\mathbf{c}''$ at P_c'' and finishing with a velocity $\mathbf{c}'$ at P_c'.

Because P_c'' is a convected point, we may assume that the penultimate collisions on $\mathcal{S}$ are uniformly distributed over it—at least to the accuracy required in the present model—and that as they converge on the point P_c', they have an average acceleration the same as that of the centre $\overline{\mathsf{m}}$. We can think of $\overline{\mathsf{m}}$ as a pseudo-molecule representing the average dynamic behaviour of all the molecules that converge on P_c' and then are collided into the stream $\mathcal{B}$.

5.9.3 *Relative acceleration of particles and phase-space elements*

Properties X say, at points P_c'', P_c', and P_c will be indicated by X'', X', and X. By (5.77) the acceleration of the phase space elements in a convected frame consists of a pressure gradient term and a fluid shear term:

$$\mathbf{a} = -\mathbf{P} - \mathbf{c} \cdot \mathbf{e}. \tag{5.79}$$

We shall denote the acceleration of $\overline{\mathsf{m}}$ at P_c' (just prior to the collision) by $\mathbf{a}_m'$ and at P_c'' by $\mathbf{a}_m''$. The acceleration of $d\boldsymbol{\nu}$ at P_c' is $\mathbf{a}'$, and hence by definition, at this point $\boldsymbol{\alpha}$ has the value

$$\boldsymbol{\alpha} = \mathbf{a}_m' - \mathbf{a}' \qquad \big(\boldsymbol{\alpha} = \boldsymbol{\alpha}(\mathbf{r}, \mathbf{c}, t)\big). \tag{5.80}$$

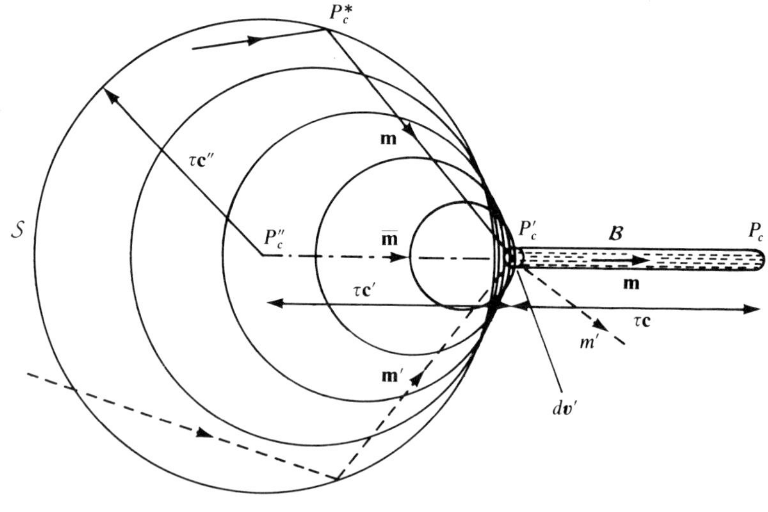

FIG. 5.2. Inverse scattering by two collisions

At P_c'', since $\overline{\mathsf{m}}$ is obtained by averaging over molecules approaching it isotropically from all directions, (5.80) is replaced by

$$\boldsymbol{\alpha}_0 = \mathbf{a}_m'' - \mathbf{a}'' \qquad \big(\boldsymbol{\alpha}_0 = \boldsymbol{\alpha}_0(\mathbf{r},\, t)\big),$$

i.e. the acceleration difference at P_c'' does not depend on $\mathbf{c}$. To relate $\mathbf{a}_m''$ and $\mathbf{a}_m'$ we must allow for the fact that the reference frames P_c'' and P_c' in which these accelerations are measured are also accelerating under the local pressure gradient and body forces. Free flight of the molecules on their penultimate trajectories thus gives

$$\mathbf{a}_m' + \mathbf{P}' + \mathbf{F}' = \mathbf{a}_m'' + \mathbf{P}'' + \mathbf{F}'', \tag{5.81}$$

where $\mathbf{F}$ is the body force per unit mass.

From the above equations it follows that

$$\boldsymbol{\alpha} = \mathbf{c}' \cdot \mathbf{e}' - \mathbf{c}'' \cdot \mathbf{e}'' + \mathbf{F}'' - \mathbf{F}' + \boldsymbol{\alpha}_0.$$

The points P_c'' and P_c' are swept along with the fluid, accelerating with it. Hence the appropriate time derivative following this motion is $\mathbb{D}_S$, defined in (5.75). Therefore, since these points are a collision interval apart,

$$\begin{aligned}\boldsymbol{\alpha} &= \mathbf{c}' \cdot \mathbf{e}' - \mathbf{c}'' \cdot \mathbf{e}'' + \mathbf{F}'' - \mathbf{F}' + \boldsymbol{\alpha}_0 \\ &= \big\{(\mathbf{c} \cdot \mathbf{e})'' + \tau \mathbb{D}_S(\mathbf{c} \cdot \mathbf{e})\big\} - (\mathbf{c} \cdot \mathbf{e})'' + \mathbf{F}'' - \big\{\mathbf{F}'' + \tau \mathbb{D}_S \mathbf{F}\big\} + \boldsymbol{\alpha}_0 + \mathrm{O}(\epsilon^2).\end{aligned}$$

Thus we arrive at the approximate expression for $\boldsymbol{\alpha}$:

$$\boldsymbol{\alpha} = \tau \mathbb{D}_S(\mathbf{c} \cdot \mathbf{e}) - \tau \mathbf{c} \cdot \nabla \mathbf{F} + \boldsymbol{\alpha}_0 + \mathrm{O}(\epsilon^2), \tag{5.82}$$

where a term $\tau \mathrm{D}^* \mathbf{F}$, independent of $\mathbf{c}$ has been included in $\boldsymbol{\alpha}_0$.

To complete the calculation of $\boldsymbol{\alpha}$ we need the value of $\boldsymbol{\alpha}_0$; this can be deduced from constraints on the inverse streaming term.

5.10 Constraints on the relative acceleration

By (5.27), (5.30), and (5.78), $\boldsymbol{\alpha}$ must satisfy the three conditions

$$\int \big(1,\, \mathbf{c},\, c^2\big)\, \boldsymbol{\alpha} \cdot \frac{\partial f_0}{\partial \mathbf{c}} = 0,$$

and since (5.38) gives $\partial f_0 / \partial \mathbf{c} = -2\mathbf{c} f_0 / \mathcal{C}^2$, these are equivalent to

$$\int \big(\mathbf{c} \cdot \boldsymbol{\alpha},\, \mathbf{c}\mathbf{c} \cdot \boldsymbol{\alpha},\, c^2 \mathbf{c} \cdot \boldsymbol{\alpha}\big)\, f_0 \, d\mathbf{c} = 0. \tag{5.83}$$

5.10.1 *Modifying the acceleration*

From (5.77) and (5.78) the kinetic equation can be written in the form

$$f - f_0 = -\tau \mathbb{D} f + \tau \mathbf{P} \cdot \frac{\partial}{\partial \mathbf{c}}(f_0 - f) + \tau \boldsymbol{\alpha} \cdot \frac{\partial f_0}{\partial \mathbf{c}} + \mathrm{O}(\epsilon^3), \tag{5.84}$$

where from (5.37),

$$\boldsymbol{\alpha} \cdot \frac{\partial f_0}{\partial \mathbf{c}} = -\frac{2}{\mathcal{C}^2} f_0 \boldsymbol{\alpha} \cdot \mathbf{c}. \tag{5.85}$$

By the discussion following (5.30), we note the liberty to add terms of the type $\{\boldsymbol{\beta}(\mathbf{r},t)\cdot\mathbf{c} + \gamma(\mathbf{r},t)c^2\}f_0$ to $(f - f_0)$ to ensure that (5.30) is satisfied. By (5.84) and (5.85) this is equivalent to adjusting $\boldsymbol{\alpha}$ to satisfy (5.83) by a replacement of the form

$$\boldsymbol{\alpha} \longrightarrow \boldsymbol{\alpha} + A(\mathbf{r},t)\mathbf{c} + \mathbf{B}(\mathbf{r},t). \tag{5.86}$$

From the first and third of the constraints in (5.83) we find that $\boldsymbol{\alpha}$ cannot include a term of the type $A\mathbf{c}$ unless it forms part of a deviator (e.g. as in $\mathbf{c}\cdot\overset{\circ}{\mathbf{e}} = \mathbf{c}\cdot\mathbf{e} - \frac{1}{3}\mathbf{c}\nabla\cdot\mathbf{v}$). Equation (5.86) allows us to remove such non-deviatoric terms.

By (5.75) and (5.82),

$$\mathbb{D}_S(\mathbf{c}\cdot\mathbf{e}) = \mathbf{c}\cdot\mathrm{D}^*\mathbf{e} + \mathbf{cc}:\nabla\mathbf{e} - \mathbf{c}\cdot\mathbf{e}\cdot\mathbf{e},$$

and using

$$\mathbf{e} = \overset{\circ}{\mathbf{e}} + \tfrac{1}{3}\mathbf{1}\,\nabla\cdot\mathbf{v},$$

we get

$$\mathbb{D}_S(\mathbf{c}\cdot\mathbf{e}) = \mathbf{c}\cdot\mathrm{D}^*\overset{\circ}{\mathbf{e}} + \mathbf{cc}:\nabla\overset{\circ}{\mathbf{e}} - \mathbf{c}\cdot\overset{\circ}{\overline{\overset{\circ}{\mathbf{e}}\cdot\overset{\circ}{\mathbf{e}}}} - \tfrac{2}{3}\nabla\cdot\mathbf{v}\,\mathbf{c}\cdot\overset{\circ}{\mathbf{e}} + \tfrac{1}{3}\mathbf{cc}\cdot\nabla\nabla\cdot\mathbf{v} - A_0\mathbf{c},$$

where

$$A_0 \equiv \tfrac{1}{3}\overset{\circ}{\mathbf{e}} : \overset{\circ}{\mathbf{e}} + \tfrac{1}{9}(\nabla\cdot\mathbf{v})^2 - \tfrac{1}{3}\mathrm{D}\nabla\cdot\mathbf{v}.$$

By (5.86) we omit the term $A_0\mathbf{c}$, and from (5.82) obtain

$$\boldsymbol{\alpha} = \boldsymbol{\alpha}_c + \boldsymbol{\alpha}_0, \tag{5.87}$$

where

$$\boldsymbol{\alpha}_c \equiv \tau\Big\{\mathbf{c}\cdot\mathrm{D}^*\overset{\circ}{\mathbf{e}} + \mathbf{cc}:\nabla\overset{\circ}{\mathbf{e}} - \mathbf{c}\cdot\overset{\circ}{\overline{\overset{\circ}{\mathbf{e}}\cdot\overset{\circ}{\mathbf{e}}}} - \tfrac{2}{3}\nabla\cdot\mathbf{v}\,\mathbf{c}\cdot\overset{\circ}{\mathbf{e}} + \tfrac{1}{3}\mathbf{cc}\cdot\nabla\nabla\cdot\mathbf{v} - \mathbf{c}\cdot\overset{\star}{\nabla\mathbf{F}}\Big\},$$

and

$$\overset{\star}{\nabla\mathbf{F}} \equiv \nabla\mathbf{F} - \tfrac{1}{3}\mathbf{1}\,\nabla\cdot\mathbf{F}.$$

Noting that for any tensor A, the integral of $\mathbf{c}\cdot\overset{\circ}{\mathsf{A}}\cdot\mathbf{c}\,f_0$ yields the result $\frac{1}{2}\mathcal{C}^2\mathbf{1}:\overset{\circ}{\mathsf{A}} = 0$, and that integrals odd in $\mathbf{c}$ also vanish, we deduce from its definition that $\boldsymbol{\alpha}_c$ satisfies

$$\int \mathbf{c}\cdot\boldsymbol{\alpha}_c\,f_0\,d\mathbf{c} = 0, \qquad \int c^2\mathbf{c}\cdot\boldsymbol{\alpha}_c\,f_0\,d\mathbf{c} = 0.$$

Substituting $\boldsymbol{\alpha} = \boldsymbol{\alpha}_0 + \boldsymbol{\alpha}_c$ into (5.83) and removing terms that are obviously zero, we get

$$\int \mathbf{cc}\cdot(\boldsymbol{\alpha}_c + \boldsymbol{\alpha}_0)\,f_0\,d\mathbf{c} = 0. \tag{5.88}$$

Hence by (2.40),

$$\boldsymbol{\alpha}_0 = -\frac{2}{n}\int \boldsymbol{\nu\nu}\cdot\boldsymbol{\alpha}_c\,f_0\,d\mathbf{c} \qquad (\boldsymbol{\nu} \equiv \mathbf{c}/\mathcal{C}). \tag{5.89}$$

One other adjustment to (5.82) concerns the collision interval τ. The component $\overset{\circ}{\mathbf{e}}$ of $\mathbf{e}$ is associated with momentum transport, for which $\tau = \tau_1$, while the component containing $\nabla \cdot \mathbf{v}$ is related to energy transport (cf. $(5.42)_2$), for which $\tau = \tau_2$. This allocation will be confirmed in §8.2.5 by comparison of like transport terms.

5.10.2 *Final expression for the acceleration*

The method of determining $\boldsymbol{\alpha}$ described in the previous section leads to the correct value for $\boldsymbol{\alpha}_c$, but not directly to the complete expression for $\boldsymbol{\alpha}$. Our stratagem is therefore to exploit the constraint on $\boldsymbol{\alpha}_0$, and write

$$\boldsymbol{\alpha} = \boldsymbol{\alpha}_c - \frac{2}{n}\int \boldsymbol{\nu\nu}\cdot\boldsymbol{\alpha}_c\, f_0\, d\mathbf{c}, \tag{5.90}$$

and then to offer physical explanations of the terms in $\boldsymbol{\alpha}_0$. For this we need the integrals

$$\frac{2}{n}\int \boldsymbol{\nu\nu}\cdot\mathbf{cc}\, f_0\, d\mathbf{c} = 2\mathcal{C}^2\int \nu^4\hat{\mathbf{c}}\hat{\mathbf{c}}\,\frac{f_0}{n}\, d\mathbf{c} = \tfrac{5}{2}\mathcal{C}^2\mathbf{1},$$

and $$\frac{2}{n}\int \boldsymbol{\nu\nu}\mathbf{cc}\, f_0\, d\mathbf{c} = 2\mathcal{C}^2\int \nu^4\hat{\mathbf{c}}\hat{\mathbf{c}}\hat{\mathbf{c}}\hat{\mathbf{c}}\,\frac{f_0}{n}\, d\mathbf{c} = \mathcal{C}^2(\mathsf{K} + \tfrac{5}{6}\mathbf{11}),$$

the first of which follows from (2.37), (2.40), and (2.41), and the second from (2.41) and (5.58). From these formulae, (5.89) and the remarks about τ made at the end of §5.10.1, we obtain

$$\boldsymbol{\alpha}_0 = -\tau_1\mathcal{C}^2\nabla\cdot\overset{\circ}{\mathbf{e}} - \tfrac{5}{6}\tau_2\mathcal{C}^2\nabla\nabla\cdot\mathbf{v}, \tag{5.91}$$

so that by (5.87),

$$\begin{aligned}\boldsymbol{\alpha} \;=\; & \tau_1\Big\{\mathbf{c}\cdot\mathrm{D}^*\overset{\circ}{\mathbf{e}} + \mathbf{cc}:\nabla\overset{\circ}{\mathbf{e}} - \mathcal{C}^2\nabla\cdot\overset{\circ}{\mathbf{e}} - \overline{\mathbf{c}\cdot\overset{\circ}{\mathbf{e}}\cdot\overset{\circ}{\mathbf{e}}}^{\,\circ} - \tfrac{2}{3}\nabla\cdot\mathbf{v}\,\mathbf{c}\cdot\overset{\circ}{\mathbf{e}}\Big\}\\ & + \tfrac{1}{3}\tau_2(\mathbf{cc} - \tfrac{5}{2}\mathcal{C}^2\mathbf{1})\cdot\nabla\nabla\cdot\mathbf{v} - \tau_1\mathbf{c}\cdot\overset{\star}{\nabla}\mathbf{F} + \mathrm{O}(\epsilon^2).\end{aligned} \tag{5.92}$$

5.10.3 *Physical explanation*

The terms in (5.91) are adjustments to $\boldsymbol{\alpha}_c$ due to fluid pressure that arise as follows. In deriving the inverse streaming impulse giving rise to the term $\mathbf{P}\cdot\partial f_0/\partial\mathbf{c}$ in (5.78), we assumed that the pressure gradient force $\mathbf{P}$ should include the effects of fluid shear. By (5.59),

$$\mathbf{P} = -\frac{1}{\rho}\nabla\cdot\mathsf{p} = -\frac{1}{\rho}\nabla p + \tau_1\mathcal{C}^2\nabla\cdot\overset{\circ}{\mathbf{e}} + \frac{2}{\rho}\nabla\mu\cdot\overset{\circ}{\mathbf{e}}, \tag{5.93}$$

from which it follows that the first right-hand term in (5.91) eliminates the effect of fluid shear on the impulsive force at the inverse collisions. This is necessary because the molecules approach $\overline{\mathsf{m}}$ almost isotropically from all directions, with the result that the fluid shear from one side of $d\boldsymbol{\nu}$ (in physical space) cancels that from the opposite side, giving zero net effect.

The second term in (5.91) is a little more elusive. By (5.42) it can be rewritten in terms of the scalar pressure:

$$\boldsymbol{\alpha}_{0p} = \tau_2 \frac{p}{\rho} \nabla \mathrm{D} \ln p.$$

From $$-\frac{1}{\rho}\nabla(p'' - p') = \frac{1}{\rho}\nabla(\tau_2 \mathrm{D}p) = \tau_2 \frac{p}{\rho}\nabla \mathrm{D} \ln p + \frac{1}{\rho}\nabla(\tau_2 p)\,\mathrm{D} \ln p$$

and (5.93) it follows that $\boldsymbol{\alpha}_{0p}$ can be interpreted as resulting from a delay of one collision interval in the scalar pressure effective at the inverse collision, i.e. it is the average pressure on $\mathcal{S}$ that is transmitted to the collisions at P'_c.

Summarizing the above, our kinetic equation, correct to $\mathrm{O}(\epsilon^2)$, is contained in (5.77), (5.78), (5.92), and (5.93). The importance of having a kinetic equation correct to second-order in ϵ will become clear when we deal with transport in a magnetoplasma across strong magnetic fields (see Chapter 11). In this situation the second-order transport terms prove to be much larger than the first-order terms, a consequence of the latter terms being almost completely suppressed by the magnetic field. In neutral gases the $\mathrm{O}(\epsilon^2)$ terms are less important than the $\mathrm{O}(\epsilon)$ terms, but as we shall show in Chapter 8, in ultrasonic sound waves and shock waves they make a significant difference to the results obtained.

5.11 Generalization to a non-synchronous model

5.11.1 *Integral form for the distribution function*

Finally, we shall outline how the above theory can be deduced from a more general model of molecular behaviour. We start from a more accurate form of equation (5.35):

$$f(\mathbf{r}, \mathbf{c}, t) = \int_{-\infty}^{t} e^{-(t-t')/\tau} f_0(\mathbf{r}', \mathbf{c}', t') \frac{dt'}{\tau}, \tag{5.94}$$

where $$\mathbf{r}' = \mathbf{r} - \int_{t'}^{t} \mathbf{c}''\, dt'', \tag{5.95}$$

and $$\mathbf{c}' = \mathbf{c} - \int_{t'}^{t} (\mathbf{a}'' + \boldsymbol{\alpha}'')\, dt''. \tag{5.96}$$

The integral in (5.94) is a generalization of the right-hand side of (5.28)—and its justification is the same—but it has the advantage that synchronous collisions are avoided. A molecule m is collided into the bunch of particles $\mathcal{B}$ in the range $(t', t' + dt')$, with probability dt'/τ; it then has the probability $\exp\{-(t - t')/\tau\}$ of reaching the current point $(\mathbf{r}, \mathbf{c}, t)$ without further collision.

The expression for $\mathbf{c}'$ in (5.96) is justified as follows. The molecules entering the stream $\mathcal{B}$ at time t'' change their acceleration impulsively from $\mathbf{a}''$ to $\mathbf{a}'' + \boldsymbol{\alpha}''$. According to the remarks following (5.75), to eliminate streaming losses from $d\boldsymbol{\nu}_S$, the typical molecule m must have an acceleration at $\mathbb{P}(\mathbf{r}, \mathbf{c}, t)$ that matches as closely as possible the acceleration $\dot{\mathbf{c}}$ of $d\boldsymbol{\nu}_S$.

This element has an acceleration $\mathbf{P}$ relative to $d\boldsymbol{\nu}$, so for there to be no streaming losses, m's acceleration relative to $d\boldsymbol{\nu}$ should equal $\dot{\mathbf{c}} - \mathbf{P} = \mathbf{a}$. But the only forces that can affect m's collision-free acceleration during its transit from $(\mathbf{r}', \mathbf{c}', t')$ to $(\mathbf{r}, \mathbf{c}, t)$ are those that enter through the motion of the reference frame P_c, which affect m and $d\boldsymbol{\nu}$ alike. It follows that streaming losses are absent from $d\boldsymbol{\nu}_S$ only if

$$\mathbf{a}(t'') + \boldsymbol{\alpha}(t'') = \mathbf{a}(t),$$

which enables (5.96) to be simplified to

$$\mathbf{c}' = \mathbf{c} - (t - t')\mathbf{a}(\mathbf{r}, \mathbf{c}, t). \tag{5.97}$$

5.11.2 *Taking the swept derivative*

The typical point $(\mathbf{r}', \mathbf{c}', t')$ at which a mean free path $\boldsymbol{\lambda}$ starts, and the point $(\mathbf{r}'+\boldsymbol{\lambda}, \mathbf{c}', t'+\tau)$ where it finishes, are points that are convected with their local fluid elements, convection including fluid acceleration and spin. The collisional pattern and the range $(-\infty, t)$ are thus *swept* along with the fluid, which allows us to make the interchange*

$$\mathbb{D}_S \int_{-\infty}^{t} \cdots = \int_{-\infty}^{t} \mathbb{D}_S \cdots, \tag{5.98}$$

where
$$\mathbb{D}_S = \mathrm{D}^* + \mathbf{c}\cdot\nabla + \dot{\mathbf{c}}\cdot\frac{\partial}{\partial\mathbf{c}} \qquad (\dot{\mathbf{c}} = -\mathbf{c}\cdot\mathbf{e}), \tag{5.99}$$

is the swept derivative.

Taking the derivative of (5.94), we get

$$\mathbb{D}_S f = \frac{1}{\tau} f_0 - \frac{1}{\tau} f + \int_{-\infty}^{t} e^{-(t-t')/\tau} \mathbb{D}_S f_0(\mathbf{r}', \mathbf{c}', t') \frac{dt'}{\tau}, \tag{5.100}$$

where f_0/τ comes from differentiating the upper limit of the integral and $-f/\tau$ from differentiating the exponential.

By (5.97),

$$\mathbb{D}_S f_0' = \frac{df_0'}{dt} + \mathbf{c}\cdot\frac{\partial f_0'}{\partial\mathbf{r}} + \dot{\mathbf{c}}\cdot\frac{\partial f_0'}{\partial\mathbf{c}},$$

whcrc

*The situation is similar to that obtaining with surface layers separating two regions of flow with different properties, say $\boldsymbol{\phi}_1$ and $\boldsymbol{\phi}_2$. If the layer has a velocity $\mathbf{V}$, then to obtain the volume integral of $\partial\boldsymbol{\phi}/\partial t$ over a thin disc on the surface it is necessary to convert to a time derivative, say $\delta/\delta t$ *following* the surface,

$$\frac{\partial\boldsymbol{\phi}}{\partial t} = \frac{\delta\boldsymbol{\phi}}{\delta t} - \mathbf{V}\cdot\nabla\boldsymbol{\phi},$$

then
$$\int \frac{\partial\boldsymbol{\phi}}{\partial t}\cdot\mathbf{n}\,dS = \frac{\delta}{\delta t}\int \boldsymbol{\phi}\cdot\mathbf{n}\,dS - \mathbf{V}\cdot(\boldsymbol{\phi}_2 - \boldsymbol{\phi}_1), \qquad \text{etc.}$$

$$\frac{df_0'}{dt} = \frac{d\mathbf{r}'}{dt} \cdot \frac{\partial f_0'}{\partial \mathbf{r}'} + \frac{d\mathbf{c}'}{dt} \cdot \frac{\partial f_0'}{\partial \mathbf{c}'}$$
$$= -\mathbf{c} \cdot \frac{\partial f_0'}{\partial \mathbf{r}'} - \left\{\mathbf{a} + (t-t')\mathrm{D}\mathbf{a}\right\} \cdot \frac{\partial f_0'}{\partial \mathbf{c}'},$$
$$\mathbf{c} \cdot \frac{\partial f_0'}{\partial \mathbf{r}} = \mathbf{c} \cdot \frac{\partial f_0'}{\partial \mathbf{r}'} - (t-t')\mathbf{c} \cdot \nabla \mathbf{a} \cdot \frac{\partial f_0'}{\partial \mathbf{c}'},$$
$$\dot{\mathbf{c}} \cdot \frac{\partial f_0'}{\partial \mathbf{c}} = \dot{\mathbf{c}} \cdot \frac{\partial f_0'}{\partial \mathbf{c}'} - (t-t')\dot{\mathbf{c}} \cdot \frac{\partial}{\partial \mathbf{c}}\mathbf{a} \cdot \frac{\partial f_0'}{\partial \mathbf{c}'}.$$

Hence
$$\mathbb{D}_S f_0' = \left\{\mathbf{P} - (t-t')\mathbb{D}_S\mathbf{a}\right\} \cdot \frac{\partial f_0'}{\partial \mathbf{c}'}. \tag{5.101}$$

Since f_0' is the same function of $\mathbf{c}'$ as is f_0 of $\mathbf{c}$, the dashes can be removed from the partial derivative. Equation (5.81) can be written as a relation between swept derivatives:

$$\mathbb{D}_S\mathbf{P} + \mathbb{D}_S\mathbf{F} = -\mathbb{D}_S\mathbf{a}_m = \mathrm{O}(\epsilon).$$

We approximate $-\mathbb{D}_S\mathbf{a}_m$ by its average over the converging molecules, introducing a function of $\mathbf{r}$, t for this average. Thus

$$\mathbb{D}_S\mathbf{a} = \mathbb{D}_S\dot{\mathbf{c}} - \mathbb{D}_S\mathbf{P} \approx \mathbb{D}_S\dot{\mathbf{c}} + \mathbb{D}_S\mathbf{F} + \boldsymbol{\beta}(\mathbf{r}, t).$$

It follows from (5.100) and (5.101) that

$$\mathbb{D}_S f = \frac{1}{\tau} f_0 + (\mathbf{P} + \boldsymbol{\alpha}) \cdot \frac{\partial f_0}{\partial \mathbf{c}} - \frac{1}{\tau} f + \mathrm{O}(\epsilon^3), \tag{5.102}$$

where
$$\boldsymbol{\alpha} = \tau\mathbb{D}_S(\mathbf{c} \cdot \mathbf{e}) - \tau\mathbf{c} \cdot \nabla\mathbf{F} + \boldsymbol{\alpha}_0, \tag{5.103}$$

in agreement with (5.82) deduced from the synchronous model. To complete the theory it remains to calculate $\boldsymbol{\alpha}_0$ as in §5.10.

5.11.3 *Role of collisions*

Collisions enter kinetic theory in two distinct roles—singly as in binary collisions, and collectively as in fluid pressure and fluid shear accelerations. Binary collisions transfer molecules in and out of an element $d\boldsymbol{\nu}$ of phase space at a rate that can be expressed via a collision operator of the form $(f - \bar{f})/\tau$, whereas collective effects enter by producing streaming losses (or gains) from $d\boldsymbol{\nu}$. If all that is required is a theory correct to first order in the Knudsen number, it is sufficient to consider the binary collisions alone. For a more accurate theory the collisional 'environment' cannot be neglected. This interaction between macroscopic and microscopic phenomena is very important, especially in plasma physics, as we shall see in Chapter 11. It also means that we cannot reduce fluid dynamics to kinetic theory, since the former is required to complete the specification of the latter.

5.12 Summary of intermediate kinetic theory

The essential features of intermediate kinetic theory are as follows:

(i) In a fully convected frame, i.e. one swept along with the fluid, spinning and accelerating with it, the collisional losses from the phase-space element $d\boldsymbol{\nu}_S$ (see 5.2.3) can be due only to scattering collisions, since relative to $d\boldsymbol{\nu}_S$, all collective motions of the particles vanish. In any other frame the loss rate cannot be expressed in the simple form f/τ. This yields the kinetic equation $\mathbb{D}_S f = -f/\tau + \mathsf{g}$, where g is the gain rate due to inverse collisions.

(ii) While particles that *leave* $d\boldsymbol{\nu}_S$, move into an ambient fluid $\mathcal{F}$ that is relatively stationary, those that enter $d\boldsymbol{\nu}_S$ belonged to $\mathcal{F}$, i.e. were experiencing the fluid accelerations, one collision interval earlier than the current time t. During the time interval $t-\tau$, t, particles p that are collided into $d\boldsymbol{\nu}$ at time t, are moving freely and independently of all the other particles comprising $\mathcal{F}$. Thus p are not accelerated with the fluid acceleration $\mathrm{D}\mathbf{v}$ and hence during the collision interval τ they acquire a velocity $-\tau\mathrm{D}\mathbf{v}$ relative to $d\boldsymbol{\nu}$.

The equation of fluid motion is (see 5.2.3) $\mathrm{D}\mathbf{v} = \mathbf{F} + \mathbf{P}$. Since the body force $\mathbf{F}$ acts continuously on both the fluid and the particles, its contribution to $\mathrm{D}\mathbf{v}$ may be removed. Then at the time of their inverse collisions, the particles p have a velocity deficiency $\tau(\mathrm{D}\mathbf{v} - \mathbf{F}) = \tau\mathbf{P}$ relative to $d\boldsymbol{\nu}_S$. This is eliminated impulsively when the inverse collisions inject p into $d\boldsymbol{\nu}_S$. Inverse scattering brings particles into $d\boldsymbol{\nu}_S$ at the rate f_0/τ and the simultaneous streaming impulse $\tau\mathbf{P}$ changes f_0 into $f_0 + \tau\mathbf{P}\cdot(\partial f_0/\partial\mathbf{c})$, giving $\mathsf{g} = f_0/\tau + \mathbf{P}\cdot(\partial f_0/\partial\mathbf{c})$. The additional term containing $\mathbf{P}$ is required in order to match the acceleration of the particles p with that of the phase-space element $d\boldsymbol{\nu}_S$.

(iii) But p are not only free of the influence of the pressure force during their flight prior to their inverse collisions, they also last 'sampled' the acceleration $\dot{\mathbf{c}} = -\mathbf{c}\cdot\mathsf{e}$ a collision time earlier than the current instant. Thus their average acceleration matches that of $d\boldsymbol{\nu}_S$ at the time $t-\tau$, i.e. matches $\dot{\mathbf{c}}(t-\tau) \approx \dot{\mathbf{c}}(t) - \tau\mathbb{D}_S\dot{\mathbf{c}} + \mathrm{O}(\tau^2)$. It follows that at the time of the inverse collision, p have a (further) velocity deficiency of $\tau\boldsymbol{\alpha}_c$, where $\boldsymbol{\alpha}_c = -\tau\mathbb{D}_S\dot{\mathbf{c}} = \tau\mathbb{D}_S(\mathbf{c}\cdot\mathsf{e})$. When this is removed impulsively, the gain rate is increased by a term $\boldsymbol{\alpha}_c\cdot(\partial f_0/\partial\mathbf{c})$.

But this expression is not complete; as explained in §5.10.2, $\boldsymbol{\alpha}_c$ needs to be modified to $\boldsymbol{\alpha}$, where

$$\boldsymbol{\alpha} = \boldsymbol{\alpha}_c - \frac{2}{n}\int \boldsymbol{\nu}\boldsymbol{\nu}\cdot\boldsymbol{\alpha}_c f_0\, d\mathbf{c},$$

in order to satisfy the conservation restrictions on the collision operator.

(iv) Correct to $O(\epsilon^2)$, the kinetic equation is

$$\mathbb{D}f = \mathbb{C}(f), \tag{5.104}$$

where

$$\begin{aligned}
\mathbb{D}f &= \left\{\mathrm{D}^* + \mathbf{c}\cdot\nabla - (\mathbf{P} + \mathbf{c}\cdot\mathbf{e})\cdot\frac{\partial}{\partial\mathbf{c}}\right\}f,\\
\mathbb{C}(f) &= \frac{1}{\tau}(\bar{f} - f) + \hat{\mathbf{P}}\cdot\frac{\partial\bar{f}}{\partial\mathbf{c}} - \mathbf{P}\cdot\frac{\partial f}{\partial\mathbf{c}},\\
\mathbf{P} &= -\frac{1}{\rho}\nabla\cdot\mathsf{p} = -\frac{1}{\rho}\nabla p - \frac{1}{\rho}\nabla\cdot\boldsymbol{\pi},\\
\hat{\mathbf{P}} &= \mathbf{P} + \boldsymbol{\alpha}, \quad \mathbf{e} = \tfrac{1}{2}(\nabla\mathbf{v} + \widetilde{\nabla\mathbf{v}}),\\
\boldsymbol{\alpha} &= \tau_1\left\{\mathbf{c}\cdot\mathrm{D}^*\mathring{\mathbf{e}} + \mathbf{c}\mathbf{c}:\nabla\mathring{\mathbf{e}} - \mathcal{C}^2\nabla\cdot\mathring{\mathbf{e}} - \mathbf{c}\cdot\overline{\mathring{\mathbf{e}}\cdot\mathring{\mathbf{e}}}^{\circ}\right\}\\
&\quad -\tfrac{2}{3}\tau_1\nabla\cdot\mathbf{v}\,\mathbf{c}\cdot\mathring{\mathbf{e}} + \tfrac{1}{3}\tau_2(\mathbf{c}\mathbf{c} - \tfrac{5}{2}\mathcal{C}^2\mathbf{1})\cdot\nabla\nabla\cdot\mathbf{v} - \tau_1\mathbf{c}\cdot\overset{\star}{\nabla}\mathbf{F},\\
\mathrm{D}^*\boldsymbol{\Phi}_n &= \mathrm{D}\boldsymbol{\Phi}_n - n\boldsymbol{\Omega}\times\boldsymbol{\Phi}_n \qquad (n = 0, 1),\\
\mathrm{D}^*\boldsymbol{\Phi}_2 &= \mathrm{D}\boldsymbol{\Phi}_2 - \boldsymbol{\Omega}\times\boldsymbol{\Phi}_2 + \boldsymbol{\Phi}_2\times\boldsymbol{\Omega},\\
\mathrm{D} &= \frac{\partial}{\partial t} + \mathbf{v}\cdot\nabla, \quad \boldsymbol{\Omega} = \tfrac{1}{2}\nabla\times\mathbf{v}, \quad \mathcal{C}^2 = 2kT/m,\\
\overset{\star}{\nabla}\mathbf{F} &= \nabla\mathbf{F} - \tfrac{1}{3}\mathbf{1}\nabla\cdot\mathbf{F},
\end{aligned}$$

$\boldsymbol{\Phi}_n$ is a tensor of order n, $\mathbf{F}$ is the body force per unit mass, and $\bar{f}$ denotes an average distribution taken over all directions. The BGK approximation to this average is the equilibrium distribution, f_0. In §7.2.3 will be found the expressions for $\bar{f}$, and τ resulting from Boltzmann's advanced kinetic theory.

Exercises 5

5.1 Let

$$\int \{m,\, m\mathbf{c},\, \tfrac{1}{2}mc^2\}(f - f_0)\,d\mathbf{c} = \{A,\, \mathbf{C},\, B\}.$$

Show that the substitution $f_0 \longrightarrow f_0[1 + \alpha + \boldsymbol{\beta}\cdot\mathbf{c} + \gamma c^2]$ reduces the right-hand side to zero if

$$\alpha = 5A/2\rho - 4B/p, \ \boldsymbol{\beta} = \mathbf{c}/p, \ \gamma = 4\rho B/3p^2 - A/2p.$$

5.2 In a magnetoplasma the body force acting on a charged species, the particles of which have mass m and charge Q, is given by $Q\{\mathbf{E} + (\mathbf{v} + \mathbf{c})\times\mathbf{B}\}$, where $\mathbf{E}$ is the electric field strength and $\mathbf{B}$ is the magnetic induction. Show that the velocity distribution function satisfies

$$\mathbb{D}f = -\frac{1}{\tau}(f - f_0) + \frac{Q}{m}\mathbf{B}\times\mathbf{c}\cdot\frac{\partial}{\partial\mathbf{c}}(f - f_0) + O(\epsilon^2),$$

where $\mathbb{D}$ is the operator defined in (5.17).

5.3 Show that

$$T\nabla\nabla\cdot\mathbf{v} = -\tfrac{3}{2}\left\{\mathrm{D}^*\nabla T + \overset{\circ}{\mathbf{e}}\cdot\nabla T + \nabla\cdot\mathbf{v}\nabla T\right\}$$

and deduce that the dilatation gradient adds a term

$$f_0\left\{-\tau^2\left(\tfrac{5}{2} - \nu^2\right)\mathbf{c}\cdot\overset{\circ}{\mathbf{e}}\cdot\nabla \ln T\right\} \qquad (\nu \equiv c/\mathcal{C})$$

to f and increments the heat flux vector by replacing ∇T in $\mathbf{q} = -\kappa\nabla T$ by $(\mathbf{1} + \tau_2\overset{\circ}{\mathbf{e}})\cdot\nabla T$.

5.4 Standard BGK kinetic theory (see §5.4.2) leads to the following equation for the second-order heat flux vector:

$$\mathbf{q}_2 = \frac{k}{m}p\tau_1^2\Big\{\theta_1\nabla\cdot\mathbf{v}\,\nabla T \;+\; \theta_2\,\overbrace{(\mathrm{D}\nabla T - \nabla\mathbf{v}\cdot\nabla T)}^{(i)} + \theta_3\frac{T}{p}\overbrace{\nabla p\cdot\overset{\circ}{\mathbf{e}}}^{(ii)}$$

$$+\;\overbrace{\theta_4 T\nabla\cdot\overset{\circ}{\mathbf{e}}}^{(iii)} + 3\theta_5\nabla T\cdot\overset{\circ}{\mathbf{e}}\Big\}, \qquad (5.105)$$

with $\quad\theta_1 = \frac{15}{4}\left(\frac{7}{2} - s\right),\ \ \theta_2 = \frac{45}{8},\ \ \theta_3 = -3,\ \ \theta_4 = 3,\ \ \theta_5 = \frac{35}{4} + s.$

Discuss the terms identified as (i), explaining why they should not appear in a *diffusion* equation, at least in their particular combination.

5.5 Referring to equation (5.105), why should you be surprised to see a term containing ∇p in an equation for the heat flux?

5.6 Consider the case of a uniform, steady temperature gradient in a medium supporting a sheared flow. Would you expect it to be possible to transfer internal energy through the medium in this case? Compare your answer with the prediction of equation (5.105).

6

ADVANCED KINETIC THEORY

6.1 Introduction

6.1.1 *Maxwell–Chapman theory*

There are two forms for the advanced theory, both of which make use of the actual distribution of velocities, $f(\mathbf{r}, \mathbf{c}, t)$. The first of these was initiated by Maxwell only six years after his pioneering work in elementary kinetic theory (see account in §3.1). It is based on his equation (3.5) for the transfer of a molecular property ϕ:

$$\frac{\partial}{\partial t}(n\overline{\phi}) + \nabla \cdot (n\mathbf{v}\overline{\phi} + n\overline{\mathbf{c}\phi}) = n\,\frac{\delta \phi}{\delta t}, \tag{6.1}$$

where by (2.2) expressed in an altered notation,

$$\overline{\phi}(\mathbf{r}, t) = \frac{1}{n}\int \phi(\mathbf{c}) f(\mathbf{r}, \mathbf{c}, t)\, d\mathbf{c}. \tag{6.2}$$

The right-hand term in (6.1) is the rate at which ϕ is changed by particle dynamics, i.e. by body force accelerations like gravity and accelerations due to particle collisions. An exact, finite expression for $\delta\phi/\delta t$ does not exist, but a good approximation to it will be given in §7.3.1. Maxwell was able to complete the calculation of expressions for the viscosity and conductivity only for molecules having inverse fifth-power repulsion (see §3.1.3).

Forty years were to pass before a means of treating a more general force law by Maxwell's method was discovered by Chapman (1916). This is an approximate method that takes the Maxwellian equilibrium distribution function f_0 as the dominant part of f and adds successive corrections in the form of a power series in the peculiar velocity:

$$f = f_0\{1 + \mathbf{A}_1 \cdot \mathbf{c} + \mathsf{A}_2 : \mathbf{cc} + \mathsf{A}_3 \,\vdots\, \mathbf{ccc} + \cdots\}, \tag{6.3}$$

where the coefficients $\mathbf{A}_1$, A_2 ... are assumed to be first order in smallness, and have the tensorial order indicated by their subscripts. They are functions of the speed $c = |\mathbf{c}|$ and the fluid variables ρ (or n), T and $\mathbf{v}$. To find first approximations for the transport coefficients, as indicated by (6.3), it is sufficient to take the expansion only as far as the cubic term in $\mathbf{c}$. Also, since these coefficients depend on integrals involving f, it proves unnecessary to find f explicitly, so that the kinetic equation for f is not involved.

6.1.2 *Boltzmann–Enskog theory*

Boltzmann's (1872) integro-differential equation for f provides the basis for the second form of the advanced theory. Despite attempts by Hilbert and others, this equation remained unsolved until 1917, when Enskog devised a successful iterative scheme. This led to almost the same expressions for the transport coefficients as achieved by the Maxwell–Chapman theory. But having the advantage of being more systematic and deductive, it is the favoured approach in most advanced treatises, including the classic by Chapman himself (Chapman and Cowling 1970). We shall give an outline of the Boltzmann–Enskog theory in Chapter 7. The Maxwell–Chapman theory is more intuitive and in accord with the style of this book; it will be presented below. For a brief history of the progress of kinetic theory from the days of Maxwell and Boltzmann, the reader is referred to the Historical Summary in the 1958 edition of Chapman and Cowling's treatise.

In §5.5.2 we found from intermediate kinetic theory that the distribution function is given by

$$f = f_0\Big\{1 - \tau_2\mathcal{C}(\nu^2 - \tfrac{5}{2})\boldsymbol{\nu}\cdot\nabla\ln T - 2\tau_1\boldsymbol{\nu}\cdot\overset{\circ}{\mathbf{e}}\cdot\boldsymbol{\nu} + \mathrm{O}(\epsilon^2)\Big\}, \tag{6.4}$$

where $\boldsymbol{\nu} \equiv \mathbf{c}/\mathcal{C}$, $\mathcal{C} \equiv \sqrt{2kT/m}$, and τ_2, τ_1 are microscopic times, whose values are either to be assigned from experiment or to be determined from advanced kinetic theory. In §7.6.2 it will be shown that a sufficiently general form for the $\mathrm{O}(\epsilon)$ terms in f is contained in an expression similar to (6.4), viz.

$$f = f_0\left\{1 - \frac{\mathcal{C}}{n}A(\nu)\boldsymbol{\nu}\cdot\nabla\ln T - \frac{2}{n}B(\nu)\boldsymbol{\nu}\cdot\overset{\circ}{\mathbf{e}}\cdot\boldsymbol{\nu} + \mathrm{O}(\epsilon^2)\right\}, \tag{6.5}$$

where the scalar functions $A(\nu)$ and $B(\nu)$ depend on the local thermodynamic state, say n, T, as well as on the peculiar speed ratio ν.

The Boltzmann–Enskog theory leads to integral equations for $A(\nu)$ and $B(\nu)$, while the Maxwell–Chapman method proceeds by a flanking attack, determining the coefficients in (6.3) step by step, each step improving the accuracy of the results obtained for the coefficients of viscosity (μ) and conductivity (κ). The first step in this iterative process amounts to assuming that A and B have values in agreement with (6.4), so intermediate theory can be viewed as being the first iteration in advanced kinetic theory. It is fortunately the case that the first stage yields expressions for μ and κ nearly equal to their exact values, so we shall be content to leave the details of the iterative scheme to the more mathematical works quoted in the references. In principal, quantum theory provides a means of determining the force laws between molecules, but except with the simplest molecules, little progress has be made. For our purposes there is little point in seeking great accuracy in the formalism.

6.2 Collision dynamics

6.2.1 *The relative motion of interacting particles*

We start by considering the relative motion of particles P_1 and P_2 moving in each other's field of force. Let the particles be at position vectors $\mathbf{r}_1$, $\mathbf{r}_2$, have masses m_1, m_2, and be subject to forces $\mathbf{F}_1$ and $\mathbf{F}_2$, where by the law of action and reaction, $\mathbf{F}_1 = -\mathbf{F}_2$. We shall assume that the forces are 'central', i.e. are parallel to $\mathbf{r} \equiv \mathbf{r}_1 - \mathbf{r}_2$ and depend only on the distance $r = |\mathbf{r}_1 - \mathbf{r}_2|$. Thus $m_1\ddot{\mathbf{r}}_1 = \mathbf{F}_1$, $m_2\ddot{\mathbf{r}}_2 = \mathbf{F}_2$, so that $m_1 m_2(\ddot{\mathbf{r}}_1 - \ddot{\mathbf{r}}_2) = m_2\mathbf{F}_1 - m_1\mathbf{F}_2$, or

$$M\ddot{\mathbf{r}} = \mathbf{F} \quad \big(M \equiv m_1 m_2/(m_1 + m_2), \quad \mathbf{F} = \mathbf{F}_1\big), \tag{6.6}$$

where M is known as the 'reduced' mass. Thus the motion of P_1 relative to P_2 is the same as the motion of a particle of mass M about a fixed centre of force $\mathbf{F}$.

Specifying the position $\mathbf{r}$ by polar coordinates (r, θ) in the plane of the orbit and introducing unit vectors $\hat{\mathbf{r}}, \hat{\boldsymbol{\theta}}$, we have $\dot{\hat{\mathbf{r}}} = \dot{\theta}\hat{\boldsymbol{\theta}}$, $\dot{\hat{\boldsymbol{\theta}}} = -\dot{\theta}\hat{\mathbf{r}}$, so that (6.6) becomes

$$M(\ddot{r} - r\dot{\theta}^2)\hat{\mathbf{r}} + M(2\dot{r}\dot{\theta})\hat{\boldsymbol{\theta}} = -\frac{\partial V}{\partial r}\,\hat{\mathbf{r}},$$

where $V(r)$ is the potential energy—taken to be zero at $r = \infty$—of the force $\mathbf{F}$. With some straightforward algebra we arrive at the conservation laws for angular momentum and energy:

$$r^2\dot{\theta} = \text{const.} = gb \tag{6.7}$$

and

$$\tfrac{1}{2}M(\dot{r}^2 + r\dot{\theta}^2) + V(r) = \text{const.} = \tfrac{1}{2}Mg^2, \tag{6.8}$$

where b is the distance known as the 'impact parameter' (see Fig. 6.1) and g is the constant relative velocity. Eliminating time from (6.7) and (6.8) we find

$$\frac{g^2 b^2}{2r^4}\left\{\left(\frac{dr}{d\theta}\right)^2 + r^2\right\} = \tfrac{1}{2}g^2 - V(r)/M,$$

whence

$$\frac{d\theta}{dr} = \pm\frac{b}{r^2}\left\{1 - \frac{b^2}{r^2} - \frac{2V(r)}{Mg^2}\right\}^{-1/2}. \tag{6.9}$$

6.2.2 *The scattering angle*

Figure 6.1 shows the trajectory of P_1 in a force field centred on P_2. The trajectory has two asymptotes, one along the initial direction of approach at infinity and the other along the final direction of motion when the particle has receded to infinity. The *scattering angle* χ is defined to be the angle between these asymptotes, as indicated in the figure. The reference frame is chosen so that initially—i.e. prior to the interaction between the particles—P_1 is at $\theta = 0$ and its final position is at $\theta = \pi - \chi$. The trajectory is

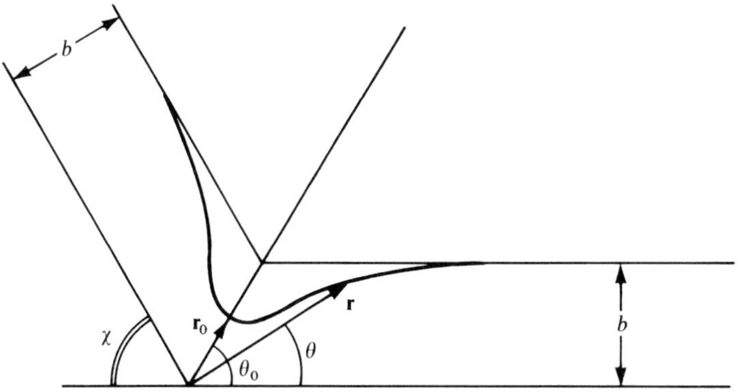

FIG. 6.1. Scattering by a central force, combining attraction with a shorter-range repulsion

symmetric about the apse line, namely a line joining P_2 to P_1 at their point of closest approach, (r_0, θ_0) say. At this point $dr/d\theta = 0$. Thus from (6.9),

$$r_0^2 - b^2 - \frac{2r_0^2}{Mg^2}V(r_0) = 0. \tag{6.10}$$

It is also clear from the figure that $\chi = \pi - 2\theta_0$. To find θ_0 we integrate (6.9) over (r_0, ∞), taking the negative square root since $dr/d\theta$ is negative along the incoming trajectory. Hence

$$\chi(b, g) = \pi - 2b\int_{r_0}^{\infty} \frac{r^{-2}dr}{\left\{1 - b^2/r^2 - 2V(r)/(Mg^2)\right\}^{1/2}}. \tag{6.11}$$

6.3 Cross sections

6.3.1 *Differential cross section*

Let d_1, d_2 denote the diameters of hard-core molecules of the first and second species, then their mutual cross section for contact collisions is*

$$\sigma = \pi d_{12}^2 \qquad \left(d_{12} \equiv \tfrac{1}{2}(d_1 + d_2)\right). \tag{6.12}$$

Suppose that a homogeneous beam of P_1 molecules has a velocity g relative to a single target molecule P_2 placed in the path of the beam. The beam intensity I is the rate at which the P_1 molecules cross unit area orthogonal to their motion. This is ng, where n is the number density of the P_1 molecules. The probability of a collision is the ratio of the target area σ to the cross sectional area of the beam. In the present case this is unity. Hence if N is the number scattered by collisions from the beam in unit time, $N = \sigma I$.

*In the remainder of this book σ will denote a collision cross section, not a diameter.

The ratio $\sigma \equiv N/I$ can be adopted to provided a more general definition of the collision cross section than that given in (6.12)—one that is independent of the force law between the molecules. Hence the generalized scattering cross section is the ratio of the number of beam molecules scattered per unit time by one fixed molecule, to the intensity of the beam. Because of the symmetry of the force field, the deflection pattern will have an axis of symmetry along the line of approach and continuing through the centre of force, as shown in Fig. 6.2. Now suppose that instead of precise

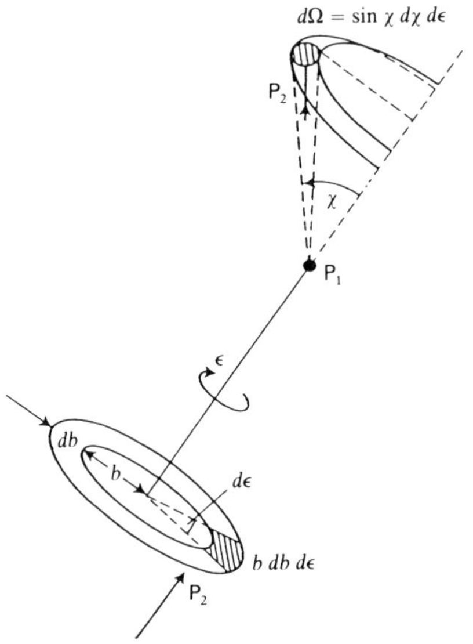

FIG. 6.2. Scattering of a homogeneous beam by a centre of force

knowledge of the impact parameter b of particle P_2, we know only that P_2 is incident on an element of area $b\,db\,d\epsilon$ (Fig. 6.2). Let $\alpha\, d\Omega$ denote the probability that it is deflected into the solid angle $d\Omega = \sin\chi\, d\chi\, d\epsilon$. By the symmetry just mentioned, the distribution α is independent of the angle ϵ. Of N incident particles per unit area per second, $|b\,db\,d\epsilon|N$ are scattered into $d\Omega$. By definition this number also equals $\alpha\, d\Omega\, N$. It follows that

$$\alpha(g, \chi) = \frac{b}{\sin\chi}\left|\frac{db}{d\chi}\right|, \tag{6.13}$$

where χ is given as a function of b and g by (6.11). Thus the *differential cross section* α can be determined when the potential $V(r)$ is known. It is usually the case that the force decreases with increasing distance, making $d\chi/db$ negative, so that $|db/d\chi|$ in (6.13) can be replaced by $-db/d\chi$.

6.3.2 *Transport cross sections*

The total scattering cross section σ is obtained by integrating $\alpha\, d\Omega$ over the complete solid angle:

$$\sigma(g) = 2\pi \int_0^\pi \alpha(g, \chi) \sin\chi\, d\chi. \tag{6.14}$$

However, this integral diverges for the usual force fields, e.g. inverse power laws. With these fields there is no upper limit to the impact parameter—deflections will occur no matter how large the distance b—making the effective area of the molecule infinite. Fortunately the *total* scattering cross section does not arise in transport theory. As we shall see in §§6.6 and 6.7, the theory leads to the so-called *transport cross sections*:

$$\begin{aligned} Q^{(\ell)}(g) &= 2\pi \int_0^\infty \Big\{1 - \cos^\ell \chi(b, g)\Big\} b\, db \qquad (\ell = 1, 2) \\ &= 2\pi \int_0^\pi \alpha(g, \chi) \Big\{1 - \cos^\ell \chi(b, g)\Big\} \sin\chi\, d\chi. \end{aligned} \tag{6.15}$$

The extra factor in the integrand vanishes like χ^2 for infinitesimal deflections and renders the integrals convergent for the usual (classical) intermolecular force fields.

6.3.3 *Rutherford scattering cross section*

An important exception to the convergence just mentioned occurs when particles P_1 and P_2 carry electric charges, giving rise to a Coulomb force field. Suppose the charges are $Z_1 e$ and $Z_2 e$, where $-e$ denotes the electron charge and Z_1, Z_2 are the (integer) charge numbers for the interacting particles. In Fig. 6.3 we show the orbit of an electron ($Z_1 = -1$) in the Coulomb field of an ion. The electric field at a vector distance $\mathbf{r}$ from a

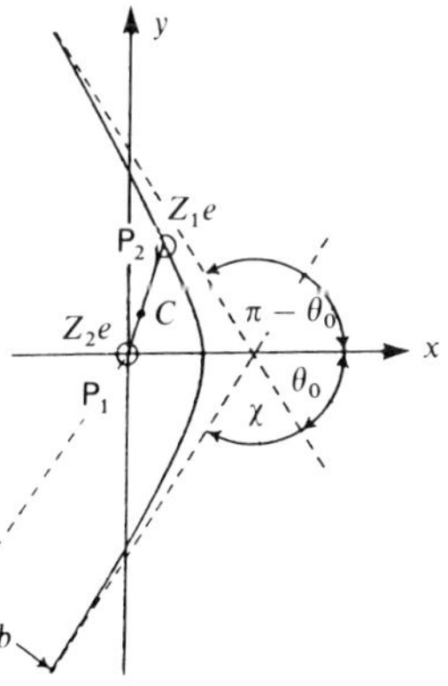

FIG. 6.3. Hyperbolic orbit of an electron in the coulomb field of an ion

charge eZ_2 is

$$\mathbf{E} = \frac{eZ_2}{4\pi\epsilon_0}\frac{\mathbf{r}}{r^3},$$

where ϵ_0 is the free space permittivity. Hence the force on P_1 due to P_2's electric field is

$$\mathbf{F} = Z_1 e\mathbf{E} = Z_1 Z_2 e^2 \mathbf{r}/(4\pi\epsilon_0 r^3), \tag{6.16}$$

whence

$$V(r) = Z_1 Z_2 e^2/(4\pi\epsilon_0 r).$$

We introduce the impact parameter b_0 defined by

$$b_0 \equiv |Z_1 Z_2| e^2/(4\pi\epsilon_0 M g^2). \tag{6.17}$$

To be definite in the following we shall assume that Z_1 and Z_2 have opposite signs—for the other case change the sign of b_0 in the final expressions. Now (6.9) becomes

$$\frac{d\theta}{dr} = \frac{\pm b\, dr}{r^2[1 - b^2/r^2 + 2b_0/r]^{\frac{1}{2}}}.$$

Choosing OX to be an axis of symmetry for the orbit of P_1 (see Fig. 6.3), we obtain the integral

$$\frac{b^2}{rb_0} = 1 + \epsilon\cos\theta, \tag{6.18}$$

where

$$\epsilon \equiv \left(1 + (b/b_0)^2\right)^{\frac{1}{2}}.$$

Equation (6.18) describes a conic with eccentricity ϵ and focus at the origin. Since $\epsilon > 1$, it is a hyperbola, as illustrated for the case of an electron being scattered by an ion in Fig. 6.3. Let θ_0 be the angle between OX and the asymptotes of (6.18), then for the upper branch of the conic, $\theta \to \pi - \theta_0$ as $r \to \infty$, whence $\cos\theta_0 = \epsilon^{-1}$ or

$$\tan\theta_0 = b/b_0. \tag{6.19}$$

Therefore

$$b = b_0 \cot\tfrac{1}{2}\chi, \qquad b\frac{db}{d\chi} = -\tfrac{1}{2}b_0^2\frac{\cos\frac{1}{2}\chi}{\sin^3\frac{1}{2}\chi},$$

so by (6.13),

$$\alpha = -\frac{b}{\sin\chi}\frac{db}{d\chi} = \frac{b_0^2}{2\sin\chi}\frac{\cos\frac{1}{2}\chi}{\sin^3\frac{1}{2}\chi},$$

or

$$\alpha(g,\chi) = \frac{b_0^2}{4\sin^4\frac{1}{2}\chi} = \left(\frac{Z_1 Z_2 e^2}{8\pi\epsilon_0 M g \sin^2\frac{1}{2}\chi}\right)^2. \tag{6.20}$$

This function is called the 'Coulomb' or sometimes the 'Rutherford' scattering cross section. It is evident that small-angle scattering is more probable than large deflections. When (6.20) is substituted into (6.15), the result is an integral divergent at both $\chi = 0$ and $\chi = \pi$. A method of dealing with this problem will be described in 6.8.4.

6.4 Transfers during a single encounter

6.4.1 *Momentum transfer per collision*

The Maxwell–Chapman theory of transport is based on (6.1), with ϕ having the values $\mathbf{cc}$ for momentum transport and $\frac{1}{2}c^2\mathbf{c}$ for energy transport. The first step in the theory is to determine how these quantities are changed by a single collision. Prior to a collision between class '1' and class '2' molecules, by (2.52) their peculiar velocities can be expressed as

$$\mathbf{c}_1 = \mathbf{G} + \mu_2\mathbf{g}, \qquad \mathbf{c}_2 = \mathbf{G} - \mu_1\mathbf{g}, \tag{6.21}$$

where $\mu_1 = m_1/(m_1+m_2)$, $\mu_2 = m_2/(m_1+m_2)$, $\mathbf{G}$ is the velocity of their centre of mass and $\mathbf{g}$ is their relative velocity. It was shown in §2.7.1 that after the collision their velocities become

$$\mathbf{c}_1' = \mathbf{G} + \mu_2\mathbf{g}', \qquad \mathbf{c}_2' = \mathbf{G} - \mu_1\mathbf{g}', \tag{6.22}$$

where
$$|\mathbf{g}'| = |\mathbf{g}|, \qquad \mathbf{g}'\cdot\mathbf{g} = \cos\chi, \tag{6.23}$$

χ being the scattering angle (cf. Figs 6.1 and 6.4).

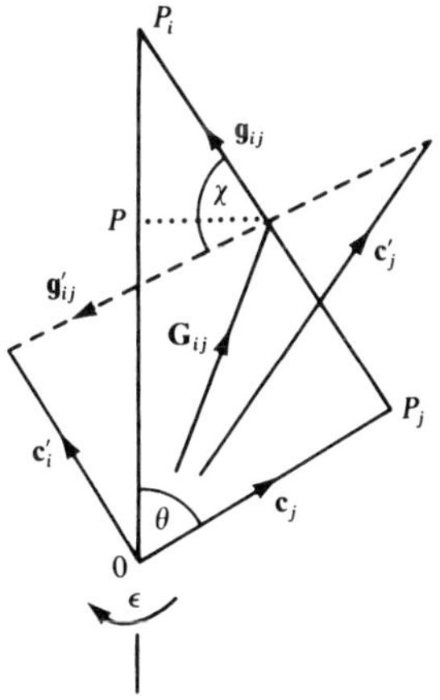

FIG. 6.4. Relative velocities of colliding particles

Let $\delta(m_1\mathbf{c}_1)$ denote the change in the momentum of P_1, i.e. the momentum transferred from P_1 to P_2, then from (6.21) and (6.22)

$$\delta(m_1\mathbf{c}_1) = m_1\mathbf{c}_1' - m_1\mathbf{c}_1 = M(\mathbf{g}'-\mathbf{g}) \qquad \big(M \equiv m_1m_2/(m_1+m_2)\big).$$

Let $\hat{\mathbf{g}}$, $\hat{\mathbf{h}}$ denote unit vectors parallel and orthogonal to $\mathbf{g}$, then

$$\hat{\mathbf{g}}\cdot\delta(m_1\mathbf{c}_1) = -Mg(1-\cos\chi), \quad \hat{\mathbf{h}}\cdot\delta(m_1\mathbf{c}_1) = Mg\sin\chi, \tag{6.24}$$

and therefore
$$\mathbf{g}' = g(\hat{\mathbf{h}}\sin\chi + \hat{\mathbf{g}}\cos\chi). \tag{6.25}$$

We shall require these expressions in §6.5 when calculating the momentum transfer occurring in mutual diffusion.

6.4.2 *Transfer in a single species*

For momentum and energy transfer within a single species we put $m_1 = m_2 = m$, whence $\mu_1 = \mu_2 = 1/2$. Now define

$$\delta\phi \equiv \phi_1' + \phi_2' - (\phi_1 + \phi_2), \tag{6.26}$$

i.e. $\delta\phi$ is the change in the dynamical quantity ϕ for both molecules involved in the encounter.

Let $\phi = \mathbf{cc}$, then by (6.21) and (6.22)

$$\delta(\mathbf{cc}) = \mathbf{c}_1'\mathbf{c}_1' + \mathbf{c}_2'\mathbf{c}_2' - \mathbf{c}_1\mathbf{c}_1 - \mathbf{c}_2\mathbf{c}_2 = \tfrac{1}{2}(\mathbf{g}'\mathbf{g}' - \mathbf{gg}). \tag{6.27}$$

By (6.23) this can be written

$$\delta(\mathbf{cc}) = \tfrac{1}{2}\Big\{g^2 \sin^2\chi(\hat{\mathbf{h}}\hat{\mathbf{h}} - \hat{\mathbf{g}}\hat{\mathbf{g}}) + g^2 \sin\chi\cos\chi(\hat{\mathbf{g}}\hat{\mathbf{h}} + \hat{\mathbf{h}}\hat{\mathbf{g}})\Big\}. \tag{6.28}$$

Now we average this expression over all directions $\hat{\mathbf{h}}$. This amounts to holding χ and $\mathbf{g}$ constant and averaging over the azimuthal angle ϵ shown in Fig. 6.2. By the symmetry of the deflection pattern about the line of approach, all vectors $\hat{\mathbf{h}}$ are equally probable. Let $\widetilde{\phi}$ denote the average of ϕ over ϵ, then writing $\hat{\mathbf{h}} = \mathbf{i}\sin\epsilon + \mathbf{j}\cos\epsilon$, where $\mathbf{i}$ and $\mathbf{j}$ are fixed unit vectors in a plane orthogonal to $\mathbf{g}$, and averaging over $0 \le \epsilon \le 2\pi$, we find that (cf. the similar calculation in 2.6.2)

$$\widetilde{\hat{\mathbf{h}}} = 0, \qquad \widetilde{\hat{\mathbf{h}}\hat{\mathbf{h}}} = \tfrac{1}{2}(\mathbf{ii} + \mathbf{jj}) = \tfrac{1}{2}(\mathbf{1} - \hat{\mathbf{g}}\hat{\mathbf{g}}), \tag{6.29}$$

where $\mathbf{1}$ is the unit tensor. It follows that the average of (6.28) is

$$\widetilde{\delta(\mathbf{cc})} = (\tfrac{1}{4}g^2\mathbf{1} - \tfrac{3}{4}\mathbf{gg})\sin^2\chi. \tag{6.30}$$

6.4.3 *Energy transfer per collision*

Next let $\phi = \mathbf{c}c^2$, then

$$\delta(\mathbf{c}c^2) = \mathbf{c}_1'c_1'^2 + \mathbf{c}_2'c_2'^2 - \mathbf{c}_1c_1^2 - \mathbf{c}_2c_2^2.$$

By (6.21), (6.22) and (6.28) this reduces to

$$\delta(\mathbf{c}c^2) = \mathbf{G}\cdot(\mathbf{g}'\mathbf{g}' - \mathbf{gg}) = 2\mathbf{G}\cdot\delta(\mathbf{cc}). \tag{6.31}$$

Hence from (6.30)

$$\widetilde{\delta(\mathbf{c}c^2)} = (\tfrac{1}{2}g^2\mathbf{G} - \tfrac{3}{2}\mathbf{gg}\cdot\mathbf{G})\sin^2\chi. \tag{6.32}$$

By (6.11) the scattering angle χ depends on the force-law potential V, on the impact parameter b and on the relative velocity $\mathbf{g}$. With b and $\mathbf{g}$ fixed, equation (6.24) determines the average momentum transfer per collision between distinguishable particles, while equations (6.30) and (6.32) determine the average total loss in momentum and energy *fluxes* per collision between like particles. Notice that if we further average $\widetilde{\delta(\mathbf{cc})}$ and $\widetilde{\delta(\mathbf{c}c^2)}$ over all directions $\hat{\mathbf{g}}$—provided these are isotropically distributed so that $\langle\hat{\mathbf{g}}\hat{\mathbf{g}}\rangle = \frac{1}{3}\mathbf{1}$—we obtain the conservation laws

$$\langle\widetilde{\delta(\mathbf{cc})}\rangle = 0, \qquad \langle\widetilde{\delta(\mathbf{c}c^2)}\rangle = 0. \tag{6.33}$$

6.5 Particle diffusion

6.5.1 *Mutual diffusion*

In §4.3.3 we presented the theory of mutual diffusion for the special case of hard-core molecules. We can now extend that theory to the general case of central molecular forces. As in §4.3.1, we first replace the peculiar velocities $\mathbf{c}_i$ and $\mathbf{c}_j$ by $\mathbf{w}_i$ and $\mathbf{w}_j$, that is we refer all velocities to the laboratory frame.

Consider encounters of class 1 molecules with class 2 molecules, as depicted in Fig. 6.2. Let $f_1 = f(\mathbf{r}, \mathbf{w}_1, t)$ and $f_2 = f(\mathbf{r}, \mathbf{w}_2, t)$ denote the phase-space number densities of these molecules. In unit volume there are $f_2\, d\mathbf{w}_2$ 'target' molecules, collectively presenting an impact area of $b\, db\, d\epsilon f_2\, d\mathbf{w}_2$ to $f_1\, d\mathbf{w}_1$ incident molecules, moving at a speed $g = |\mathbf{w}_1 - \mathbf{w}_2|$ relative to the target molecules. The differential collision rate per unit volume is therefore

$$d^8 N = \{b\, db\, d\epsilon f_2\, d\mathbf{w}_2\} \times g \times \{f_1\, d\mathbf{w}_1\}.$$

By (6.13), $\alpha \sin\chi\, d\chi = \alpha \sin\chi |d\chi/db|\, db = b\, db$; also $d\mathbf{k}' = \sin\chi\, d\chi\, d\epsilon$ is the element of solid angle into which the incident molecules are scattered. Hence there are three ways we can express the collision rate:

$$\begin{aligned} d^8 N &= f_1 f_2 g\, b\, db\, d\epsilon\, d\mathbf{w}_1\, d\mathbf{w}_2 \\ &= f_1 f_2 \alpha\, g \sin\chi\, d\chi\, d\epsilon\, d\mathbf{w}_1\, d\mathbf{w}_2 \\ &= f_1 f_2 \alpha\, g\, d\mathbf{k}'\, d\mathbf{w}_1\, d\mathbf{w}_2. \end{aligned} \tag{6.34}$$

With the assumption that $\mathbf{w}_1 - \mathbf{w}_2 = (\mathbf{c}_1 - \mathbf{c}_2) + (\mathbf{v}_1 - \mathbf{v}_2) \approx \mathbf{c}_1 - \mathbf{c}_2$, we may take the momentum transferred per encounter to be given by (6.24). By the first of (6.24), the average over ϵ is $-M\mathbf{g}(1 - \cos\chi)$. Hence from (6.34) the momentum transferred from class 1 to class 2 molecules per second per unit volume is

$$d^7 \mathbf{R}_{12} = 2\pi M \alpha(g, \chi) g \mathbf{g} (1 - \cos\chi) \sin\chi\, d\chi\, f_1 f_2\, d\mathbf{w}_1\, d\mathbf{w}_2 \tag{6.35}$$

in which χ can be varied independently of $\mathbf{w}_1$, $\mathbf{w}_2$ and $\mathbf{g}$. Hence we may integrate over χ to find

$$d^6 \mathbf{R}_{12} = M g \mathbf{g} \sigma_D(g)\, f_1 f_2\, d\mathbf{w}_1\, d\mathbf{w}_2, \tag{6.36}$$

where σ_D is the transport cross section $Q^{(1)}(g)$ defined in (6.15), i.e.

$$\sigma_D = 2\pi \int_0^{2\pi} \alpha(g, \chi)\{1 - \cos\chi\} \sin\chi\, d\chi. \tag{6.37}$$

6.5.2 *Friction force between diffusing species*

The total transfer rate is the friction force $\mathbf{R}_{12}$ between the fluids. Therefore

$$\mathbf{R}_{12} = M n_1 n_2 \langle g \mathbf{g} \sigma_D(g) \rangle, \tag{6.38}$$

where $$\langle g\mathbf{g}\sigma(g)\rangle = \int\!\!\int g\mathbf{g}\sigma(g)\,\frac{f_1}{n_1}\frac{f_2}{n_2}\,d\mathbf{w}_1\,d\mathbf{w}_2.$$

Now we can follow the treatment given in §4.3.2, from equation (4.15) to that following (4.16), to find

$$\begin{aligned}\langle g\mathbf{g}\sigma(g)\rangle &= \frac{M}{3kT}\frac{1}{(\pi C_1 C_2)^3}(\mathbf{v}_1-\mathbf{v}_2)\int g^3\sigma(g)\exp\left\{-\frac{M}{2kT}g^2\right\}d\mathbf{g}\\ &\qquad\times\int\exp\left\{-\frac{m_1+m_2}{2kT}G^2\right\}d\mathbf{G}.\end{aligned}$$

With the help of (2.67) we find that

$$\langle g\mathbf{g}\sigma_D(g)\rangle = \tfrac{16}{3}\Omega_D, \tag{6.39}$$

where

$$\Omega_D = \left(\frac{kT}{2\pi M}\right)^{1/2}\int_0^\infty y^5\sigma_D(y)e^{-y^2}\,dy \qquad \left(y^2=\frac{Mg^2}{2kT}\right). \tag{6.40}$$

Hence from (6.38), $$\mathbf{R}_{12} = Mn_1n_2\tfrac{16}{3}\Omega_D. \tag{6.41}$$

Equations (4.5), (4.7), and (4.9) yield

$$\mathbf{R}_{12} = \frac{\rho_1\rho_2}{\rho}\frac{(\mathbf{v}_1-\mathbf{v}_2)}{\tau_{12}} = \frac{n_1n_2}{n}kT\frac{(\mathbf{v}_1-\mathbf{v}_2)}{D_{12}},$$

from which it follows that the coefficient of mutual diffusion is given by

$$D_{12} = \frac{3kT}{16Mn\Omega_D}. \tag{6.42}$$

Notice that this formula for D_{12} shows that it is not dependent on the proportions of the mixture, i.e. n_1 and n_2 do not appear separately, but only in the combination $n = n_1 + n_2$.

Equation (6.42) agrees with the first approximation obtained in the Chapman–Enskog theory (see §7.10.1). Higher approximations do show a small but complicated dependence on the composition (see Chapman and Cowling 1970, p.259); for neutral gases the resulting variation in D_{12} does not exceed a few per cent and can usually be ignored. The variation is more important when the masses of the molecules are very unequal—for the Lorentz gas (see §§4.5.1 and 6.8.4) it can be up to 13 per cent.

6.6 Viscosity

6.6.1 *Total change in momentum flux due to collisions*

The first step is to find the connection between the total change in the momentum flux due to collisions, say $\Delta(\mathbf{cc})$ per unit mass, and τ_1. By (6.34) the collision rate per unit volume between class 1 and class 2 molecules is

$$d^7N = -2\pi\alpha g\sin\chi\,d\chi\,f_1f_2\,d\mathbf{c}_1\,d\mathbf{c}_2.$$

To find $\Delta(\mathbf{cc})$ we multiply this rate by the change in $\mathbf{cc}$ (for both molecules) per collision, given by (6.30), and then integrate it over $\mathbf{c}_1$ and $\mathbf{c}_2$. Thus

$$\Delta(\mathbf{cc}) = -\tfrac{1}{2} \iiint (\tfrac{1}{4}g^2\mathbf{1} - \tfrac{3}{4}\mathbf{gg})2\pi\alpha g \sin^3\chi\, d\chi\, f_1 f_2\, d\mathbf{c}_1\, d\mathbf{c}_2.$$

The factor $\frac{1}{2}$ introduced here allows for the fact that now we are concerned with collisions between *like* molecules. To amplify this point: with unlike particles, a collision between molecules with velocities $\mathbf{c}_a$, $\mathbf{c}_b$ will appear once with $\mathbf{c}_1 = \mathbf{c}_a$, $\mathbf{c}_2 = \mathbf{c}_b$ and again with $\mathbf{c}_1 = \mathbf{c}_b$, $\mathbf{c}_2 = \mathbf{c}_a$, whereas with like molecules these two events are indistinguishable and must be counted as one.

Now denote by σ_μ the transport cross section $Q^{(2)}(g)$ defined in (6.15):

$$\sigma_\mu = 2\pi \int_0^\pi \alpha(g,\chi)\sin^3\chi\, d\chi. \tag{6.43}$$

Then the expression for $\Delta(\mathbf{cc})$ becomes

$$\Delta(\mathbf{cc}) = -\tfrac{1}{8} \iint g^3\sigma_\mu(g)(\mathbf{1} - 3\hat{\mathbf{g}}\hat{\mathbf{g}})\, f_1 f_2\, d\mathbf{c}_1\, d\mathbf{c}_2, \tag{6.44}$$

where $\hat{\mathbf{g}}$ denotes unit vector parallel to $\mathbf{g}$.

By (6.4,)

$$\begin{aligned} f_1 f_2 \ &= \ f_{10}f_{20}\Big\{1 - \tau_2\big[(\nu_1^2 - \tfrac{5}{2})\mathbf{c}_1\cdot\nabla\ln T + (\nu_2^2 - \tfrac{5}{2})\mathbf{c}_2\cdot\nabla\ln T\big] \\ &\quad -2\tau_1(\boldsymbol{\nu}_1\cdot\overset{\circ}{\mathbf{e}}\cdot\boldsymbol{\nu}_1 + \boldsymbol{\nu}_2\cdot\overset{\circ}{\mathbf{e}}\cdot\boldsymbol{\nu}_2) + \mathrm{O}(\epsilon^2)\Big\}, \end{aligned} \tag{6.45}$$

where f_{10} and f_{20} denote the equilibrium distributions. The terms odd in the vectors $\mathbf{c}_1$ and $\mathbf{c}_2$ contribute nothing to the integral in (6.44). Also, because $\langle\hat{\mathbf{g}}\hat{\mathbf{g}}\rangle = \frac{1}{3}\mathbf{1}$, the factor $(\mathbf{1} - 3\hat{\mathbf{g}}\hat{\mathbf{g}})$ ensures that $\Delta(\mathbf{cc})$ vanishes in equilibrium conditions. We are left with

$$\Delta(\mathbf{cc}) = \frac{\tau_1 n^2}{4C^2} \iint g^3\sigma_\mu(g)(\mathbf{1} - 3\hat{\mathbf{g}}\hat{\mathbf{g}})(\mathbf{c}_1\mathbf{c}_1 + \mathbf{c}_2\mathbf{c}_2) : \overset{\circ}{\mathbf{e}}\ \frac{f_{10}}{n}\frac{f_{20}}{n}\, d\mathbf{c}_1\, d\mathbf{c}_2. \tag{6.46}$$

To evaluate this integral we follow the method of 2.8.1, that is we transform to variables $\mathbf{G}$, $\mathbf{g}$, using (cf. (6.21))

$$\mathbf{c}_1 = \mathbf{G} + \tfrac{1}{2}\mathbf{g}, \quad \mathbf{c}_2 = \mathbf{G} - \tfrac{1}{2}\mathbf{g}, \quad \mathbf{c}_1\mathbf{c}_1 + \mathbf{c}_2\mathbf{c}_2 = 2\mathbf{GG} + \tfrac{1}{2}\mathbf{gg}. \tag{6.47}$$

An equation like (2.65) is obtained, with g $(= g_{ij})$ in the integrand replaced by $g^3\sigma_\mu(\mathbf{1} - 3\hat{\mathbf{g}}\hat{\mathbf{g}})(2\mathbf{GG} + \frac{1}{2}\mathbf{gg}) : \overset{\circ}{\mathbf{e}}$. Hence

$$\Delta(\mathbf{cc}) = \frac{\tau_1 n^2}{8\pi^3 C^8} \int g^3\sigma_\mu(g)(\mathbf{1} - 3\hat{\mathbf{g}}\hat{\mathbf{g}})\mathbf{gg} : \overset{\circ}{\mathbf{e}}\ e^{-mg^2/4kT}\, d\mathbf{g} \int e^{-mG^2/kT}\, d\mathbf{G}, \tag{6.48}$$

where the term containing $\mathbf{GG} : \overset{\circ}{\mathbf{e}}$ in the integrand has been omitted—it adds nothing to the integral because of the presence of the factor $(\mathbf{1} - 3\hat{\mathbf{g}}\hat{\mathbf{g}})$.

From (2.40), (2.41), and (2.67) we have

$$\iint e^{-mG^2/kT}\, d\mathbf{G} = 4\pi \int_0^\infty G^2\, e^{-mG^2/kT}\, dG = \left(\frac{\pi}{2}\right)^{3/2} \mathcal{C}^3.$$

Also by (5.54) and (5.58),

$$\int_\Omega (\mathbf{1} - 3\hat{\mathbf{g}}\hat{\mathbf{g}})\hat{\mathbf{g}}\hat{\mathbf{g}}\, d\Omega : \mathring{\mathbf{e}} = \left\{\tfrac{1}{3}\mathbf{11} - 3(\tfrac{2}{15}\,\mathsf{K} + \tfrac{1}{9}\mathbf{11})\right\} : \mathring{\mathbf{e}} = -\tfrac{2}{5}\,\mathring{\mathbf{e}}\,.$$

Then with the definition

$$\Omega_\mu \equiv \left(\frac{kT}{\pi m}\right)^{1/2} \int_0^\infty y^7 \sigma_\mu(y)\, e^{-y^2}\, dy \qquad \left(y^2 = \frac{mg^2}{4kT}\right), \tag{6.49}$$

we reduce (6.48) to

$$\Delta(\mathbf{cc}) = -\frac{16p\tau_1}{5m} n\Omega_\mu\, \mathring{\mathbf{e}}\,. \tag{6.50}$$

6.6.2 *The coefficient of viscosity*

The next step in the Maxwell–Chapman theory is to set $\phi = \mathbf{cc}$ in (6.1) and obtain a relationship between $n\,\delta(\mathbf{cc})/\delta t$ and the gradient of the fluid velocity.* An equivalent and more direct approach is to use the kinetic equation in the collision integral:

$$n\,\frac{\delta}{\delta t}(\mathbf{cc}) = \int \mathbf{cc}\mathbb{C}f\, d\mathbf{c} = \int \mathbf{cc}\mathbb{D}f\, d\mathbf{c}, \tag{6.51}$$

and then adopt the approximation $f = f_0 + \mathrm{O}(\epsilon)$. By (5.46), i.e.

$$\mathbb{D} \ln f_0 = (\nu^2 - \tfrac{5}{2})\mathbf{c} \cdot \nabla \ln T + 2\boldsymbol{\nu} \cdot \mathring{\mathbf{e}} \cdot \boldsymbol{\nu} \qquad (\boldsymbol{\nu} \equiv \mathbf{c}/\mathcal{C}), \tag{6.52}$$

(2.37), (2.41), and (5.58) we get

$$\begin{aligned} n\,\frac{\delta}{\delta t}(\mathbf{cc}) &= 2\int \mathbf{cc}\boldsymbol{\nu\nu} : \mathring{\mathbf{e}}\, f_0\, d\mathbf{c} = \frac{8n}{\sqrt{\pi}}\mathcal{C}^2 \int_0^\infty \nu^6\, e^{-\nu^2}\, d\nu \int_\Omega \hat{\mathbf{c}}\hat{\mathbf{c}}\hat{\mathbf{c}}\hat{\mathbf{c}}\, d\Omega : \mathring{\mathbf{e}} \\ &= n\mathcal{C}^2\, \tfrac{15}{2} \left(\tfrac{2}{15}\,\mathsf{K} + \tfrac{1}{9}\mathbf{11}\right) : \mathring{\mathbf{e}} = \frac{2p}{m}\, \mathring{\mathbf{e}}\,. \end{aligned} \tag{6.53}$$

Now $\Delta(\mathbf{cc})$ is the total *loss* of momentum flux in a volume element $d\mathbf{r}$ of physical space, whereas $n\delta(\mathbf{cc})/\delta t$ is the total gain. These expressions are different forms for the same collisional changes. Hence $\Delta(\mathbf{cc}) = -n\delta(\mathbf{cc})/\delta t$, and it follows from (6.50) and (6.53) that

$$\tau_1 = \frac{5}{8n\Omega_\mu}. \tag{6.54}$$

The coefficient of viscosity now follows from (see (5.59))

$$\mu = p\tau_1 = \frac{5kT}{8\Omega_\mu}, \tag{6.55}$$

*See account in Present (1958), p. 217.

which is the first approximation in the Chapman–Enskog theory. We shall apply this general formula to some important special cases in §6.8. The higher approximations will be discussed in §6.8.3.

6.7 Thermal conductivity

6.7.1 *Total rate of change of energy flux due to collisions*

The first step is to calculate $\Delta(\mathbf{c}c^2)$. We can shorten the account by making use of the equations of the previous section. We use (6.31) and (6.44) to deduce that

$$\Delta(\mathbf{c}c^2) = -\tfrac{1}{4}\iint g^3\sigma_\mu(g)(\mathbf{1}-3\hat{\mathbf{g}}\hat{\mathbf{g}})\cdot\mathbf{G}\, f_1 f_2\, d\mathbf{c}_1\, d\mathbf{c}_2.$$

An argument similar to that giving (6.46) now leads to

$$\Delta(\mathbf{c}c^2) = \frac{\tau_2 n^2}{4\mathcal{C}^2}\iint g^3\sigma_\mu(g)(\mathbf{1}-3\hat{\mathbf{g}}\hat{\mathbf{g}})\cdot\mathbf{GG}\cdot\mathbf{gg}\cdot\nabla\ln T\,\frac{f_{10}}{n}\frac{f_{20}}{n}\, d\mathbf{c}_1\, d\mathbf{c}_2,$$

in which we used (6.47) to write

$$(c_1^2-\tfrac{5}{2}\mathcal{C}^2)\mathbf{c}_1+(c_2^2-\tfrac{5}{2}\mathcal{C}^2)\mathbf{c}_2 = 2\big(G^2-\tfrac{5}{2}\mathcal{C}^2+\tfrac{1}{4}g^2\big)\mathbf{G}+\mathbf{G}\cdot\mathbf{gg},$$

and then omitted the term not containing the vector $\mathbf{g}$, since the factor $(\mathbf{1}-3\hat{\mathbf{g}}\hat{\mathbf{g}})$ in the integrand eliminates its contribution. The next step is to complete the transformation to $\mathbf{G}$, $\mathbf{g}$ as the variables of integration, which gives

$$\begin{aligned}\Delta(\mathbf{c}c^2) &= \frac{\tau_2 n^2}{4\pi^3\mathcal{C}^8}\int g^3\sigma_\mu(g)(\mathbf{1}-3\hat{\mathbf{g}}\hat{\mathbf{g}})\cdot \\ &\qquad \int \mathbf{GG}\, e^{-mG^2/kT}\, d\mathbf{G}\cdot\mathbf{gg}\cdot\nabla\ln T\, e^{-mg^2/4kT}\, d\mathbf{g}.\end{aligned}$$

By (2.40), (2.41), and (2.67),

$$\int \mathbf{GG}\, e^{-mG^2/kT}\, d\mathbf{G} = \tfrac{1}{2}\pi^{3/2}\left(\frac{kT}{m}\right)^{5/2}\mathbf{1};$$

so $$\Delta(\mathbf{c}c^2) = \frac{\tau_2 n^2}{8\pi^{3/2}2^{5/2}\mathcal{C}^3}\int g^5\sigma_\mu(\mathbf{1}-3\hat{\mathbf{g}}\hat{\mathbf{g}})\cdot\hat{\mathbf{g}}\hat{\mathbf{g}}\, e^{-mg^2/4kT}\, d\mathbf{g}\,\cdot\,\nabla\ln T.$$

By (6.49) this reduces to

$$\Delta(\mathbf{c}c^2) = -\tfrac{16}{3}\tau_2(nk/m)^2\,\Omega_\mu T\nabla T. \tag{6.56}$$

6.7.2 *Coefficient of thermal conductivity*

From

$$n\frac{\delta}{\delta t}(\mathbf{c}c^2) = \int \mathbf{c}c^2\mathbb{C}f\, d\mathbf{c} = \int \mathbf{c}c^2\,\mathbb{D}f\, d\mathbf{c} \approx \int \mathbf{c}c^2\mathbb{D}f_0\, d\mathbf{c}$$

and (6.52), we obtain

$$n\frac{\delta}{\delta t}(\mathbf{c}c^2) = \frac{5k^2}{m^2}nT\nabla T,$$

a rate equal to $-\Delta(\mathbf{c}c^2)$ (see final paragraph of §6.6.2). We conclude that

$$\tau_2 = \frac{15}{16n\Omega_\mu}. \tag{6.57}$$

It follows from (5.62) that the coefficient of thermal conductivity is given by

$$\kappa = \frac{75}{32}\frac{k^2T}{m\Omega_\eta} = \frac{25}{16}c_v\frac{kT}{\Omega_\eta} = \tfrac{5}{2}c_v\mu. \tag{6.58}$$

Also by (6.55),

$$\tau_2 = \tfrac{3}{2}\tau_1, \tag{6.59}$$

a relation that we deduced in §3.3.3 by using mean-free-path arguments and velocity persistence. The formulae in (6.58) agree with the first approximations of the Chapman–Enskog theory; some special cases will now be considered.

6.8 Formulae for special molecular models

6.8.1 *List of basic equations*

The sequence of equations from which the first approximations to μ, κ, and D_{12} can be determined are (6.11), (6.10), (6.13), (6.15), (6.37), (6.43), (6.40), (6.49), (6.42), (6.55), and (6.58), which for the reader's convenience are collected below:

$$\chi(b,g) = \pi - 2b\int_{r_0}^{\infty}\frac{r^{-2}dr}{\{1 - b^2/r^2 - 2V(r)/(Mg^2)\}^{1/2}},$$

$$0 = r_0^2 - b^2 - \frac{2r_0^2}{Mg^2}V(r_0),$$

$$\alpha(g,\chi) = \frac{b}{\sin\chi}\left|\frac{db}{d\chi}\right|,$$

$$Q^{(\ell)} = 2\pi\int_0^{2\pi}\alpha(g,\chi)\{1-\cos^\ell\chi\}\sin\chi\, d\chi,$$

$$\Omega_D \equiv \left(\frac{kT}{2\pi M}\right)^{1/2}\int_0^\infty y^5 Q^{(1)}(y)e^{-y^2}\,dy \qquad \left(y^2 = \frac{Mg^2}{2kT}\right),$$

$$\Omega_\mu \equiv \left(\frac{kT}{\pi m}\right)^{1/2}\int_0^\infty y^7 Q^{(2)}(y)\,e^{-y^2}\,dy \qquad \left(y^2 = \frac{mg^2}{4kT}\right),$$

$$D_{12} = \frac{3kT}{16Mn\Omega_D}, \quad \mu = p\tau_m = \frac{5kT}{8\Omega_\mu}, \quad \kappa = \tfrac{5}{2}c_v\mu.$$

6.8.2 *Hard-core molecules*

The simplest special case of the above theory is a gas of hard-core molecules. Consider the collision of two spheres of diameters d_1 and d_2, as illustrated in Fig. 6.5 below. The line joining the centres of the spheres at the instant of contact has length $d_{12} = \frac{1}{2}(d_1 + d_2)$, and it is evident from the figure that the impact parameter b is related to the scattering angle χ by

$$b = d_{12} \sin\theta = d_{12} \sin \tfrac{1}{2}(\pi - \chi) = d_{12} \cos \tfrac{1}{2}\chi.$$

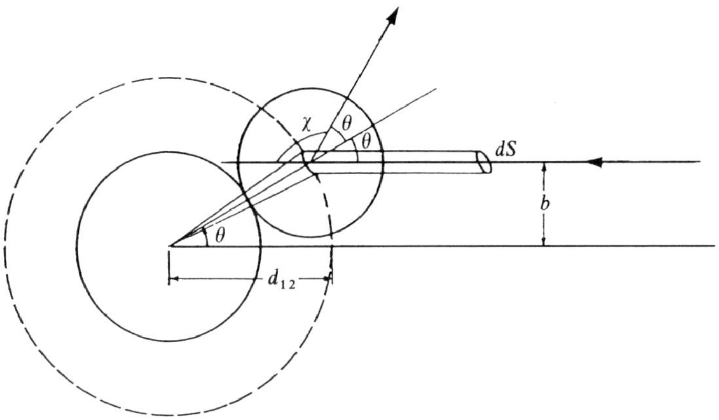

FIG. 6.5. Collision between rigid elastic spheres

Hence $b\,db = -\frac{1}{4}d_{12}^2 \sin\chi\, d\chi$, and $\alpha = \frac{1}{4}d_{12}^2$. It follows from the definitions given above that $Q^{(1)} = \pi d_{12}^2$, $Q^{(2)} = \frac{2}{3}\pi d_{12}^2$. Therefore the above equations and the standard integrals in (2.41) yield

$$\Omega_D = d_{12}^2 \left(\frac{\pi kT}{2M}\right)^{1/2}, \qquad \Omega_\mu = 4\pi d^2 \left(\frac{\pi kT}{m}\right)^{1/2},$$

where for Ω_μ we have taken the case of identical molecules. Finally, we arrive at

$$D_{12} = \frac{3}{8\pi d_{12}^2 n}\left(\frac{\pi kT}{2M}\right)^{1/2}, \qquad \mu = \frac{5}{16\, d^2}\left(\frac{mkT}{\pi}\right)^{1/2}, \tag{6.60}$$

results in agreement with those already derived by more direct arguments—see (4.20) and (3.50).

6.8.3 *Inverse power-law force*

Next let us assume that a class '1' molecule repels a class '2' molecules, a distance r away, with the force $K_{12}r^{-\nu}$, where K_{12} and ν are constants (see §3.2.1). The potential energy is $V(r) = K_{12}r^{1-\nu}/(\nu - 1)$, so that with the substitutions

$$x = b/r, \quad x_0 = b/r_0, \quad z = b(Mg^2/K_{12})^{1/(\nu-1)}, \quad y^2 = Mg^2/2kT, \tag{6.61}$$

the scattering angle can be written

$$\chi(b,\, g) = \pi - 2b \int_0^{x_0} \left\{ 1 - x^2 - \frac{2}{\nu - 1} \left(\frac{x}{z} \right)^{\nu - 1} \right\}^{-1/2} dx. \tag{6.62}$$

As x_0 is the (unique) positive square root of

$$1 - x^2 - \frac{2}{\nu - 1} \left(\frac{x}{z} \right)^{\nu - 1} = 0,$$

χ depends only on ν and z. Replacing the variable b by z, we find

$$Q^{(\ell)} \equiv 2\pi \int_0^{\infty} (1 - \cos^{\ell} \chi) b\, db = 2\pi \left(\frac{K_{12}}{2kTy^2} \right)^{2/(\nu - 1)} A_{\ell}(\nu),$$

where

$$A_{\ell} \equiv \int_0^{\infty} (1 - \cos^{\ell} \chi) z\, dz. \tag{6.63}$$

Then with the gamma function, i.e.

$$\Gamma(s) \equiv \int_0^{\infty} e^{-t}\, t^{s-1}\, dt$$

we obtain the expressions

$$\Omega_D = \left(\frac{\pi kT}{2M} \right)^{1/2} \left(\frac{K_{12}}{2kT} \right)^{2/(\nu - 1)} A_1(\nu)\, \Gamma\left(3 - \frac{2}{\nu - 1} \right), \tag{6.64}$$

and

$$\Omega_\mu = \left(\frac{\pi kT}{m} \right)^{1/2} \left(\frac{K}{2kT} \right)^{2/(\nu - 1)} A_2(\nu)\, \Gamma\left(4 - \frac{2}{\nu - 1} \right). \tag{6.65}$$

The functions $A_{\ell}(\nu)$ have to be obtained by numerical analysis; the values in Table 6.1 were taken from Chapman and Cowling (1970). It

Table 6.1 *Values of $A_1(\nu)$ and $A_2(\nu)$*

ν	5	7	9	11	15	21	25	∞
$A_1(\nu)$	0.422	0.385	0.382	0.383	0.393	—	—	0.5
$A_2(\nu)$	0.436	0.357	0.332	0.319	0.309	0.307	0.306	0.333

follows from the above theory that the first approximations to μ and κ are given by

$$[\mu]_1 = 5(kmT/\pi)^{1/2}(2kT/K)^{2/(\nu-1)} \Big/ 8\Gamma\left(4 - \frac{2}{\nu - 1} \right) A_2, \quad [\kappa]_1 = \tfrac{5}{2} c_v [\mu]_1. \tag{6.66}$$

To obtain higher approximations, it is necessary to replace the constants τ_e and τ_m by functions of c as shown in (6.5). Chapman's method of correcting the first approximations was equivalent to expanding these functions in powers of c^2, substituting the resulting form for f in (6.46) and calculating the collisional rate of destruction of the various functions so obtained. He thereby found an infinite number of linear equations to determine the coefficients in his expansions, which generalize the single relations we obtained for τ_m and τ_e in (6.54) and (6.57). In most cases the corrections are found to be quite small. They can be represented thus:

$$[\mu]_n = (1 + \delta_n^\mu)[\mu]_1, \qquad [\kappa]_n = (1 + \delta_n^\kappa)[\kappa]_1,$$

where the small numbers δ_n^μ and δ_n^κ depend only on the force index ν. For example

$$\delta_2^\mu = \frac{3(\nu - 5)^2}{2(\nu - 1)(101\nu - 113)}, \qquad \delta_2^\kappa = \frac{(\nu - 5)^2}{4(\nu - 1)(11\nu - 13)}. \tag{6.67}$$

Thus for Maxwellian molecules (see discussion in §3.1.3), $\nu = 5$, and the first approximation is unchanged, a result that holds for all the higher approximations. As ν varies from 5 to ∞, the numerical multiplier of $[\mu]_1$ varies from unity to 1.0160.., and that of $[\kappa]_1$, from unity to 1.0251.. . The Eucken ratio (see §3.7.3), $\kappa/c_v\mu$ varies from 2.5 at $\nu = 5$ to 2.522 at $\nu = \infty$. For practical applications the first approximation is entirely sufficient, since the force law itself is not known with any accuracy.

There are other more complicated force laws that give better agreement with experiment over a wider range of temperatures, but these depend on an empirical input of two or more parameters. Two of these laws have been discussed in §3.2.2. For an extensive account of the higher-order corrections and other force laws, the reader is referred to the work by Ferziger and Kaper (1972).

6.8.4 *Inverse-square or Coulomb force law*

By (6.16), for the force between particles '1' carrying charges $Z_1 e$ and particles '2' carrying charges $Z_2 e$, the parameters are

$$K_{12} = Z_1 Z_2 e^2 / 4\pi\epsilon_0, \qquad \nu = 2, \tag{6.68}$$

where ϵ_0 is the free-space permittivity. For this case (6.62) becomes

$$\chi(b, g) = \pi - 2b \int_0^{x_0} \left\{1 - x^2 - 2\frac{x}{z}\right\}^{-1/2} dx,$$

where x_0 is the positive root of $1 - x^2 - 2x/z = 0$. The integral is analytic and yields

$$\cos\chi = \frac{z^2 - 1}{z^2 + 1} \qquad (z = bMg^2/K_{12}). \tag{6.69}$$

Then from its definition in §6.8.1,

$$Q^{(\ell)} = 2\pi \left(\frac{K_{12}}{2kTy^2}\right)^2 \int_0^{z_{\max}} \left\{1 - \left(\frac{z^2 - 1}{z^2 + 1}\right)^\ell\right\} z\, dz,$$

where, since the integral over $(0, \infty)$ is divergent, we have replaced the upper limit by $z_{\max}$, the maximum value of z that makes physical sense.

We shall give the value of $z_{\max}$ usually adopted shortly. It follows that

$$Q^{(1)} = 2\pi \left(\frac{K_{12}}{2kTy^2}\right)^2 \ln(1 + z_{\max}^2), \tag{6.70}$$

$$Q^{(2)} = 4\pi \left(\frac{K_{12}}{2kTy^2}\right)^2 \left\{\ln(1 + z_{\max}^2) - \frac{z_{\max}^2}{1 + z_{\max}^2}\right\}. \tag{6.71}$$

Notice from (6.69) that when $z = 1$, i.e. when

$$b = b_0 \equiv \frac{|K_{12}|}{Mg^2} = \frac{|Z_1 Z_2|e^2}{4\pi\epsilon_0 M g^2}, \tag{6.72}$$

$\chi = \pi/2$. This scattering angle serves to separate the range of b into 'close' collisions ($b \leq b_0$) and 'grazing' collisions ($b_0 \leq b$). It remains to integrate over y to complete the theory, but as $Mg^2 = 2kTy^2$, the functions $Q^{(\ell)}(y)$ are rather complicated. Satisfactory approximations can be obtained as follows.

First it can be shown (see §9.2.3) that the movement of electrons in a plasma masks the Coulomb potential due to a fixed charge almost completely within a distance (known as the Debye length)

$$\lambda_D = (\epsilon_0 k T_e / n_e e^2)^{1/2}, \tag{6.73}$$

from the charge. It follows that impact parameters greater than λ_D will have a negligible effect on the trajectories of the passing particles. Therefore it is physically sound to set $z = z_{\max}$ at $b = \lambda_D$. Secondly, because $z_{\max} = \lambda_D / b_0 \gg 1$, the terms containing $z_{\max}$ in (6.70) and (6.71) can be reduced to $2 \ln z_{\max}$ and $2 \ln z_{\max} - 1$ respectively; these expressions are insensitive to variations of g^2 about its mean value $\overline{g^2}$. We therefore define an average impact parameter by

$$\bar{b}_0 \equiv |Z_1 Z_2| e^2 / (4\pi\epsilon_0 M \overline{g^2}) \tag{6.74}$$

and adopt the approximation $z_{\max} = \Lambda$, where Λ is the constant

$$\Lambda \equiv \lambda_D / \bar{b}_0. \tag{6.75}$$

Then from equations given in §6.8.1 we find

$$\Omega_D = 2\pi \ln \Lambda \left(\frac{kT}{2\pi M}\right)^{1/2} \left(\frac{K_{12}}{2kT}\right)^2, \tag{6.76}$$

and

$$\Omega_\mu = 2\pi(2 \ln \Lambda - 1) \left(\frac{kT}{\pi m}\right)^{1/2} \left(\frac{K_{12}}{2kT}\right)^2, \tag{6.77}$$

from which the first approximations $[D_{12}]_1$ and $[\mu]_1$ can be found.

There are two difficulties with this theory. First, the values of $[D_{12}]_1$ and $[\mu]_1$ are appreciably smaller than $[D_{12}]_2$ and $[\mu]_2$, making higher approximations essential for accuracy. Secondly, Coulomb fields have such an extended range, that grazing encounters involve the *simultaneous* interactions of a great number of particles, and the assumption of binary collisions needs reconsideration. We shall return to this topic in §9.4.4 when dealing with particle trajectories in plasmas.

Exercises 6

6.1 Weakly attracting rigid-sphere molecules of diameters d_1, d_2 have a reduced mass M and a relative velocity g. By considering only those impact parameters for which a contact collision can occur, show that the corresponding cross section can be written

$$\sigma = \pi d_{12}^2 \left(1 - \frac{V(d_{12})}{\frac{1}{2}Mg^2}\right) \qquad \left(d_{12} \equiv \tfrac{1}{2}(d_1 + d_2)\right).$$

6.2 A particle is moving along $y = b$ in the positive direction along OX with velocity c_x and energy E. A central force $F(r)$, $r = (x^2 + y^2)^{1/2}$ deflects it through a small angle χ. Starting from $c_y = \int F_y \, dt$, show that

$$\chi \approx \frac{b}{2E} \int_{-\infty}^{\infty} F(r) \frac{dr}{r}.$$

Given that the potential energy is $V = Kr^{-S}$, where s is a positive integer and the constant K can have either sign, show that $\chi \approx s|K|A_S/Eb^S$, where

$$A_S = \int_0^{\pi/2} \cos^S \theta \, d\theta = \frac{\pi^{1/2}}{2} \frac{\Gamma\left(\frac{1}{2}(s+1)\right)}{\Gamma\left(\frac{1}{2}s + 1\right)}.$$

6.3 In exercise 6.1 let $V(r) = -Kr^{-S}$, where K and s are positive numbers. Show that to first order in K, the scattering angle in a contact collision is given by

$$\chi = 2\cos^{-1} z_0 + \frac{K}{\varepsilon b^S} \int_0^{z_0} \frac{z^S \, dz}{(1 - z^2)^{3/2}} \qquad (z_0 \leq 1),$$

where $z = b/r$, $z_0 = b/d_{12}$ and $\varepsilon = \frac{1}{2}Mg^2$. Why is it sufficient to restrict attention to contact collisions when using this expression to calculate transport coefficients?

6.4 Show that for the preceding exercise

$$\sigma_D = \pi d_{12}^2 + 2\pi K I_1(s)/\varepsilon d_{12}^{S-2},$$

and

$$\Omega_D = \Omega_D^0\left\{1 + S_{12}/T\right\}, \quad D_{12} = D_{12}^0\left\{1 + S_{12}/T\right\},$$

where

$$I_1(s) = 2\int_0^1 dz_0 \, z_0^{2-S}\left(1 - z_0^2\right)^{1/2} \int_0^{z_0} dz \, z^S\left(1 - x^2\right)^{-3/2},$$

$$S_{12} = KI_1(s)/kd_{12}^S,$$

and Ω_D^0, D_{12}^0 denote the rigid sphere values. (The function $I_1(s+1) \equiv i_1(s)$ is tabulated by Chapman and Cowling (1970), p. 183.)

6.5 Use the result in exercise 6.2 to show that

$$\sigma_\mu = 2\pi d^2/3 + 2\pi K I_2(s)/\varepsilon d^{S-2},$$

where

$$I_2(s) = 4\int_0^1 dz_0\, z_0^{2-S}\big(2z_0^2 - 1\big)\big(1 - z_0^2\big)^{1/2}\int_0^{z_0} dz\, z^S\big(1 - z^2\big)^{-3/2}.$$

Prove that the coefficient of viscosity is given by (cf. (3.10)) $\mu = \mu^0/\big(1 + S/T\big)$, where $S = KI_2(s)/kd^S$ and μ^0 denotes the rigid-sphere value.

6.6 Consider the diffusion of an electron gas through a background of ions of charge number Z. Using expressions given in exercise 4.4, show that the electron–ion collision frequency is given by

$$\tau_{ei}^{-1} = \frac{1}{3}\frac{Zn_e}{m_e^2}\frac{e^4}{\epsilon_0^2}\left(\frac{m_e}{2\pi kT}\right)^{3/2}\ln\Lambda.$$

6.7 Show that if the scattering angle χ is small, allowing terms $O(\chi^3)$ to be ignored,

$$\iint \delta\mathbf{g}\, b\, db\, d\epsilon = -gQ^{(1)}(g), \qquad \iint \delta\mathbf{g}\delta\mathbf{g}\, b\, db\, d\epsilon = (g^2\mathbf{1} - \mathbf{gg})Q^{(1)}(g).$$

7

BOLTZMANN'S KINETIC EQUATION

7.1 Introduction

7.1.1 *Irreversibility*

In this chapter we shall present and apply Boltzmann's fundamental equation for the evolution of the velocity distribution function f. He derived it in order to explain how f approached the equilibrium distribution f_0 discovered by Maxwell a few years earlier. His principal objective was to *deduce* the second law of thermodynamics by a purely mechanical argument. He managed this—at least for the case of a dilute gas—by proving a remarkable theorem, now known as the 'H-theorem'. What is surprising about this theorem is that it establishes irreversibility analytically, using what seems to be a very innoxious (statistical) assumption about the behaviour of colliding molecules. Yet Newton's laws of motion are reversible and the molecules in a tenuous gas of elastic spheres obey these laws. This apparent conflict was the cause of considerable scepticism amongst Boltzmann's peers about the merit of his theorem, some of whom even questioned the reality of atoms.

The irreversibility of macroscopic descriptions of the gas is evident in the simplest of transport equations. For example in the Fourier law $\mathbf{q} = -\kappa\nabla T$, the heat flux vector $\mathbf{q}$ is related to the particle motion by (3.55), viz. $\mathbf{q} = \frac{1}{2}\rho\langle \mathbf{c}c^2\rangle^*$, whereas the temperature is proportional to $\langle c^2\rangle$. Thus if the molecules abruptly reverse direction, $\mathbf{q}$ will change sign but T will not, which means that Fourier's law is not invariant under time reversal, i.e. it represents an *irreversible* phenomenon. Boltzmann's kinetic equation has the same property—it is not invariant under time reversal, a property that can be traced to the statistical elements he introduced into the equation. Although the statistical nature of the second law had already been identified by Maxwell, it was left to Boltzmann to find an analytic formula relating entropy and probability.

7.1.2 *Solving Boltzmann's equation*

In 1872 Boltzmann published the paper *Further Studies on the Thermal Equilibrium of Gas Molecules* that contains his famous integro-differential equation. Although the work runs to 96 pages, only seven deal with the calculation of the transport properties of gases. He showed how the coefficients of viscosity, heat conduction and diffusion could be found in the particular case of Maxwellian molecules, but apart from some cautionary observations about the complexity with other force laws, he did not attempt the general case. In an effort some years later to calculate the viscosity for the case of elastic spheres, after struggling with a very complicated approximating procedure, he arrived at no simple result, and remarked that one must almost despair of the general solution of his equation. It was not until 1905 that Lorentz made some progress with the equation for the special case of a mixture of heavy and light molecules, now termed a 'Lorentzian gas' (see §4.5.1).

Enskog wrote two papers in 1911, following the method of expansion in powers of the peculiar velocity initiated by Boltzmann, but found unwieldy expressions. Hilbert (1910) studied the mathematical structure of Boltzmann's equation and proved a uniqueness theorem by adopting a formal expansion for the distribution function. But his expansion was incomplete since it ignored the fact that the expansion parameter was also implicit in the equations of fluid motion. This lead to the paradoxical conclusion that f is uniquely determined by the value of its first five velocity moments at any initial instant. These moments specify ρ, $\mathbf{v}$ and T, which according to Hilbert's method, are sufficient to determine the values of all higher moments for all time. While the approach gave the Euler (non-dissipative) equations of fluid mechanics, it failed to yield the Navier–Stokes equations. The problem lay with Hilbert's treatment of the time derivatives in the moment equations.

In 1916, using Maxwell's equations of transfer, together with an expansion in powers of velocity—as briefly described in §6.8.3—Chapman overcame the problem of the time derivatives, and was able to obtain simple expressions for transport in a gas, with general force laws. A more formal and direct treatment based on Boltzmann's equation was published in 1917 by Enskog, whose approach differed from Hilbert's in that he expanded *both* the solution (i.e. f) and the equations determining ρ, $\mathbf{v}$, and T. As Enskog's method gave identical results to those found by Chapman, it is now called the 'Chapman–Enskog theory'. It is this version that we shall describe below.

A significant improvement to the mathematical aspects of the theory was made in 1935 by Burnett, who showed that instead of the power series used by Chapman and Enskog, the use of Sonine-polynomial expansions both speeded up the convergence of the series and somewhat simplified the accurate evaluation of the transport coefficients. Burnett also obtained transport equations correct to second-order in the Knudsen number, work started by Maxwell some sixty years earlier in an effort to explain the curious phenomenon of the Crookes radiometer. Another significant contribution to the second-order theory was made by H. Grad in 1949, whose work is described in the text by Chapman and Cowling (1970). We shall deal with second-order theory for neutral gases in the next chapter.

7.2 The classical derivation of Boltzmann's equation

7.2.1 *General form of the kinetic equation*

In §5.2 we gave an account of the general form of the kinetic equation. This was expressed in terms of a convected coordinate system, whereas Boltzmann's equation is usually formulated in the laboratory frame, with the particle velocity $\mathbf{w} = \mathbf{v} + \mathbf{c}$ taken to be the independent velocity space variable. Strictly $\mathbf{w}$ is not the *particle* velocity, but the velocity of a phase space element, $d\boldsymbol{\nu}$ say. The acceleration of this element is chosen to be $\mathbf{F}$, the body force per unit mass, with the expectation that molecules with the velocity $\mathbf{w}$ will remain in it until scattered out by collisions. Of course the reference frame in which we chose to formulate the kinetic equation should not matter at all—it is merely a question of the ease with which we can model the collisional processes involved. We shall start by re-expressing the results of §5.2 in the laboratory frame coordinates.

The differential of a scalar function $\psi(\mathbf{r}, \mathbf{w}, t)$ is

$$d\psi = \left(\frac{\partial \psi}{\partial t}\right)_{\mathbf{r},\,\mathbf{w}} dt + \left(\frac{\partial \psi}{\partial \mathbf{r}}\right)_{\mathbf{w},\,t} \cdot d\mathbf{r} + \left(\frac{\partial \psi}{\partial \mathbf{c}}\right)_{\mathbf{r},\,t} \cdot d\mathbf{w}, \tag{7.1}$$

from which it follows that the rate of change of ψ following an element $d\boldsymbol{\nu} = d\mathbf{r}\,d\mathbf{w}$ of phase space moving with velocity $\mathbf{w}$ and acceleration $\mathbf{F}$, is given by

$$\mathbb{D}\psi = \frac{\partial \psi}{\partial t} + \mathbf{w} \cdot \frac{\partial \psi}{\partial \mathbf{r}} + \mathbf{F} \cdot \frac{\partial \psi}{\partial \mathbf{w}}. \tag{7.2}$$

It is important to remember that the vectors $\mathbf{w}$ and $\mathbf{F}$ describe the motion of a *phase-space element*, and do not necessarily relate to particular molecules. Once $\mathbf{w}$ has been chosen, then molecules that happen to have the same velocity will remain in $d\boldsymbol{\nu}$ until removed either by a mismatch of their acceleration with $\mathbf{F}$, or by being randomly scattered out of $d\boldsymbol{\nu}$ by collisions. Although the body force alone accelerates the particles during their free flight between collisions, they do receive impulsive accelerations at these collisions that are not purely random.

The number of particles in an element $d\boldsymbol{\nu}$ is $f\,d\boldsymbol{\nu}$, and its rate of change is $\mathbb{D}(f\,d\boldsymbol{\nu})$. Now we assume that the losses and gains experienced by the element are due only to collisions, and let $(df/dt)_{col}\,d\boldsymbol{\nu}$ denote this rate. Then by conservation of particle numbers,

$$\mathbb{D}(f\,d\boldsymbol{\nu}) = (df/dt)_{col}\,d\boldsymbol{\nu}. \tag{7.3}$$

In phase space the six-dimensional 'velocity' is $(\mathbf{w},\ \mathbf{F})$, and the corresponding divergence is $(\partial/\partial\mathbf{r},\ \partial/\partial\mathbf{w})$, so by a generalization of the argument given in §1.10.1, we obtain

$$\mathbb{D}(d\boldsymbol{\nu}) = (\frac{\partial}{\partial \mathbf{r}} \cdot \mathbf{w} + \frac{\partial}{\partial \mathbf{w}} \cdot \mathbf{F})\,d\boldsymbol{\nu},$$

in which the term $(\partial/\partial\mathbf{r}) \cdot \mathbf{w}$ vanishes because $\mathbf{r}$ and $\mathbf{w}$ are independent variables. Thus $\mathbb{D}(d\boldsymbol{\nu})$ is zero provided either $\mathbf{F}$ is independent of $\mathbf{w}$ or more generally, $(\partial/\partial\mathbf{w})\cdot\mathbf{F}$ is zero. Whereupon (7.3) reduces to the general form of the kinetic equation,

$$\mathbb{D}f = \mathbb{C}(f) \qquad \big(\mathbb{C}(f) \equiv (df/dt)_{col}\big), \tag{7.4}$$

where
$$\mathbb{D} = \frac{\partial}{\partial t} + \mathbf{w}\cdot\frac{\partial}{\partial \mathbf{r}} + \mathbf{F}\cdot\frac{\partial}{\partial \mathbf{w}}. \tag{7.5}$$

7.2.2 *Boltzmann's collision integral*

Boltzmann's collision integral, $\mathbb{C}_B$, is based on several assumptions:

1. Only binary collisions are taken into account.
2. The walls and other boundary discontinuities are sufficiently 'distant' (several mean free paths), to have negligible effects.
3. The velocity of a particle is uncorrelated, both with its position and with the initial velocity of any particle with which it is about to collide. Thus the probability that a class '1' molecule collides with a class '2' molecule is assumed to be proportional to the product of their phase-space number densities, $f_1 f_2$. The neglect of correlations between the particles is termed the 'hypothesis of molecular chaos'.
4. Collective effects are ignored, in other words the kinetic equation is independent of the macroscopic variables, the latter being determined by the former.

Consider class 1 particles (P_1) moving with velocity $\mathbf{g} = \mathbf{w}_1 - \mathbf{w}_2 = g\mathbf{k}$ relative to class 2 particles and colliding with them. Suppose that the collisions scatter P_1 into the element $d\mathbf{k}'$ of solid angle, then by (6.34) the collision rate is

$$d^8N = f_1 f_2\, \alpha(\mathbf{k}\,|\,\mathbf{k}'; g)\, d\mathbf{k}'\, d\mathbf{w}_1\, d\mathbf{w}_2,$$

where $\alpha(\mathbf{k}\,|\,\mathbf{k}'; g)$ is the differential cross section defined in §6.3.1. The total number of particles lost per second from the element $d\boldsymbol{\nu}_1 = d\mathbf{w}_1\, d\mathbf{r}$ due to such collisions is therefore

$$d\mathbf{w}_1\, d\mathbf{r} \iint f_1 f_2\, \alpha(\mathbf{k}\,|\,\mathbf{k}'; g)\, d\mathbf{k}'\, d\mathbf{w}_2. \tag{7.6}$$

In addition to the direct collisions $\mathbf{w}_1 \to \mathbf{w}_1'$ that scatter the particles P_1 out of $d\boldsymbol{\nu}_1$, there are others, namely the inverse collisions $-\mathbf{w}_1' \to -\mathbf{w}_1$ that scatter class '1' particles *into* $d\boldsymbol{\nu}_1$. On reversing the direction of the collisions in (7.6), we find that the gain of particles per second due to such encounters is

$$d\mathbf{w}_1'\, d\mathbf{r} \iint f_1' f_2'\, \alpha(-\mathbf{k}'\,|\,-\mathbf{k}; g')\, d\mathbf{k}\, d\mathbf{w}_2'. \tag{7.7}$$

By (2.59), modified to the present notation, we have

$$d\mathbf{k}'\, d\mathbf{w}_1\, d\mathbf{w}_2 = g^2 dg\, d\mathbf{G}\, d\mathbf{k}\, d\mathbf{k}' = d\mathbf{k}\, d\mathbf{w}_1'\, d\mathbf{w}_2', \tag{7.8}$$

where $\mathbf{G}$ is the velocity of the centre of mass of the colliding particles and we have expressed the volume element $d\mathbf{g}$ in spherical polar coordinates.

The particle equations of motion are invariant under motion reversal, so the scattering cross section is unchanged by the transformation $\mathbf{k} \to -\mathbf{k}'$. Further, as the forces acting between the particles are assumed to have central symmetry, α must be invariant to an inversion of the coordinates. Hence

$$\alpha(\mathbf{k} \,|\, \mathbf{k}'; g) = \alpha(-\mathbf{k}' \,|\, -\mathbf{k}; g) = \alpha(\mathbf{k}' \,|\, \mathbf{k}; g). \tag{7.9}$$

7.2.3 *Boltzmann's integro-differential equation*

Equations (7.8) and (7.9) enable us to combine (7.6) and (7.7) into the form $d\mathbf{w}_1\, d\mathbf{r}\, \mathbb{C}_B$, where $\mathbb{C}_B$ is Boltzmann's collision integral:

$$\mathbb{C}_B(f_1 f_2) = \iint (f_1' f_2' - f_1 f_2) g\alpha(\mathbf{k} \,|\, \mathbf{k}'; g)\, d\mathbf{k}'\, d\mathbf{w}_2. \tag{7.10}$$

Boltzmann's integro-differential equation for f_1 therefore reads

$$\frac{\partial f_1}{\partial t} + \mathbf{w} \cdot \frac{\partial f_1}{\partial \mathbf{r}} + \mathbf{F} \cdot \frac{\partial f_1}{\partial \mathbf{w}} = \iint (f_1' f_2' - f_1 f_2) g\alpha(\mathbf{k} \,|\, \mathbf{k}'; g)\, d\mathbf{k}'\, d\mathbf{w}_2. \tag{7.11}$$

The first point to note is that Boltzmann's equation possesses the irreversibility associated with dissipation—the left-hand side has negative parity* under time reversal, while the right hand side has positive parity. A second point that will be of interest later (§7.11.3) is that the equation can be put into a form resembling the BGK kinetic equation:

$$\mathbb{D} f_1 = \frac{1}{\tau}(\overline{f} - f_1), \tag{7.12}$$

where

$$\tau^{-1} \equiv \iint f_2 g\alpha(\mathbf{k} \,|\, \mathbf{k}'; g)\, d\mathbf{k}'\, d\mathbf{w}_2, \tag{7.13}$$

and

$$\overline{f} \equiv \tau \iint f_1' f_2' g\alpha(\mathbf{k} \,|\, \mathbf{k}'; g)\, d\mathbf{k}'\, d\mathbf{w}_2. \tag{7.14}$$

In §5.8.1 we gave a physical interpretation of (7.12), describing $\overline{f}$ as an average of the distributions of the particles approaching the inverse collision and τ as being the time interval between successive collisions. Boltzmann's first important contribution was to find an accurate expression for this average. Using this, he was able to deduce a special case of the second law of thermodynamics (see §7.4.3). Maxwell might well have anticipated Boltzmann, for his equation of transfer (3.5) embodies the same principle as (7.11), namely a balance between streaming and collisional changes. He also appreciated the physical significance of the factor $(f_1' f_2' - f_1 f_2)$ appearing in the collision integral (see (2.28)).

*Negative parity means that its sign changes when all the velocities and the time is reversed in sign; with positive parity there is no change in sign. See discussion in §8.3.3.

7.3 The equilibrium distribution function

7.3.1 *Rate of change of a dynamic property*

In §2.5 we deduced the form for the Maxwellian distribution function f_0 from the principle of maximum entropy. Another treatment employs Boltzmann's collision integral as follows.

The first step is to derive a general symmetrical expression for the collisional rate of change of a dynamic property $\varphi(\mathbf{r}, \mathbf{w}, t)$. We follow a similar method to that of §5.3, leading to equation (5.23), except that now we start from the form of the kinetic equation given by (7.11). Thus

$$\int \varphi \mathbf{F} \cdot \frac{\partial f}{\partial \mathbf{w}} \, d\mathbf{w} = \int \frac{\partial}{\partial \mathbf{w}} \cdot (\mathbf{F}\varphi f) \, d\mathbf{w} - \int f \frac{\partial}{\partial \mathbf{w}} \cdot (\mathbf{F}\varphi) \, d\mathbf{w}.$$

By the divergence theorem the first integral becomes a surface integral over the infinite sphere $|\mathbf{w}| = R \to \infty$, that vanishes provided that as $R \to \infty$, $|\varphi \mathbf{F} f| \to 0$ rapidly enough. Assuming that this is so, and using $(\partial/\partial \mathbf{w}) \cdot \mathbf{F} = 0$, we have

$$\int \varphi \mathbf{F} \cdot \frac{\partial f}{\partial \mathbf{w}} \, d\mathbf{w} = -n \overline{\mathbf{F} \cdot \frac{\partial \varphi}{\partial \mathbf{w}}}.$$

Also, since $\mathbf{r}$ and $\mathbf{w}$ are independent variables,

$$\int \varphi \mathbf{w} \cdot \frac{\partial f}{\partial \mathbf{r}} \, d\mathbf{w} = \frac{\partial}{\partial \mathbf{r}} \cdot (n \overline{\mathbf{w}\varphi}) = \nabla \cdot (n \mathbf{v} \overline{\varphi}) + \nabla \cdot (n \overline{\mathbf{c}\varphi}),$$

on replacing $\mathbf{w}$ by $\mathbf{v} + \mathbf{c}$ and $(\partial/\partial \mathbf{r})$ by ∇. Hence if (7.11) is multiplied by φ and then integrated over velocity space, the outcome can be written

$$\frac{\partial}{\partial t}(n\overline{\varphi}) + \nabla \cdot (n\mathbf{v}\overline{\varphi} + n\overline{\mathbf{c}\varphi}) - n\overline{\mathbf{F} \cdot \frac{\partial \varphi}{\partial \mathbf{w}}} = \int \varphi \mathbb{C}_B \, d\mathbf{w}. \tag{7.15}$$

With the collision integral in (7.10), the right-hand member of (7.15) is

$$\left(\frac{\partial \varphi}{\partial t}\right)_{col} \equiv \int \varphi_1 \mathbb{C}_B \, d\mathbf{w}_1 = \iiint \varphi_1 (f_1' f_2' - f_1 f_2) g \alpha(\mathbf{k} \,|\, \mathbf{k}'; g) \, d\mathbf{k}' \, d\mathbf{w}_2 \, d\mathbf{w}_1,$$

where $\varphi_1 = \varphi(\mathbf{r}, \mathbf{w}_1, t)$, and we shall shortly introduce $\varphi_2 = \varphi(\mathbf{r}, \mathbf{w}_2, t)$. As $(\partial \varphi/\partial t)_{col}$ is clearly independent of the class of particle chosen in the integrand, by (7.8) we can write

$$\left(\frac{\partial \varphi}{\partial t}\right)_{col} = \frac{1}{2} \iiiint (\varphi_1 + \varphi_2)(f_1' f_2' - f_1 f_2) g^3 \alpha \, dg \, d\mathbf{G} \, d\mathbf{k} \, d\mathbf{k}'.$$

7.3.2 *Collisional invariants*

Let F, G, and φ be any functions of $\mathbf{r}$, $\mathbf{w}$, and t; then $\varphi_1 F_1' G_2'$ in the integrand of

$$\iiiint \varphi_1 F_1' G_2' g^3 \alpha \, dg \, d\mathbf{G} \, d\mathbf{k} \, d\mathbf{k}'$$

can be replaced by $\varphi_1' F_1 G_2$, and similarly $\varphi_2 F_1' G_2'$ could be replaced by $\varphi_2' F_1 G_2$. To see this, observe that the interchange of the roles of direct and inverse scattering cannot affect either the cross-section α because of (7.9) or the integration because of (7.8). This result allows us to write the integral for $(\partial\varphi/\partial t)_{col}$ in the symmetrical form

$$\left(\frac{\partial\varphi}{\partial t}\right)_{col} = \frac{1}{4}\iiiint(\varphi_1+\varphi_2-\varphi_1'-\varphi_2')(f_1'f_2'-f_1f_2)g^3\alpha\,dg\,d\mathbf{G}\,d\mathbf{k}\,d\mathbf{k}'. \quad (7.16)$$

It follows that

$$(\partial\varphi/\partial t)_{col} = 0 \quad \text{if} \quad \varphi(\mathbf{w}_1) + \varphi(\mathbf{w}_2) = \varphi(\mathbf{w}_1') + \varphi(\mathbf{w}_2'); \quad (7.17)$$

functions satisfying this relation are known as collisional (or summational) invariants. For a monatomic gas, i.e. a gas with only translatory energy, there are just five independent scalar invariants, since during a collision only the mass, momentum, and energy of the colliding particles are conserved (see (5.27)). Any linear combination of these invariants is also an invariant.

7.3.3 *The equilibrium distribution*

Turning now to the equilibrium distribution f_0, we define this to be that solution of the kinetic equation (7.11) which is independent of time. A *sufficient* condition for $f = f_0(\mathbf{r}, \mathbf{w})$ to satisfy the kinetic equation is that $(\partial\varphi/\partial t)_{col} = 0$, that is $f_1'f_2' = f_1f_2$, or

$$\ln f(\mathbf{r}, \mathbf{w}_1') + \ln f(\mathbf{r}, \mathbf{w}_2') = \ln f(\mathbf{r}, \mathbf{w}_1) + \ln f(\mathbf{r}, \mathbf{w}_2).$$

Hence $\ln f_0$ must be a collisional invariant, and from the observation of the previous paragraph it follows that its most general form is

$$\ln f_0 = k_1 + \mathbf{k}_2 \cdot \mathbf{w} + k_3 w^2,$$

where k_1, $\mathbf{k}_2$, and k_3 are independent of $\mathbf{w}$ but may depend on $\mathbf{r}$. On replacing these coefficients by α, $\mathbf{v}$, and β, we can rearrange the equilibrium distribution in the form

$$f_0 = \exp\{-\alpha - \beta(\mathbf{w} - \mathbf{v})\},$$

where α, $\mathbf{v}$, and β are to be determined from

$$n = \int f_0\,d\mathbf{w}, \qquad n\mathbf{v} = \int \mathbf{w} f_0\,d\mathbf{w}, \qquad \tfrac{3}{2}nkT = \int \tfrac{1}{2}(\mathbf{w} - \mathbf{v})^2 f_0\,d\mathbf{w}.$$

The calculation proceeds as in §2.5, with the same outcome:

$$f_0 = n\left(\frac{m}{2\pi kT}\right)^{\frac{3}{2}} \exp\left\{-\frac{m(\mathbf{w} - \mathbf{v})^2}{2kT}\right\}. \quad (7.18)$$

To prove that (7.18) is a *necessary* condition for equilibrium, we need Boltzmann's H-theorem.

7.4 The H-theorem

7.4.1 *Entropy production rate*

In §2.2 we showed that in kinetic theory the entropy density is given by

$$\rho s = -k \int f(\ln f - 1)\, d\mathbf{w}. \tag{7.19}$$

Hence by (7.10) and (7.11),

$$\begin{aligned}\frac{\partial}{\partial t}(\rho s) &= -k \int \frac{\partial f}{\partial t} \ln f \, d\mathbf{w} \\ &= -k \int \mathbb{C}_B \ln f \, d\mathbf{w} + k \int \left(\mathbf{w} \cdot \frac{\partial f}{\partial \mathbf{r}} + \mathbf{F} \cdot \frac{\partial f}{\partial \mathbf{w}} \right) \ln f \, d\mathbf{w}.\end{aligned}$$

As $(\partial/\partial\mathbf{w}) \cdot \mathbf{F} = 0$,
$$\int \mathbf{F} \cdot \frac{\partial f}{\partial \mathbf{w}} \ln f \, d\mathbf{w} = \int \frac{\partial}{\partial \mathbf{w}} \cdot \{\mathbf{F} f (\ln f - 1)\}\, d\mathbf{w}.$$

By the divergence theorem in velocity space this becomes a surface integral over $|\mathbf{w}| = R \to \infty$ that equals zero, since f tends to zero exponentially as $|\mathbf{w}| \to \infty$. Also defining the entropy flux vector by (see (1.69) and exercise 7.2)

$$\mathbf{J} \equiv -k \int (\mathbf{w} - \mathbf{v}) f (\ln f - 1)\, d\mathbf{w}, \tag{7.20}$$

we have
$$\begin{aligned}k \int \mathbf{w} \cdot \frac{\partial f}{\partial \mathbf{r}} \ln f \, d\mathbf{w} &= \frac{\partial}{\partial \mathbf{r}} \cdot \left\{ k \int \mathbf{w} f (\ln f - 1)\, d\mathbf{w} \right\} \\ &= -\frac{\partial}{\partial \mathbf{r}} \cdot \mathbf{J} - \frac{\partial}{\partial \mathbf{r}} \cdot (\rho s \mathbf{v}),\end{aligned}$$

the first step of which results from the independence of $\mathbf{r}$ and $\mathbf{w}$.

It follows from the above that the expression for $\partial(\rho s)/\partial t$ may be written

$$\frac{\partial(\rho s)}{\partial t} + \nabla \cdot (\rho s \mathbf{v}) + \nabla \cdot \mathbf{J} = -k \int \mathbb{C}_B \ln f \, d\mathbf{w}, \tag{7.21}$$

or by conservation of mass (cf. (1.49))

$$\rho \mathrm{D} s + \nabla \cdot \mathbf{J} = \sigma, \tag{7.22}$$

where

$$\sigma \equiv -k \int \mathbb{C}_B \ln f \, d\mathbf{w}. \tag{7.23}$$

7.4.2 *Local form of the second law of thermodynamics*

We could at this stage appeal to the second law, viz. $\sigma \geq 0$, which we established in §1.11.2, and impose this as a *restriction* on $\mathbb{C}_B$ through (7.23). Boltzmann proceeded in the opposite direction. He was seeking to *prove* the second law, at least for the case of a dilute gas. Therefore accepting (7.23) as the definition of σ and ignoring previous knowledge of the second law, we proceed as follows.

From (7.23) and (7.10),

$$\sigma = -k \iiint (f_1' f_2' - f_1 f_2) \ln f g\alpha \, d\mathbf{k}' \, d\mathbf{w}_2 \, d\mathbf{w}_1$$
$$= -\tfrac{1}{4} k \iiiint (f_1' f_2' - f_1 f_2) \ln(f_1 f_2 / f_1' f_2') g^3 \alpha \, dg \, d\mathbf{G} \, d\mathbf{k} \, d\mathbf{k}', \quad (7.24)$$

where we have symmetrized the integrand in a manner similar to that used to derive (7.16). As the function $\ln(f_1 f_2/f_1' f_2')$ is always opposite in sign to $(f_1' f_2' - f_1 f_2)$,* their product is always negative or zero, i.e.

$$\sigma \geq 0, \quad (7.25)$$

which is the local form of the second law of thermodynamics.

7.4.3 *Boltzmann's theorem*

Boltzmann's H-theorem is the statement that in a closed system S the functional

$$H \equiv \iint f \ln f \, d\mathbf{w} \, d\mathbf{r}, \quad (7.26)$$

where the integrals are over the volume of S and all of velocity space, steadily decreases until it reaches a minimum value, corresponding to an equilibrium state. First we shall establish the relation between H and the total entropy S of S. By (7.19),

$$S = \int \rho s \, d\mathbf{r} = -k \iint f(\ln f - 1) \, d\mathbf{w} \, d\mathbf{r} = -kH + kN,$$

where N is the total number of particles in S. Hence

$$H = -\frac{1}{k} S + N. \quad (7.27)$$

The rate at which the entropy in S increases is obtained by integrating the local rate of increase per unit volume:

$$\frac{dS}{dt} = \int \sigma \, d\mathbf{r},$$

whence by (7.25) and (7.27),

$$\frac{dH}{dt} = -\frac{1}{k} \int \sigma \, d\mathbf{r} \leq 0, \quad (7.28)$$

and the theorem is proved.

The important conclusion is that the integral in (7.24) can vanish if and only if

$$f_1' f_2' = f_1 f_2, \quad (7.29)$$

*The logarithmic function is convex, i.e. $y = \ln x$ always lies below its tangent; this follows from $d^2y/dx^2 = -1/x^2 \leq 0$ $(0 \leq x \leq \infty)$. The tangent at $x = 0$ is $y = x - 1$, whence $\ln x \leq x - 1$, with equality if, and only if, $x = 1$. Hence $0 \leq (\ln x)^2 \leq (x - 1) \ln x$, so $(1 - x)$ and $\ln x$ have opposite signs.

which is thus the necessary condition for equilibrium. It is now proved that, subject to the conditions under which Boltzmann's collision integral is valid, Maxwell's distribution function is both necessary and sufficient for equilibrium.

7.4.4 *Loschmidt's paradox*

Shortly after Boltzmann published the H-theorem in 1872, Loschmidt objected that it introduced irreversibility into the kinetic model, whereas the individual particles certainly obeyed the reversible equations of dynamics. It took many years before this paradox was clearly resolved. The essential point is that irreversible processes are merely *statistically* much more likely than reversible ones. When equilibrium is attained, the theorem requires H to have a constant minimum value for an isolated system, whereas small variations in f due to thermodynamic noise will cause H to fluctuate above this value. However, on a large enough time scale, such fluctuations will be unimportant. Boltzmann's collision integral is itself statistical, depending as it does on the hypothesis of molecular chaos. For a system in a state of chaos, the binary interaction represented in the collision term does cause H to decrease, but on a dynamical view rather than a statistical one, the collision causes a correlation between the velocities of the two particles, thereby destroying the initial state of molecular chaos. Hence on the dynamic theory H no longer necessarily decreases in the next instant, and a gas near the equilibrium state may experience an *increase* in its H function. For a gas very far from the equilibrium state, the statistical view predicts an overwhelmingly large probability that H will decrease.

7.5 The Chapman–Enskog series

7.5.1 *Knudsen number power series*

Chapman and Enskog independently found equivalent methods of solving Boltzmann's equation by successive approximation, the first step of which amounted to expanding the distribution function *and* the time derivative as a power series in the Knudsen number.* What follows is a modified version of their method; in §7.11 we shall discuss the validity of the procedure.

We shall modify the notation of Boltzmann's equation by replacing f_1 by f and f_2 by $\bar{f}$, then (7.11) reads

$$\frac{\partial f}{\partial t} + \mathbf{w} \cdot \frac{\partial f}{\partial \mathbf{r}} + \mathbf{F} \cdot \frac{\partial f}{\partial \mathbf{w}} = \mathbb{C}_B(f\bar{f}), \tag{7.30}$$

where from (7.10)

$$\mathbb{C}_B(f\bar{f}) = \iint (f'\bar{f}' - f\bar{f}) g\alpha(\mathbf{k} \,|\, \mathbf{k}'; g) \, d\mathbf{k}' \, d\bar{\mathbf{w}}. \tag{7.31}$$

*Hilbert's similar expansion procedure did not include the time derivative.

Notice that $\mathbb{C}_B$ is a function of f and a functional of $\bar{f}$. The next step is to change the independent variable from $\mathbf{w}$ to $\mathbf{c} = \mathbf{w} - \mathbf{v}(\mathbf{r}, t)$, where the time derivatives of $\mathbf{c}$ are specified in the spinning frame P_c, described in Mathematical note 5.2. To effect this transformation we need the following vector algebra.

Mathematical note 7.5 *Transformation of the velocity reference frame*

Let L, P and P_c denote the laboratory, fluid, and fully convected reference frames, then both P and P_c move with the velocity $\mathbf{v}$ relative to L and in addition P_c rotates with the angular velocity $\boldsymbol{\Omega}$ $(\equiv \frac{1}{2}\nabla \times \mathbf{v})$ relative to P. Let ψ represent a function that is either a scalar ψ_0, a vector $\boldsymbol{\psi}_1$ or a second-order tensor $\boldsymbol{\psi}_2$, then the time derivatives of ψ in the three frames are: (see Mathematical note 5.2)

$$L: \quad \frac{\partial \psi}{\partial t}, \qquad P: \quad \mathrm{D}\psi = \frac{\partial \psi}{\partial t} + \mathbf{v} \cdot \nabla \psi,$$

P_c :

$$\mathrm{D}^*\boldsymbol{\psi}_n \equiv \mathrm{D}\boldsymbol{\psi}_n - n\boldsymbol{\Omega}\times\boldsymbol{\psi}_n \quad (n = 0, 1), \quad \mathrm{D}^*\boldsymbol{\psi}_2 \equiv \mathrm{D}\boldsymbol{\psi}_2 - \boldsymbol{\Omega}\times\boldsymbol{\psi}_2 + \boldsymbol{\psi}_2\times\boldsymbol{\Omega}. \tag{7.32}$$

Let $\mathbf{a}(\mathbf{r}, \mathbf{w}, t)$ be a vector specified in the L-frame, and let $\mathbf{A}(\mathbf{r}, \mathbf{c}, t)$ be the same vector specified in the P_c-frame. Denoting differentials in L by d and in P_c by d^*, then with $\mathbf{a} = \mathbf{w} - \mathbf{v}$, $\mathbf{A} = \mathbf{c}$, we have $d(\mathbf{w} - \mathbf{v}) = d^*\mathbf{c}$, i.e.

$$d\mathbf{w} = d^*\mathbf{c} + d\mathbf{v} = \big(d\mathbf{c} - \boldsymbol{\Omega} \times \mathbf{c}\big)dt + \left(\frac{\partial \mathbf{v}}{\partial t} + \mathbf{w} \cdot \nabla \mathbf{v}\right) dt,$$

$$= d\mathbf{c} + \big(\mathrm{D}\mathbf{v} + \mathbf{c} \cdot \nabla \mathbf{v} - \boldsymbol{\Omega} \times \mathbf{c}\big)\, dt,$$

or since

$$\mathbf{c} \cdot \nabla \mathbf{v} - \boldsymbol{\Omega} \times \mathbf{c} = \mathbf{c} \cdot (\nabla \mathbf{v} + \mathbf{1} \times \boldsymbol{\Omega}) = \mathbf{c} \cdot (\overset{\circ}{\nabla} \mathbf{v} + \tfrac{1}{3}\nabla \cdot \mathbf{v}\, \mathbf{1}) = \mathbf{c} \cdot \mathbf{e},$$

we have

$$d\mathbf{w} = d\mathbf{c} + \big(\mathrm{D}\mathbf{v} + \mathbf{c} \cdot \mathbf{e}\big)\, dt.$$

Hence

$$d\mathbf{a} = dt\, \frac{\partial \mathbf{a}}{\partial t} + d\mathbf{r} \cdot \frac{\partial \mathbf{a}}{\partial \mathbf{r}} + d\mathbf{w} \cdot \frac{\partial \mathbf{a}}{\partial \mathbf{w}},$$

$$= dt \left\{ \frac{\partial \mathbf{a}}{\partial t} + (\mathrm{D}\mathbf{v} + \mathbf{c} \cdot \mathbf{e}) \cdot \frac{\partial \mathbf{a}}{\partial \mathbf{w}} \right\} + d\mathbf{r} \cdot \frac{\partial \mathbf{a}}{\partial \mathbf{r}} + d\mathbf{c} \cdot \frac{\partial \mathbf{a}}{\partial \mathbf{w}}.$$

This will equal the change in the other representation of the same vector, namely

$$d^*\mathbf{A} = d\mathbf{A} - \boldsymbol{\Omega} \times \mathbf{A}\, dt = dt \left(\frac{\partial \mathbf{A}}{\partial t} - \boldsymbol{\Omega} \times \mathbf{A}\right) + d\mathbf{r} \cdot \frac{\partial \mathbf{A}}{\partial \mathbf{r}} + d\mathbf{c} \cdot \frac{\partial \mathbf{A}}{\partial \mathbf{c}},$$

if

$$\frac{\partial \mathbf{a}}{\partial \mathbf{r}} = \frac{\partial \mathbf{A}}{\partial \mathbf{r}}, \quad \frac{\partial \mathbf{a}}{\partial \mathbf{w}} = \frac{\partial \mathbf{A}}{\partial \mathbf{c}}, \quad \frac{\partial \mathbf{a}}{\partial t} = \frac{\partial \mathbf{A}}{\partial t} - \boldsymbol{\Omega} \times \mathbf{A} - (\mathrm{D}\mathbf{v} + \mathbf{c} \cdot \mathbf{e}) \cdot \frac{\partial \mathbf{A}}{\partial \mathbf{c}}.$$

It follows that the derivative

$$\frac{d\mathbf{a}}{dt} = \frac{\partial \mathbf{a}}{\partial t} + \mathbf{w} \cdot \frac{\partial \mathbf{a}}{\partial \mathbf{r}} + \mathbf{F} \cdot \frac{\partial \mathbf{a}}{\partial \mathbf{w}} \qquad (\mathbf{w} = \mathbf{v} + \mathbf{c})$$

transforms into

$$\frac{d^*\mathbf{A}}{dt} = \mathrm{D}^*\mathbf{A} + \mathbf{c} \cdot \nabla \mathbf{A} + (\mathbf{F} - \mathrm{D}\mathbf{v} - \mathbf{c} \cdot \mathbf{e}) \cdot \frac{\partial \mathbf{A}}{\partial \mathbf{c}}. \tag{7.33}$$

The result contained in (7.33) is easily generalized to the function ψ defined above.

We have now established that the rate of change of ψ in the L-frame, namely $\mathbb{D}\psi$, equals its rate of change $\mathbb{D}^*\psi$ in the P_c-frame, provided that*

$$\mathbb{D}^*\psi = \mathrm{D}^*\psi + \mathbf{c} \cdot \nabla \psi + (\mathbf{F} - \mathrm{D}\mathbf{v} - \mathbf{c} \cdot \mathbf{e}) \cdot \frac{\partial \psi}{\partial \mathbf{c}}, \tag{7.34}$$

where $\mathrm{D}^*\psi$ is the frame-indifferent time derivative defined in **5.2**. Notice that in the P_c-frame the phase space element acquires an acceleration $-\mathbf{c} \cdot \mathbf{e}$ due to the fluid velocity field. We have now formally derived the result contained in (5.17) and (5.18) (also see (5.15)).

Returning to (7.30) and replacing $\mathbf{w}$ by $\mathbf{c}$, we have

$$\mathbb{C}_B(f\bar{f}) = \mathbb{D}^* f, \tag{7.35}$$

the collision integral being unaffected. It is not possible to solve this non-linear, integro-differential equation exactly. A numerical method of treating it will be described in §7.12. Provided the gas is sufficiently collisional, we may proceed as follows: we enlist the time scale τ defined in (7.13), assume that

$$\epsilon \equiv |\tau \mathbb{D}^* \ln f| \ll 1, \tag{7.36}$$

and adopt ϵ as the expansion parameter. To zero order (7.35) gives

$$\mathbb{C}_B(f_0\bar{f}_0) = 0, \tag{7.37}$$

where by the theory of §7.3, f_0 is the Maxwellian distribution.

Shortly we shall show that not only f, but also $\mathbb{D}^*$ must be expanded in a series in ϵ. Hence

$$\left.\begin{aligned} f &= f_0 + f_1 + f_2 + \dots \qquad (f_r = \mathrm{O}(\epsilon^r)) \\ \text{and} \qquad \mathbb{D}^* &= \mathbb{D}_0^* + \mathbb{D}_1^* + \mathbb{D}_2^* + \dots \quad (\tau\mathbb{D}_r^* = (\epsilon^{r+1})) \end{aligned}\right\} \tag{7.38}$$

are substituted into (7.35), terms of the same order are equated, and the following sequence of linear equations obtained:

$$\Phi_0 = 0, \quad \Phi_1 = -\tau\mathbb{D}_0^* \ln f_0, \quad \Phi_2 = -f_0^{-1}(\tau\mathbb{D}_0^* f_1 + \tau\mathbb{D}_1^* f_0), \dots, \tag{7.39}$$

*We have added the asterisk to $\mathbb{D}$ to make the change of reference frame clear; in §5.2 the distinction was not required.

where

$$\Phi_r(f_0, f_1, f_2, \ldots, f_r) \equiv -f_0^{-1}\tau\mathbb{C}_B(f_0\bar{f}_r + f_1\bar{f}_{r-1} + \ldots + f_r\bar{f}_0). \qquad (7.40)$$

Equations (7.39) are to be solved in turn for f_1, $f_2, \ldots$, but before we can proceed with this, expressions for $\mathbb{D}_r^*$, $r = 0, 1, \ldots$, are required.

7.5.2 *Expansion of the streaming derivative*

The pressure tensor and heat flux vector are expanded in the Chapman–Enskog series

$$\mathbf{p} = p\mathbf{1} + \boldsymbol{\pi}_1 + \boldsymbol{\pi}_2 + \ldots, \qquad \mathbf{q} = \mathbf{q}_1 + \mathbf{q}_2 + \ldots \qquad \big(\boldsymbol{\pi}_r, \mathbf{q}_r \text{ are } \mathrm{O}(\epsilon^r)\big), \quad (7.41)$$

and when these expansions are substituted into the conservation equations (1.46) to (1.49), it becomes evident that a matching expansion in D is required. Equating terms of the same order, we obtain

$$\left.\begin{aligned} \mathrm{D}_0 n &= -n\nabla\cdot\mathbf{v}, & \mathrm{D}_r n &= 0 \\ \rho\mathrm{D}_0\mathbf{v} &= -\nabla p + \rho\mathbf{F}, & \rho\mathrm{D}_r\mathbf{v} &= -\nabla\cdot\boldsymbol{\pi}_r \\ \rho c_v\mathrm{D}_0 T &= -p\nabla\cdot\mathbf{v}, & \rho c_v\mathrm{D}_r T &= -\boldsymbol{\pi}_r : \nabla\mathbf{v} - \nabla\cdot\mathbf{q}_r, \end{aligned}\right\} (r \geq 1), \qquad (7.42)$$

where D_r is $\mathrm{O}(\epsilon^r)$.

We next use (7.32) and (7.42) to define the frame-indifferent derivatives

$$\mathrm{D}_r^*\boldsymbol{\psi}_n \equiv \mathrm{D}_r\boldsymbol{\psi}_n - n\boldsymbol{\Omega}\times\boldsymbol{\psi}_n \quad (n = 0, 1), \quad \mathrm{D}_r^*\boldsymbol{\psi}_2 \equiv \mathrm{D}_r\boldsymbol{\psi}_2 - \boldsymbol{\Omega}\times\boldsymbol{\psi}_2 + \boldsymbol{\psi}_2\times\boldsymbol{\Omega}. \qquad (7.43)$$

Replacing $\rho(\mathbf{F} - \mathrm{D}\mathbf{v})$ in (7.34) by $\nabla\cdot\mathbf{p} = \nabla p + \nabla\cdot\boldsymbol{\pi}_1 + \ldots$, and D^* by $\mathrm{D}_0^* + \mathrm{D}_1^* + \ldots$, we arrive at the required expressions

$$\left.\begin{aligned} \mathbb{D}_0^* &= \mathrm{D}_0^* + \mathbf{c}\cdot\nabla + \left(\frac{1}{\rho}\nabla p - \mathbf{c}\cdot\mathbf{e}\right)\cdot\frac{\partial}{\partial\mathbf{c}}, \\ \mathbb{D}_r^* &= \mathrm{D}_r^* + \frac{1}{\rho}\nabla\cdot\boldsymbol{\pi}_r\cdot\frac{\partial}{\partial\mathbf{c}} \quad (r = 1, 2, \ldots). \end{aligned}\right\} \qquad (7.44)$$

Equations (5.30) impose the conditions

$$\int (f - f_0)\boldsymbol{\psi}\, d\mathbf{c} = \int \sum_{r=1}^{\infty} f_r\boldsymbol{\psi}\, d\mathbf{c} \qquad \big(\boldsymbol{\psi} \equiv (1, \mathbf{c}, \tfrac{1}{2}c^2)\big),$$

and since the solutions must be complete at each order, we require

$$\int f_r\boldsymbol{\psi}\, d\mathbf{c} = 0 \qquad (r = 1, 2, \ldots). \qquad (7.45)$$

Should these conditions not be satisfied for a given f_r, one is free to add a complementary function χ_r to f_r to ensure that the modified solution, $\chi_r + f_r$, does meet the constraints. Any function χ_r that satisfies the homogeneous equation,

$$\Phi_r(f_0, f_1, f_2, \ldots, f_r) = 0, \qquad (7.46)$$

will serve the purpose. It follows from (7.17) and the remarks following it that χ_r/f_0 must be a collisional invariant, i.e.

$$\chi_r = f_0(\alpha_r + \boldsymbol{\beta}_r \cdot \mathbf{c} + \gamma_r c^2), \tag{7.47}$$

where α_r, $\boldsymbol{\beta}_r$, and γ_r are functions of $\mathbf{r}$ and t.

The account given above differs in two respects from the Chapman–Enskog theory. First, in their time-derivative expansion, the order indices rderapproximationto$f$

7.6.1 *The Fredholm equation for the distribution function*

First we remark that the zeroth-order approximation to f is the Maxwellian distribution f_0, and when this is substituted into (5.60) and (5.48), viz.

$$\mathbf{q} = \tfrac{1}{2}m \int c^2 \mathbf{c} f \, d\mathbf{c}, \qquad \mathsf{p} = m \int \mathbf{c}\mathbf{c} f \, d\mathbf{c}, \tag{7.48}$$

the result is $\mathsf{p}_0 = p\mathbf{1}$ and $\mathbf{q}_0 = 0$. The corresponding fluid equations are those due to Euler, the set listed in the first column of (7.42):

$$\mathrm{D}\rho = -\rho\nabla \cdot \mathbf{v}, \quad \rho\mathrm{D}\mathbf{v} = -\nabla p + \rho\mathbf{F}, \quad \rho c_v \mathrm{D}T = -p\nabla \cdot \mathbf{v}. \tag{7.49}$$

The first-order approximation to f is determined by the case $r = 1$ in (7.39) and (7.40):

$$\Phi_1 \equiv -f_0^{-1}\tau\mathbb{C}_B(f_0\bar{f}_1 + f_1\bar{f}_0) = -\tau\mathbb{D}_0^* \ln f_0. \tag{7.50}$$

Let

$$\left.\begin{aligned} \varphi &\equiv (f - f_0)/f_0, \\ \varphi_1 = f_1/f_0, &\quad \varphi_2 = f_2/f_0, \quad \ldots, \end{aligned}\right\} \tag{7.51}$$

then

$$\begin{aligned} f_0'\bar{f}_1' + f_1'\bar{f}_0' - f_0\bar{f}_1 - f_1\bar{f}_0 &= f_0'\bar{f}_0'(\bar{\varphi}_1' + \varphi_1') - f_0\bar{f}_0(\bar{\varphi}_1 + \varphi) \\ &= f_0\bar{f}_0(\bar{\varphi}_1' + \varphi_1' - \bar{\varphi}_1 - \varphi_1), \end{aligned}$$

where we have used the equilibrium condition $f_0'\bar{f}_0' = f_0\bar{f}_0'$. From (5.46), in which $\mathbb{D}$ is replaced here by $\mathbb{D}_0^*$,

$$\mathbb{D}_0^* \ln f_0 = (\nu^2 - \tfrac{5}{2})\mathbf{c} \cdot \nabla \ln T + \boldsymbol{\nu} \cdot \mathring{\mathbf{e}} \cdot \boldsymbol{\nu} \quad (\boldsymbol{\nu} \equiv \mathbf{c}/\mathcal{C},\ \mathcal{C} \equiv \sqrt{2kT/m}). \tag{7.52}$$

Write

$$\Phi_1(f_0, f_1) = -f_0^{-1}\tau\mathbb{C}_B(f_0\bar{f}_1 + f_1\bar{f}_0) = f_0^{-1}\tau n^2 I(\varphi_1), \tag{7.53}$$

where by (7.31) I is the functional defined by

$$I(F) \equiv \frac{1}{n^2} \iint f_0 \bar{f}_0 (F + \bar{F} - F' - \bar{F}') g\alpha \, d\mathbf{k}' \, d\bar{\mathbf{w}}. \tag{7.54}$$

It now follows from (7.50) and (7.52) that φ_1 satisfies the inhomogeneous Fredholm equation

$$n^2 I(\varphi_1) = -f_0 \Big\{ (\nu^2 - \tfrac{5}{2}) \mathbf{c} \cdot \nabla \ln T + \boldsymbol{\nu} \cdot \overset{\circ}{\mathbf{e}} \cdot \boldsymbol{\nu} \Big\}. \tag{7.55}$$

7.6.2 *The general solution*

Equation (7.55) does not determine φ_1 uniquely, since to any particular solution we may add a linear combination of the solutions of the homogeneous equation, $I(\varphi_1) = 0$, or $\Phi(f_0, f_1) = 0$. By (7.46) and (7.47) this complementary function is of the form

$$\varphi_1^* = \alpha_1 + \boldsymbol{\beta}_1 \cdot \mathbf{c} + \gamma_1 c^2, \tag{7.56}$$

where α_1, $\boldsymbol{\beta}_1$ and γ_1 are functions of $\mathbf{r}$ and t. Let $\hat{\varphi}_1$ denote a particular solution of (7.55), then the general solution is $\hat{\varphi}_1 + \varphi_1^*$, where φ_1^* is uniquely determined by (7.45):

$$\int (\hat{\varphi}_1 + \varphi_1^*) \boldsymbol{\psi} \, f_0 \, d\mathbf{c} = 0 \qquad \Big(\boldsymbol{\psi} \equiv (1, \mathbf{c}, \tfrac{1}{2} c^2) \Big). \tag{7.57}$$

Now $I(\varphi_1)$ is linear in φ_1 and the right-hand side of (7.55) is likewise linear in $\nabla \ln T$ and $\overset{\circ}{\mathbf{e}}$. Hence the most general solution of (7.55) is the linear combination

$$\varphi_1 = -\frac{\mathcal{C}}{n} \mathbf{A} \cdot \nabla \ln T - \frac{2}{n} \mathbf{B} : \overset{\circ}{\mathbf{e}} + \varphi_1^*, \tag{7.58}$$

where the vector $\mathbf{A}$ and tensor $\mathbf{B}$ depend on $\mathbf{c}$ (or equivalently on $\boldsymbol{\nu}$) and on the local thermodynamic state (n, T). It is readily established from relations given in Mathematical note 1.13, that for tensors $\mathbf{A}$ and $\mathbf{B}$,

$$\mathbf{B} : \overset{\circ}{\mathbf{A}} = \overset{\circ}{\mathbf{B}} : \overset{\circ}{\mathbf{A}},$$

whence it follows that the tensor $\mathbf{B}$ in (7.58) is a deviator.

The only vector that can be formed from the elements $\mathbf{c}$, n and T is $\mathbf{c}$ itself multiplied by a scalar function of $|\mathbf{c}|$, n and T; likewise the only deviator is $\overset{\circ}{\mathbf{cc}}$ $(= \mathbf{cc} - \frac{1}{3} c^2 \mathbf{1})$ times a similar scalar function. Hence (replacing $\mathbf{c}$ by $\boldsymbol{\nu}$)

$$\mathbf{A} = \boldsymbol{\nu} A(\nu), \qquad \mathbf{B} = \overset{\circ}{\boldsymbol{\nu}\boldsymbol{\nu}} B(\nu), \tag{7.59}$$

where the scalars A and B also depend on n and T. The solution (7.58) now takes the form

$$\varphi_1 = -\frac{\mathcal{C}}{n} A(\nu) \boldsymbol{\nu} \cdot \nabla \ln T - \frac{2}{n} B(\nu) \, \overset{\circ}{\boldsymbol{\nu}\boldsymbol{\nu}} : \overset{\circ}{\mathbf{e}} + \alpha_1 + \boldsymbol{\beta}_1 \cdot \mathbf{c} + \gamma_1 c^2. \tag{7.60}$$

7.6.3 *Uniqueness of the solution*

By symmetry we find that for any function $F(c)$,

$$0 = \int F(c)\mathbf{cc}\, d\mathbf{w} - \tfrac{1}{3}\int F(c)c^2\, d\mathbf{w}\, \mathbf{1} = \int F(c)\, \overset{\circ}{\mathbf{cc}}\, d\mathbf{w}.$$

On using this and eliminating the integrands that are odd functions in $\mathbf{c}$, we find that (7.60) satisfies the constraints in (7.57) if α_1 and γ_1 are zero, and $\boldsymbol{\beta}_1$ is given by

$$\int \Big(-\frac{1}{n}A(\nu)\mathbf{cc}\cdot\nabla\ln T + \boldsymbol{\beta}_1\cdot\mathbf{cc}\Big) f_0\, d\mathbf{c} = 0.$$

Thus, provided $A(\nu)$ satisfies

$$\int A(\nu)\nu^2 f_0\, d\mathbf{c} = 0, \tag{7.61}$$

we can absorb the term containing $\boldsymbol{\beta}_1$ into the first term on the right hand side of (7.60). It follows that the integral equation (7.55) has the unique solution

$$\varphi_1 = -\frac{\mathcal{C}}{n}\mathbf{A}\cdot\nabla\ln T - \frac{2}{n}\mathbf{B} : \overset{\circ}{\mathbf{e}}, \tag{7.62}$$

where $\mathbf{A}$ and $\mathbf{B}$ are special solutions of the integral equations

$$nI(\mathbf{A}) = f_0(\nu^2 - \tfrac{5}{2})\boldsymbol{\nu}, \tag{7.63}$$

$$nI(\mathbf{B}) = f_0\, \overset{\circ}{\boldsymbol{\nu\nu}}, \tag{7.64}$$

obtained by substituting (7.62) into (7.55) and splitting the result into independent equations for $\nabla\ln T$ and $\overset{\circ}{\mathbf{e}}$.

7.7 Thermal conductivity and viscosity

7.7.1 *Thermal conductivity*

From (7.48) and (7.51),

$$\mathbf{q} = \tfrac{1}{2}m\int c^2\mathbf{c}\,\varphi f_0\, d\mathbf{c}, \tag{7.65}$$

therefore by (7.62) the first-order heat flux is

$$\mathbf{q}_1 = \tfrac{1}{2}m\int c^2\mathbf{c}\varphi f_0\, d\mathbf{c} = -\frac{m}{2n}\mathcal{C}\int c^2\mathbf{c}\mathbf{A}\cdot\nabla\ln T f_0\, d\mathbf{c},$$

since integrals over odd functions of $\mathbf{c}$ vanish (see (5.57). From (7.59) and $\mathbf{c} = \mathcal{C}\nu\hat{\mathbf{c}}$, where $\hat{\mathbf{c}}$ is the unit vector,

$$\begin{aligned}\mathbf{q}_1 &= -\frac{m}{2n}\mathcal{C}^4\int \nu^4 A(\nu)\hat{\mathbf{c}}\hat{\mathbf{c}}\cdot\nabla\ln T\, f_0\, d\mathbf{c}\\ &= -\frac{m}{6n}\mathcal{C}^4\nabla\ln T\int \nu^4 A(\nu) f_0\, d\mathbf{c}\\ &= -\frac{2m}{3n}\left(\frac{kT}{m}\right)^2\nabla\ln T\int (\nu^2 - \tfrac{5}{2})\boldsymbol{\nu}\cdot\boldsymbol{\nu}A(\nu) f_0\, d\mathbf{c},\end{aligned}$$

where in the final step we have used (7.61). Hence by (7.63),

$$\mathbf{q}_1 = -\frac{2k^2T}{3m}\nabla T \int \mathbf{A}\cdot I(\mathbf{A})\, d\mathbf{c}. \tag{7.66}$$

The coefficient of $-\nabla T$ in this expression is the thermal conductivity, κ, i.e.

$$\kappa = \frac{2k^2T}{3m}\,[\,\mathbf{A},\mathbf{A}\,], \tag{7.67}$$

where

$$[\,\mathbf{A},\mathbf{A}\,] \equiv \int \mathbf{A}\cdot I(\mathbf{A})\, d\mathbf{c}.$$

7.7.2 *Viscosity*

From (7.48) and (7.51),

$$\mathbf{p} = m\int \mathbf{cc}\, f_0\, d\mathbf{c} + \boldsymbol{\pi} = p\mathbf{1} + \boldsymbol{\pi}, \tag{7.68}$$

where

$$\boldsymbol{\pi} = m\int \mathbf{cc}\,\varphi f_0\, d\mathbf{c}. \tag{7.69}$$

Thus by (7.62), (7.59), and (5.58),

$$\begin{aligned}
\boldsymbol{\pi} &= -\frac{2m}{n}\int \mathbf{ccB} : \overset{\circ}{\mathbf{e}}\; f_0\, d\mathbf{c} \\
&= -\frac{2m}{n}\mathcal{C}^2 \int \nu^4 B(\nu)\, \hat{\mathbf{c}}\hat{\mathbf{c}}\; \overset{\circ}{\hat{\mathbf{c}}\hat{\mathbf{c}}} \;:\; \overset{\circ}{\mathbf{e}}\; f_0\, d\mathbf{c} \\
&= -\frac{4kT}{n}\left(\tfrac{2}{15}\,\mathbf{K} + \tfrac{2}{9}\mathbf{11}\right) : \overset{\circ}{\mathbf{e}}\; \frac{4}{\sqrt{\pi}}\int_0^\infty \nu^6 B(\nu) e^{-\nu^2}\, d\nu \\
&= -\frac{8kT}{15n}\,\overset{\circ}{\mathbf{e}}\; \frac{4}{\sqrt{\pi}}\int_0^\infty \tfrac{3}{2}\, \overset{\circ}{\boldsymbol{\nu\nu}} : \overset{\circ}{\boldsymbol{\nu\nu}}\; B(\nu)\, \nu^2 e^{-\nu^2}\, d\nu \\
&= -\frac{4kT}{5n}\,\overset{\circ}{\mathbf{e}} \int B(\nu)\, \overset{\circ}{\boldsymbol{\nu\nu}} : \overset{\circ}{\boldsymbol{\nu\nu}}\; f_0\, d\mathbf{c},
\end{aligned}$$

where in the final step we have used (7.59). Hence by (7.64)

$$\boldsymbol{\pi} = -\tfrac{4}{5}kT\; \overset{\circ}{\mathbf{e}} \int \mathbf{B} : I(\mathbf{B})\, d\mathbf{c}. \tag{7.70}$$

It follows that the coefficient of viscosity is given by

$$\mu = \tfrac{2}{5}kT\,[\,\mathbf{B},\mathbf{B}\,], \tag{7.71}$$

where

$$[\,\mathbf{B},\mathbf{B}\,] \equiv \int \mathbf{B} : I(\mathbf{B})\, d\mathbf{c}.$$

7.8 The maximum principle

7.8.1 *Bracket integrals*

Take F and G to be functions of the molecular velocity, then the bracket integral of F and G is defined by

$$[\,F, G\,] \equiv \int G\, I(F)\, d\mathbf{c}. \tag{7.72}$$

As I is a linear operator, $[\,F, G\,]$ is a bilinear form. By symmetry arguments similar to those used to derive (7.16), it is readily shown that

$$[\,F, G\,] = \frac{1}{4n^2} \iiint f_0 \bar{f}_0 (F + \bar{F} - F' - \bar{F}')(G + \bar{G} - G' - \bar{G}') g\alpha d\mathbf{k}'\, d\bar{\mathbf{c}}\, d\mathbf{c}. \tag{7.73}$$

Hence
$$[\,F, G\,] = [\,G, F\,]. \tag{7.74}$$

Also notice that
$$[\,F, F\,] = \int F\, I(F)\, d\mathbf{c} \geq 0, \tag{7.75}$$

since in this case the integrand in (7.73) is always positive, unless F is a linear combination of the collisional invariants and therefore vanishes.

7.8.2 *Entropy production rate*

In 7.4.2 we showed that* the entropy production rate per unit volume is

$$\sigma = \tfrac{1}{4} k \iiint (f'\bar{f}' - f\bar{f}) \ln(f'\bar{f}'/f\bar{f})\, g\alpha\, d\bar{\mathbf{k}}\, d\bar{\mathbf{c}}\, d\mathbf{c}. \tag{7.76}$$

Using (7.51) and dropping terms $O(\epsilon^2)$ and smaller, we obtain the first-order rate

$$\sigma_1 = \tfrac{1}{4} k \iiint f_0 \bar{f}_0 (\varphi_1' + \bar{\varphi}_1' - \varphi_1 - \bar{\varphi}_1)(\varphi_1' + \bar{\varphi}_1' - \varphi_1 - \bar{\varphi}_1)\, g\alpha\, d\mathbf{k}'\, d\bar{\mathbf{c}}\, d\mathbf{c},$$

i.e.
$$\sigma_1 = n^2 k \int \varphi_1\, I(\varphi_1)\, d\mathbf{c} = n^2 k\, [\,\varphi_1, \varphi_1\,]. \tag{7.77}$$

Whence by (7.75)
$$\sigma_1 \geq 0, \tag{7.78}$$

which is a special case of the general result in (7.25).

Substituting (7.62) for φ_1 into (7.77), with the aid of (7.59) we find that

$$\sigma_1 = \frac{2k^2}{3mT} [\,\mathbf{A}, \mathbf{A}\,] \nabla T \cdot \nabla T + \tfrac{4}{5} k\, [\,\mathbf{B}, \mathbf{B}\,]\, \overset{\circ}{\mathbf{e}} : \overset{\circ}{\mathbf{e}}, \tag{7.79}$$

an expression that can also be obtained more directly from (2.18), (7.66), and (7.70).

*We have changed the notation to avoid subscripts, which now refer to order in Knudsen number.

7.8.3 *Maximum rate of entropy production*

Let $\mathbf{a} = a(\nu)\boldsymbol{\nu}$ denote a trial value for the vector $\mathbf{A}$ appearing in (7.63) and suppose it satisfies the condition

$$[\,\mathbf{a}, \mathbf{a}\,] = [\,\mathbf{a}, \mathbf{A}\,]. \tag{7.80}$$

Then it follows from the relation (see (7.75)) $[\,\mathbf{a} - \mathbf{A}, \mathbf{a} - \mathbf{A}\,] \geq 0$, that

$$[\,\mathbf{a}, \mathbf{a}\,] \leq [\,\mathbf{A}, \mathbf{A}\,], \tag{7.81}$$

the inequality applying if and only if $a(\nu) = A(\nu)$.

Similarly, if the tensor $\mathbf{b} = b(\nu)\,\overset{\circ}{\boldsymbol{\nu}\boldsymbol{\nu}}$ is a trial value for the tensor $\mathbf{B}$ in (7.64) and it satisfies

$$[\,\mathbf{b}, \mathbf{b}\,] = [\,\mathbf{b}, \mathbf{B}\,],$$

then

$$[\,\mathbf{b}, \mathbf{b}\,] \leq [\,\mathbf{B}, \mathbf{B}\,], \tag{7.82}$$

with equality holding if and only if $b(\nu) = B(\nu)$.

We can now draw the following conclusion from (7.79), (7.81), and (7.82):

'In non-equilibrium systems, with given temperature and velocity gradients, the distribution function assumes the value that maximizes the rate of change of the entropy density.'

This principle plus the variational calculus is sometimes employed to solve the basic integral equations (7.63) and (7.64), to which task we now turn.

7.9 Solving the integral equations

The most efficient means of solving the integral equations for $\mathbf{A}$ and $\mathbf{B}$ is by expressing them in terms of Sonine polynomials, a method first introduced into kinetic theory by Burnett in 1935.

7.9.1 *Sonine Polynomials*

Sonine polynomials arise in the study of certain definite integrals involving Bessel functions (see Watson 1952, Chapter XII). The polynomial of order integer n and index m is defined by

$$S_m^{(n)} \equiv \sum_{p=0}^{n} \frac{(m+n)_{n-p}}{p!(n-p)!}\,(-x)^p, \tag{7.83}$$

where a_r denotes the product of the r factors $a, a-1, \ldots a-r+1$. They are generated as the coefficients in the expansion

$$(1-s)^{-m-1} \exp\left(-\frac{xs}{1-s}\right) = \sum_{n=0}^{\infty} S_m^{(n)}(x)\, s^n.$$

Note the particular values

$$S_m^{(0)}(x) = 1, \qquad S_m^{(1)} = m + 1 - x. \tag{7.84}$$

By equating the coefficients $s^p t^q$ on the two sides of the expression

$$(1-s)^{-m-1}(1-t)^{-m-1}\int_0^\infty \exp\left\{-\frac{x(1-st)}{(1-s)(1-t)}\right\} x^m\,dx = (1-st)^{-m-1}\Gamma(m+1),$$

we deduce the orthogonality relation

$$\int_0^\infty e^{-x} S_m^{(p)}(x)\, S_m^{(q)}\, x^m\,dx = \left.\begin{array}{cc} 0 & p \neq q \\ \Gamma(m+p+1)/p! & p = q \end{array}\right\}. \tag{7.85}$$

7.9.2 *Formal evaluation of A and κ*

The integral equation in (7.63), viz.

$$nI(\mathbf{A}) = f_0(\nu^2 - \tfrac{5}{2})\boldsymbol{\nu} \qquad (\mathbf{A} = \boldsymbol{\nu}A(\nu)), \tag{7.86}$$

is solved as follows. It is assumed that $A(\nu)$ can be expanded in a convergent series of the type

$$A(\nu) = \sum_{p=0}^{\infty} a_p S_{\frac{3}{2}}^{(p)}(\nu^2), \tag{7.87}$$

where the coefficients a_p are independent of ν. Substituting into (7.61), and using (2.37), i.e.

$$\frac{f_0}{n}\,d\mathbf{c} = \frac{4}{\sqrt{\pi}}\,\nu^2\exp(-\nu^2)\,d\nu\,d\Omega \qquad (0 \le \nu < \infty,\ 0 \le \Omega \le 1), \tag{7.88}$$

and $1 = S_{\frac{3}{2}}^{(0)}$, we find

$$0 = \frac{4n}{\sqrt{\pi}}\int_\Omega\int_0^\infty \sum_{p=0}^{\infty} a_p\, S_{\frac{3}{2}}^{(p)} S_{\frac{3}{2}}^{(0)}\nu^2 e^{-\nu^2}\,d\nu\,d\Omega = \frac{2n}{\sqrt{\pi}}\,\Gamma(\tfrac{5}{2})a_0,$$

where we have replaced $2\nu^4\,d\nu$ by $\nu^3\,d\nu^2$ and then applied the orthogonality relation (7.85), with $m = 3/2$ and $p = q = 0$. Thus (7.61) is satisfied if $a_0 = 0$, and it follows from (7.87) that

$$\mathbf{A} = A(\nu)\boldsymbol{\nu} = \sum_{p=1}^{\infty} a_p \mathbf{a}^{(p)}, \tag{7.89}$$

where

$$\mathbf{a}^{(p)} \equiv S_{\frac{3}{2}}^{(p)}(\nu^2)\boldsymbol{\nu}. \tag{7.90}$$

From (7.84), (7.89), and (7.86),

$$nI(\mathbf{A}) = -\mathbf{a}^{(1)} f_0. \tag{7.91}$$

Multiplying this by $\mathbf{a}^{(q)}$, integrating over $\mathbf{c}$ and using (7.89), we get

$$[\,\mathbf{a}^{(q)}, \mathbf{A}\,] = \alpha_q, \tag{7.92}$$

where by (7.88),

$$\alpha_q = -\frac{2}{\sqrt{\pi}}\int_0^\infty e^{-\nu^2}\nu^3 S_{\frac{3}{2}}^{(1)}(\nu^2) S_{\frac{3}{2}}^{(q)}(\nu^2)\,d(\nu^2).$$

Hence from (7.85),

$$\alpha_q = \begin{cases} -\frac{15}{4} & (q=1), \\ 0 & (q=0). \end{cases} \tag{7.93}$$

It follows from (7.89) that (7.92) can now be written as

$$\sum_{p=1}^{\infty} a_p a_{pq} = \alpha_q \qquad (q = 1, 2, \ldots, \infty), \tag{7.94}$$

where

$$a_{pq} \equiv [\mathbf{a}^{(p)}, \mathbf{a}^{(q)}]. \tag{7.95}$$

As the functions $\mathbf{a}^{(p)}$ are known, we can in principle determine the values of a_{pq} from the infinite set of algebraic equations in (7.94).

From (7.67) we find that the thermal conductivity is given by

$$\kappa = \frac{2k^2T}{3m}[\mathbf{A}, \mathbf{A}] = \frac{2k^2T}{3m}\sum_{p=1}^{\infty} a_p [\mathbf{a}^{(p)}, \mathbf{A}],$$

which by (7.92) and (7.93) reduces to

$$\kappa = -\frac{5k^2T}{2m} a_1. \tag{7.96}$$

7.9.3 *Formal evaluation of* B *and* μ

The integral equation in (7.64),

$$nI(\mathsf{B}) = \overset{\circ}{\boldsymbol{\nu\nu}} f_0, \tag{7.97}$$

is solved by a method similar to the used above for (7.63). We assume that B can be expanded in a series of the form

$$\mathsf{B} = \sum_{p=1}^{\infty} b_p \mathsf{b}^{(p)}, \tag{7.98}$$

where

$$\mathsf{b}^{(p)} = \overset{\circ}{\boldsymbol{\nu\nu}} S_{\frac{5}{2}}^{(p-1)}(\nu^2), \tag{7.99}$$

and the coefficients b_p are to be determined.

Multiplying (7.97) by $\mathsf{b}^{(q)}$, integrating over $\mathbf{c}$, and using (7.98), we find

$$[\mathsf{b}^{(q)}, \mathsf{B}] = \beta_p, \tag{7.100}$$

or

$$\sum_{p=1}^{\infty} b_p b_{pq} = \beta_q,$$

where

$$b_{pq} \equiv [\mathsf{b}^{(p)}, \mathsf{b}^{(q)}], \tag{7.101}$$

and

$$\beta_q = \frac{4}{3\sqrt{\pi}} \int_0^{\infty} e^{-\nu^2} \nu^5 S_{\frac{5}{2}}^{(q-1)}(\nu^2)\, d(\nu^2).$$

Hence by (7.85),

$$\beta_q = \begin{cases} \frac{5}{2} & (q = 1), \\ 0 & (q = 0). \end{cases} \tag{7.102}$$

The viscosity follows from (7.71):

$$\mu = \tfrac{2}{5}kT\ [\mathbf{B}, \mathbf{B}\ ] = \tfrac{2}{5}kT \sum_{p=1}^{\infty} b_p\ [\mathbf{b}^{(p)}, \mathbf{B}\ ],$$

whence from (7.100) and (7.102),

$$\mu = kTb_1. \tag{7.103}$$

7.10 Transport properties

7.10.1 *Truncating the series*

Burnett's method of dealing with the set of linear equations in (7.94) was to truncate the series, writing

$$\mathbf{A}^{(m)} = \sum_{p=1}^{m} a_p^{(m)}\mathbf{a}^{(p)}; \qquad \sum_{p=1}^{m} a_p^{(m)} a_{pq} = \alpha_q \quad (q = 1, 2, \ldots, m), \tag{7.104}$$

and regarding $\mathbf{A}^{(m)}$ as an m-th approximation to $\mathbf{A}$. This assumes that $\mathbf{A}^{(m)}$ and $a_p^{(m)}$ tend to $\mathbf{A}$ and a_p as m tends to infinity. With the value of α_q given by (7.93), we obtain the solution

$$a_q^{(m)} = -\tfrac{15}{4}\mathcal{A}_{1q}^{(m)}/\mathcal{A}^{(m)} \qquad (q = 1, 2, \ldots, m), \tag{7.105}$$

where $\mathcal{A}^{(m)}$ is the symmetric determinant, with elements a_{pq} $(p, q = 1, 2, \ldots, m)$ and $\mathcal{A}_{1q}^{(m)}$ is the cofactor of a_{1q} in $\mathcal{A}^{(m)}$. It follows from (7.96) and (7.105) that the m-th approximation to κ is

$$[\kappa]_m = \tfrac{25}{4}c_v kT\mathcal{A}_{11}^{(m)}/\mathcal{A}^{(m)} \qquad \left(\kappa = \lim_{m\to\infty} [\kappa]_m\right). \tag{7.106}$$

A similar treatment of (7.98), (7.100), and (7.103) leads to

$$[\mu]_m = \tfrac{5}{2}kT\mathcal{B}_{11}^{(m)}/\mathcal{B}^{(m)} \qquad \left(\mu = \lim_{m\to\infty} [\mu]_m\right), \tag{7.107}$$

where $\mathcal{B}^{(m)}$ is the determinate with elements $b_{pq}\,(p, q = 1, 2, \ldots, m)$, and $\mathcal{B}_{1q}^{(m)}$ is the cofactor of b_{1q} in $\mathcal{B}^{(m)}$.

Another approach to the task of calculation starts with a trial function

$$\mathbf{a} = \sum_{p=1}^{m} a_p^{(m)}\mathbf{a}^{(p)},$$

forms the bracket integral $g = [\mathbf{a}, \mathbf{a}\ ]$ and, taking advantage of (7.81), uses the variation criterion $\delta\{g\} = 0$ to deduce a set of algebraic equations for the $a_p^{(m)}$. This leads to exactly the same solution as obtained more directly by Burnett.

As we shall show below, the first approximations, $[\kappa]_1$ and $[\mu]_1$, are the same expressions as given in (6.66). Convergence is very rapid. For example, with rigid elastic spheres, $\kappa = 1.025\ldots[\kappa]_1$ and $\mu = 1.016\ldots[\mu]_1$.

7.10.2 *Expressions for* $[\kappa]_1$ and $[\mu]_1$

From (7.106) and (7.107),

$$[\kappa]_1 = \tfrac{25}{4} c_v kT/a_{11}, \qquad [\mu]_1 = \tfrac{5}{2} kT/b_{11}. \tag{7.108}$$

First we shall evaluate a_{11}. From (7.90), (7.95), (7.73), and (7.84):

$$\begin{aligned} a_{11} &= \left[S^{(1)}_{\frac{3}{2}}(\nu^2)\boldsymbol{\nu}, S^{(1)}_{\frac{3}{2}}(\nu^2)\boldsymbol{\nu} \right] \\ &= \frac{1}{4n^2}\iiint f_0\bar{f}_0 \Big[S^{(1)}_{\frac{3}{2}}(\nu^2)\boldsymbol{\nu} + S^{(1)}_{\frac{3}{2}}(\bar{\nu}^2)\bar{\boldsymbol{\nu}} \\ &\qquad\qquad - S^{(1)}_{\frac{3}{2}}(\nu'^2)\boldsymbol{\nu}' - S^{(1)}_{\frac{3}{2}}(\bar{\nu}'^2)\bar{\boldsymbol{\nu}}' \Big]^2 g\alpha\, d\mathbf{k}'\, d\bar{\mathbf{c}}\, d\mathbf{c}, \\ &= \frac{1}{4\pi^3}\iiint \exp\{-(\nu^2+\bar{\nu}^2)\} \big[\nu^2\boldsymbol{\nu} + \bar{\nu}^2\bar{\boldsymbol{\nu}} \\ &\qquad\qquad - \nu'^2\boldsymbol{\nu}' - \bar{\nu}'^2\bar{\boldsymbol{\nu}}'\big]^2 g\alpha\, d\mathbf{k}'\, d\bar{\boldsymbol{\nu}}\, d\boldsymbol{\nu}, \end{aligned} \tag{7.109}$$

where $[\cdots]^2$ denotes $[\cdots]\cdot[\cdots]$.

We now follow the method of §2.8.1 and transform to centre of mass and relative velocity variables. With $\mathbf{G}_* \equiv \frac{1}{2}(\boldsymbol{\nu}+\bar{\boldsymbol{\nu}})$, $\mathbf{g}_* \equiv \boldsymbol{\nu} - \bar{\boldsymbol{\nu}}$, we find

$$\begin{aligned} a_{11} &= \frac{1}{4\pi^2}\iiint e^{-2G_*^2} e^{-\frac{1}{2}g_*^2} \big[g_*^2(\mathbf{g}_*\cdot\mathbf{G}_*)^2 + g_*'^2(\mathbf{g}_*'\cdot\mathbf{G}_*)^2 \\ &\qquad\qquad -2(\mathbf{g}_*\cdot\mathbf{g}_*')(\mathbf{g}_*\cdot\mathbf{G}_*)(\mathbf{g}_*'\cdot\mathbf{G}_*)\big] g\alpha\, d\mathbf{k}'\, d\mathbf{g}_*\, d\mathbf{G}_* \\ &= \frac{2^{1/2}}{64\pi^{3/2}}\iint e^{-\frac{1}{2}g_*^2}\big[g_*^4 + g_*'^4 - 2(\mathbf{g}_*\cdot\mathbf{g}_*')^2\big] g\alpha\, d\mathbf{k}'\, d\mathbf{g}_*, \end{aligned} \tag{7.110}$$

where the integration over $\mathbf{G}_*$ is effected using expressions given in §2.6.2. From §6.4.1 we have $g_*^2 = g_*'^2$ and $\mathbf{g}_*\cdot\mathbf{g}_*' = g_*^2\cos\chi$, where $\chi = \chi(b, g)$ is the scattering angle. From §6.3.1, $\alpha\, d\mathbf{k}' = b\, db\, d\epsilon = \alpha\sin\chi\, d\chi\, d\epsilon$. Thus integrating over $0 \le \epsilon \le 2\pi$, setting $d\mathbf{g}_* = g_*^2\, dg_*$ and $y^2 = g_*^2/2$, we find

$$a_{11} = 4\left(\frac{kT}{\pi m}\right)^{1/2}\int_0^\infty e^{-y^2}y^7\left[2\pi\int_0^\pi \alpha(1-\cos^2\chi)\sin\chi\, d\chi\right] dy. \tag{7.111}$$

Hence, in terms of the integral Ω_μ, defined in 6.8.1, we arrive at $a_{11} = 4\Omega_\mu$ and therefore

$$[\kappa]_1 = \tfrac{25}{16} c_v kT/\Omega_\mu, \tag{7.112}$$

in agreement with the result obtained by the Maxwell–Chapman theory.

The integral b_{11}, is calculated in a similar way. Corresponding to (7.109), we find a like expression for b_{11}, except that in place of the term in square brackets in the integrand there is

$$\left[\boldsymbol{\nu}\boldsymbol{\nu}+\bar{\boldsymbol{\nu}}\bar{\boldsymbol{\nu}}-\boldsymbol{\nu}'\boldsymbol{\nu}'-\bar{\boldsymbol{\nu}}'\bar{\boldsymbol{\nu}}'\right]:\left[\boldsymbol{\nu}\boldsymbol{\nu}+\bar{\boldsymbol{\nu}}\bar{\boldsymbol{\nu}}-\boldsymbol{\nu}'\boldsymbol{\nu}'-\bar{\boldsymbol{\nu}}'\bar{\boldsymbol{\nu}}'\right].$$

Transforming to the centre of mass and relative velocity variables, we obtain

$$b_{11}=\frac{1}{16\pi^3}\iiint e^{-2G_*^2}\,e^{-\frac{1}{2}g_*^2}\left[\mathbf{g}'_*\mathbf{g}'_*-\mathbf{g}_*\mathbf{g}_*\right]^2 g\alpha\,d\mathbf{k}'\,d\mathbf{g}_*\,d\mathbf{G}_*.$$

Integrating this over the variable $\mathbf{G}_*$, we obtain an expression identical to that in (7.111) for a_{11}, whence $b_{11}=a_{11}$.

It follows from (7.108) that

$$[\mu]_1=\tfrac{5}{8}kT/\Omega_\mu. \tag{7.113}$$

Also notice that, regardless of the molecular force law,

$$[\kappa]_1=\tfrac{5}{2}c_v[\mu]_1. \tag{7.114}$$

For details of the higher approximations the reader may consult the works by Chapman and Cowling (1970) or Ferziger and Kaper (1972). These authors also describe an even simpler approximation due to Kihara (1949).

7.11 Boltzmann's equation and pressure gradients

7.11.1 *Pressure gradients in the Chapman–Enskog solution*

The pressure gradient plays no role in the *formulation* of Boltzmann's evolutionary equation for f, but is involved in the Chapman–Enskog method of solving the equation. Together with the divergence of the heat flux vector, it enters the theory via the time derivatives of n, $\mathbf{v}$, and T (see (7.42)). These derivatives are needed to determine the rate of change of f_0, which in turn leads to the value of f_1. The difficulty Enskog had to overcome arises because the pressure tensor, $\mathsf{p}=p\mathbf{1}+\boldsymbol{\pi}$, and the heat flux, $\mathbf{q}$, can be evaluated only when f is known. Fortunately, knowledge of p, or more precisely, of ∇p, is sufficient to enable f_1 to be found; from f_1 we can find $\mathbf{q}_1$ and $\boldsymbol{\pi}_1$, their derivatives allow us to determine f_2, and so on, the iterative process being as described in §7.5.1.

The Chapman–Enskog solution introduces the Knudsen number ϵ into the theory. Thus, as in elementary kinetic theory, mean free paths have a role. But there is an important difference from the elementary theory, namely that, at least in principle, solutions to *any* order in ϵ can be derived from Boltzmann's equation. While it is possible to press the elementary theory to yield $O(\epsilon^2)$ terms, uncertainty about both the accuracy of the coefficients and the completeness of the set of second-order derivatives obtained, would rob the work of conviction. The intermediate kinetic theory of Chapter 5, which is more secure in its treatment of second-order terms, reveals the complexity resulting when just two mean free paths in succession are taken into account. To obtain $O(\epsilon^3)$ terms, the average history of molecules over three mean free paths is required. This would present an almost impossible task for the elementary theory, and an analytically very demanding one for the related intermediate theory. While the *analytic* solution of the kinetic equation does require knowledge of the macroscopic variables, albeit one step behind at each stage of an iterative process, Boltzmann's kinetic equation itself is free of this constraint, from which it appears that, used directly to determine f, it holds for all values of the Knudsen number.

Boltzmann's collision integral is based on an accurate description of a single generic collision between class '1' and class '2' molecules, which apparently liberates it from the complexities of pursuing molecular histories over a series of previous collisions. It is not obvious why this should be so. *Defining* pressure to be momentum flux is a another key simplification needing justification, since it by-passes the problem of the imbalance of collisional forces due to pressure gradients.

7.11.2 *The correlation function*

A central assumption in the theory is that the velocity of a particle is uncorrelated with the initial velocity of any particle with which it is about to collide. The probability that a class 1 molecule collides with a class 2 molecule is proportional to the joint probability P_{12} that these molecules be in the same element of real space. With P_{12} written in the form*

$$n^2 P_{12} = f_1(\mathbf{r}, \mathbf{c}_1, t) f_2(\mathbf{r}, \mathbf{c}_2, t) + G(\mathbf{r}, \mathbf{c}_1, \mathbf{c}_2, t), \qquad (7.115)$$

the function $G(\mathbf{r}, \mathbf{c}_1, \mathbf{c}_2, t)$ is termed the 'correlation' function. The assumption above, described as the 'hypothesis of molecular chaos', amounts to setting $G = 0$.

Consider the case of a stationary atmosphere, maintained in a near-isothermal state by boundary conditions above and below. The equilibrium equation is $kT\nabla n = \rho \mathbf{g}$, where $\mathbf{g}$ is the gravitational acceleration . This zero-order relation follows from Boltzmann's equation, but only if an *additional* physical constraint concerning collisions is admitted (see exercise 7.1). Our concern here is how Boltzmann's collision operator should be modified to allow for such collisions. His operator is proportional to the product of the distribution functions for the molecules involved, i.e. G in (7.115) is zero. By (7.51) and (7.62), since ∇T and $\overset{\circ}{\mathbf{e}}$ are zero, to $O(\epsilon)$ accuracy, these are Maxwellian distributions and P_{12} is therefore independent of the directions of the colliding molecules. Were it otherwise, G would not be zero. But the atmosphere is supported in equilibrium by the upward moving molecules from the denser regions having a larger effect on a group of target molecules, than those descending from above, a collisional anisotropy that is essential for equilibrium.

*Reverting to subscripts on f identifying the particle class for the purposes of this section.

In a collision interval τ a target molecule P_2 will fall from rest to reach a speed of $\tau\mathbf{g}$ (we are ignoring P_2's random motion, which will average to zero over an ensemble of target molecules). Hence, when at the end of this time it experiences a collision, to cancel its downward motion the impact must supply an impulsive force of $-\tau\mathbf{g} = \tau g\hat{\mathbf{z}} = -\tau kT\nabla n/\rho$, where $\hat{\mathbf{z}}$ is unit vector pointing upwards. Thus the reduction in the $\hat{\mathbf{z}}$-component of the momentum flux per collided class 1 molecule is $\tau m g\hat{\mathbf{z}}\cdot\mathbf{c}_1$. The average momentum flux of class 1 molecules in the $\hat{\mathbf{z}}$-direction is $m\hat{\mathbf{z}}\cdot\langle\mathbf{c}_1\mathbf{c}_1\rangle\cdot\hat{\mathbf{z}} = \frac{1}{2}m\mathcal{C}^2$, where we have used (2.43). Of this flux, a fraction $2\tau g\hat{\mathbf{z}}\cdot\mathbf{c}_1/\mathcal{C}^2$ is expended in supporting the atmosphere, by what in §5.8.2 we termed 'streaming' collisions. Such collisions therefore deplete the number of impinging molecules by $2\tau g\hat{\mathbf{z}}\cdot\mathbf{c}_1 f_1/\mathcal{C}^2$ per unit volume of phase space.

As in the present case the viscous stress tensor is zero, the equilibrium condition can be written

$$\mathbf{g} = \frac{1}{\rho}kT\nabla n = \frac{1}{\rho}\nabla p = \frac{1}{\rho}\nabla\cdot\mathsf{p} = -\mathbf{P}, \tag{7.116}$$

where $\mathbf{P}$ is the pressure gradient force. Hence the collision rate for streaming collisions is proportional to

$$-2\tau g\hat{\mathbf{z}}\cdot\mathbf{c}_1 f_1 f_2/\mathcal{C}^2 = -2\tau\mathbf{P}\cdot\mathbf{c}_1 f_1 f_2/\mathcal{C}^2 \approx \tau\mathbf{P}\cdot\frac{\partial f_1}{\partial\mathbf{c}_1}f_2,$$

the last form following from (5.85) and the fact that to sufficient accuracy, f_1 is Maxwellian. Including the scattering collisions, we have a total collision rate proportional to

$$\left\{f_1 + \tau\mathbf{P}\cdot\frac{\partial f_1}{\partial\mathbf{c}_1}\right\}f_2 = f_1 f_2\{1+G_1\} \quad \big(G_1 \equiv \tau\mathbf{P}\cdot(\partial\ln f_1/\partial\mathbf{c}_1)\big). \tag{7.117}$$

This is the rate that should appear in Boltzmann's collision integral. Our conclusion is that the assumption of 'pure' molecular chaos is not valid in the presence of pressure gradients, although the correction required makes no difference to the Chapman–Enskog $\mathrm{O}(\epsilon)$ theory. As will be shown in the next chapter, it does make a significant change to the second-order theory.

7.11.3 *Boltzmann's integral modified*

The example given above is a special case of the more general theory of the influence of pressure gradients given in §5.8.2, where the expression $f(1+G)d\boldsymbol{\nu}/\tau$ for the rate at which collisions remove particles from the element $d\boldsymbol{\nu}$ was derived. When the derivation of the collision integral in (7.10) is modified to allow for streaming losses, it becomes

$$\mathbb{C}(f_1 f_2) = \iint\{f_1'(1+G_1')f_2' - f_1(1+G_1)f_2\}g\alpha\,d\mathbf{k}'\,d\mathbf{w}_2, \tag{7.118}$$

where G_1 is the function defined in (7.117). It may appear that we have added an $\mathrm{O}(\epsilon)$ term to the collision integral, but cancellation reduces this to an $O(\epsilon^2)$ term (cf. 5.72).

There remains the correction due to fluid shear discussed in §5.9. This replaces $\mathbf{P}$ by $\mathbf{P}+\boldsymbol{\alpha}$ in G_1', but leaves G_1 unchanged (see (5.104)). However our purpose here is not to revise Boltzmann's equation, but to show that it is not correct for terms beyond first order in the Knudsen number ϵ, at least for $\epsilon < 1$. At very large values of ϵ, the macroscopic variables $\mathbf{p}$ and $\mathbf{e}$ may have less influence on the kinetic equation, and Boltzmann's collision integral may recover accuracy. For our purposes the intermediate kinetic theory, summarized in §5.12, is much more convenient, especially for magnetoplasmas, where the presence of varying magnetic fields requires us to replace the collision interval τ by a microscopic time that does not involve particle collisions.

7.12 The direct simulation Monte Carlo approach

7.12.1 *Direct simulation*

The restriction imposed on the Knudsen number ϵ by the Chapman-Enskog series solution of Boltzmann's equation and the growing importance of flow problems in the 'transition' regime embracing $\epsilon = 1$, has lead researchers in rarefied gas dynamics to exploit the power of modern computers to obtain direct numerical solutions for specified problems. The method is assumed to be valid over the whole range of Knudsen numbers.

The gas is simulated as an assembly of thousands (even millions) of particles, each one of which is followed between collisions that are treated as statistical events with a range of outcomes. Samples of the velocity distribution f are taken at intervals and the macroscopic variables—various moments of f—deduced. The flow obtained is intrinsically unsteady since, despite the large sample of representative molecules, thermodynamics fluctuations are much larger than in a real gas. With appropriate boundary conditions, the solution may converge to a pattern consisting of small fluctuations about a steady mean.

The following outline of the method is taken from Bird's (1978) definitive text.

7.12.2 *The Monte Carlo procedure*

The simulated region of physical space is divided into a network of cells, that may either be small regions with specified boundaries or just an array of points, each representing the surrounding region. The cells have dimension Δr, typically much smaller than a mean free path λ, say $\lambda/3$, and may contain ~ 30 molecules, choices that are determined by a balance between a feasible computational time and a need to limit the amplitude of the thermodynamic fluctuations. With too few molecules, the fluctuations may swamp the flow phenomenon of interest.

The velocities of all the molecules in every cell are assigned initial values.

Time is advanced in discrete steps of magnitude Δt, where Δt is small compared with the mean collision interval τ and Δr, Δt in general depend on $(\mathbf{r}, t)$. Thus, given their initial states, after a time Δt, the new positions of the molecules (cell locations) can be found. Boundary and symmetry conditions are involved in determining the appropriate migrations of the simulated molecules.

The next step is to compute a representative set of collisions between some of the molecules occupying the same cell, then to replace the pre-collision velocity components of the collision-pairs by their post-collision values. The latter are found via Monte Carlo selection from an isotropic scattering distribution in the centre of mass frame. The probability of a collision between two hard molecules is proportional to the collision-frequency $\tau^{-1} = \pi n \sigma^2 g$ (see §1.1.1), where σ is the molecular diameter and g is their relative velocity. More generally, for the case of molecules represented by point centres of force, it follows from (3.6) that $\tau^{-1} \propto g^{(\nu-5)/(\nu-1)}$. Therefore in choosing collision-pairs, it is necessary to ensure that the probability that a given pair gets selected is proportional to $g^{(\nu-5)/(\nu-1)}$. Bird gives details of an efficient method for doing this.

It remains to determine how many collisions occur in the time interval Δt. Let N denote the number of particles in the cell of interest, then for all the $\frac{1}{2}N$ pairs to collide, a full collision interval τ is required. It follows that if a time-counter for each cell is advanced by $\Delta t_c = 2\tau/N$ per collision, and sufficient collisions are allowed in each cell to keep these counters synchronous with Δt, then the average collision frequency of the simulation will approximate that of the real gas being modelled.

Professor Bird has recently (1987) dropped the time-counter method in favour of setting the number of collision-pairs in the time interval Δt equal to $N\overline{N}f_N(\sigma^2 g)_{max}\Delta t/V_C$, where N is the number of simulated molecules in the cell, $\overline{N}$ is their time-averaged number, f_N is the number of real molecules per simulated molecule, V_C is the cell volume, and 'max' denotes the maximum value revealed by the Monte Carlo procedure. We are omitting several practical considerations in this brief survey; for these the reader is referred to Bird's text.

Bird demonstrates that the 'DSMC' method—as it is now generally labelled—is both consistent with and more general than Boltzmann's formulation of kinetic theory. He and many others have successfully applied the method to a wide range of problems in rarefied gas dynamics; agreement with experiment is usually good.

7.12.3 *Pressure gradients in DSMC*

Consider the case of a uniform, isothermal atmosphere, which we discussed for Boltzmann's formulation in 7.11.2. It is evident from the description given above of the DSMC method that during the time interval Δt, in addition to their linear translation, molecules will fall a distance $\frac{1}{2}(\Delta t)^2 g$

due to gravity. In the final equilibrium state such descents must be offset by upwards displacements caused by collisions. And although such streaming collisions *could* be incorporated as constraints on the probabilities in the Monte Carlo process, in Bird's original formulation of DSMC it appears that this is not done.

It might be argued that the increase of number density with depth, would itself be sufficient to provide the required balance, but so long as binary collisions are assumed to have isotropic outcomes in the centre of mass frame—which is accelerating under the gravitation force—there can be no steady state. Of course the density gradient *does* provide the required anisotropy, but it cannot do this through the agency of isolated binary collisions (cf. discussion in 3.9.1).

We have dealt with a special problem to make it clear that both Boltzmann's equation and the DSMC method (at least as originally proposed) have the same failing so far as pressure gradients are concerned. In *all* flow problems, regardless of whether there is a body force or not, these gradients should be represented by streaming collisions that modify the collision probabilities, as explained in 7.11.3. Similar remarks apply to fluid shear.

Exercises 7

7.1 Show that
$$\frac{\partial}{\partial t}(\rho\mathbf{v}) + \nabla\cdot(\rho\mathbf{vv} + \mathsf{M}) - \rho\mathbf{F} = \int m\mathbf{w}\mathbb{C}\,d\mathbf{w},$$
where M is the momentum flux tensor. Distinguish betweeen (i) the vanishing of $\mathbb{C}$ and (ii) the vanishing of the *integral* containing $\mathbb{C}$, writing appropriate forms for the equation for each case. Show that for a spatially uniform atmosphere, in case (i) $\rho\mathrm{D}\mathbf{v} = \rho\mathbf{g}$ and in case (ii) $\rho\mathrm{D}\mathbf{v} = -\nabla p + \rho\mathbf{g}$. Deduce that without collisions, no equilibrium state is possible.

7.2 Show that the entropy flux vector defined in (7.20) is given by
$$\mathbf{J} = k\int \mathbf{c}f_0 \ln f_0\, \tau_2(\nu^2 - \tfrac{5}{2})\mathbf{c}\cdot\nabla \ln T\, d\mathbf{c} + \mathrm{O}(\epsilon^2),$$
and hence verify that $\mathbf{J} = \mathbf{q}/T + \mathrm{O}(\epsilon^2)$.

7.3 Show that the entropy production rate per unit volume can be expressed
$$\sigma = -k\int \mathbb{D}f \ln f\, d\mathbf{w}.$$
Given that ρ, $\mathbf{v}$, u and σ have specified values, show that the entropy density is maximized by a distribution function f that satisfies (cf. §2.5)
$$\ln f + \alpha + \boldsymbol{\beta}\cdot\mathbf{c} + \gamma c^2 + \tau\mathbb{D}\ln f = 0,$$
where α, $\boldsymbol{\beta}$, γ, and τ are Lagrangian multipliers. If $\tau\mathbb{D} = \mathrm{O}(\epsilon)$, show that to first order in ϵ,
$$f = f_0(1 - \tau\mathbb{D}\ln f_0).$$

7.4 Prove that (in the notation of §7.10)
$$[\kappa]_2 = [\kappa]_1(1 - a_{12}^2/a_{11}a_{22})^{-1}, \qquad [\mu]_2 = [\mu]_1(1 - b_{12}^2/b_{11}b_{22})^{-1}.$$
Establish the relations $b_{12} = a_{12},\;\; b_{22} = a_{22} + \frac{32}{24}a_{11}$.

7.5 Use the expressions in exercise 7.4 to prove that

$$[\kappa]_2 \geq \tfrac{5}{2}c_v[\mu]_2.$$

7.6 Assuming that in an encounter between molecules P_1 and P_2 their velocities $\mathbf{c}_1$ and $\mathbf{c}_2$ alter by small amounts $\Delta\mathbf{c}_1$, $\Delta\mathbf{c}_2$, by applying Taylor's theorem to $f_1' = f_1(\mathbf{c}_1 + \Delta\mathbf{c}_1)$ and $f_2' = f_2(\mathbf{c}_2 + \Delta\mathbf{c}_2)$, show that correct to second order in these changes, Boltzmann's collision integral can be expressed as

$$\mathbb{C}_B(f_1 f_2) = \iiint \left\{\Delta\mathbf{g}\cdot\mathbf{D}f_1 f_2 + \tfrac{1}{2}\Delta\mathbf{g}\Delta\mathbf{g}:\mathbf{D}^2 f_1 f_2\right\} gb\,db\,d\epsilon\,d\mathbf{c}_2,$$

where

$$\mathbf{D} \equiv \mu_1\frac{\partial}{\partial\mathbf{c}_2} - \mu_2\frac{\partial}{\partial\mathbf{c}_1}$$

(see §§6.3.1, 6.4.1, and 7.2.3).

7.7 Use the expression in exercise 6.7 to show that

$$\iint \Delta\mathbf{g}\,gb\,db\,d\epsilon = \frac{\partial}{\partial\mathbf{g}}\cdot\iint \Delta\mathbf{g}\,\Delta\mathbf{g}\,gb\,db\,d\epsilon = \mathbf{D}\cdot\iint \Delta\mathbf{g}\,\Delta\mathbf{g}\,gb\,db\,d\epsilon,$$

where $\mathbf{D}$ is defined in exercise 7.6. Hence show that the formula for $\mathbb{C}_B$ given in exercise 7.6 can be written

$$\mathbb{C}_B = -\frac{\partial}{\partial\mathbf{c}_1}\cdot\mathbf{Q}_{12},$$

where

$$\mathbf{Q}_{12} = \tfrac{1}{2}\mu_2\iiint\left\{\mathbf{D}f_1 f_2\right\}\cdot\Delta\mathbf{g}\Delta\mathbf{g}\,gb\,db\,d\epsilon\,d\mathbf{c}_2.$$

8

SECOND-ORDER KINETIC THEORY

8.1 Introduction

8.1.1 *Knudsen number expansions*

Our purpose in this chapter is to calculate and apply the second-order terms $\mathbf{q}_2$ and $\boldsymbol{\pi}_2$ in the Knudsen number expansions for the heat flux vector $\mathbf{q}$ and the viscous stress tensor $\boldsymbol{\pi}$, namely

$$\mathbf{q} = \mathbf{q}_1 + \mathbf{q}_2 + \cdots, \qquad \boldsymbol{\pi} = \boldsymbol{\pi}_1 + \boldsymbol{\pi}_2 + \cdots. \tag{8.1}$$

One expects $(\mathbf{q}_1 + \mathbf{q}_2)$ and $(\boldsymbol{\pi}_1 + \boldsymbol{\pi}_2)$ to be more accurate representations of $\mathbf{q}$ and $\boldsymbol{\pi}$ than the first-order terms alone, but with the second-order terms more in the role of small corrections than substantial changes. Were the terms described as being 'second order' comparable in magnitude to the first-order terms, this would be a signal, either that there was some deviant flow pattern making the first-order terms singularly small, or that the theory was beyond its range of validity. The first circumstance prevails in magnetoplasmas, when strong magnetic fields strongly suppress cross-field, first-order transport, leaving second-order terms dominant (see 11.4.2). We shall test the importance of second-order terms in a neutral gas later in the chapter by considering the cases of ultrasonic sound waves and strong shock waves, examples with Knudsen numbers larger than usual and for which good experimental data is available.

The first-order transport equations for a neutral monatomic gas are (see §5.6)

$$\mathbf{q}_1 = -\kappa \nabla T \qquad \left(\kappa = \tfrac{5}{2} R p \tau_2\right), \tag{8.2}$$

and

$$\boldsymbol{\pi}_1 = -2\mu \overset{\circ}{\mathbf{e}} \qquad (\mu = p\tau_1), \tag{8.3}$$

where

$$\overset{\circ}{\mathbf{e}} = \tfrac{1}{2}(\nabla \mathbf{v} + \widetilde{\nabla \mathbf{v}}) \quad \tfrac{1}{3}\mathbf{1}\, \nabla \cdot \mathbf{v}.$$

In the following, the viscosity coefficient μ will be assumed to be related to the temperature by

$$\mu = p\tau_1 = AT^s \qquad (A, s \text{ constants}), \tag{8.4}$$

where τ_1 is the relaxation time described in §3.3.2 and defined in §6.8.1. In §3.3.3 the relation $\tau_2 = 3\tau_1/2$ was derived from mean-free-path considerations and in §6.7.2 it was confirmed by advanced kinetic theory. It follows that the thermal conductivity κ follows a law similar to that given in (8.4) for μ.

8.1.2 *Equations of intermediate kinetic theory*

We shall base our second-order transport theory on the intermediate kinetic theory developed in Chapter 5. A summary of the basic equations, given at the end of that chapter, is repeated below.

Correct to $\mathrm{O}(\epsilon^2)$, the kinetic equation is

$$\mathbb{D}f = \mathbb{C}(f), \tag{8.5}$$

where

$$\left.\begin{aligned}
\mathbb{D}f &= \left\{\mathrm{D}^* + \mathbf{c}\cdot\nabla - (\mathbf{P}+\mathbf{c}\cdot\mathbf{e})\cdot\frac{\partial}{\partial\mathbf{c}}\right\}f,\\
\mathbb{C}(f) &= \frac{1}{\tau}(f_0 - f) + \hat{\mathbf{P}}\cdot\frac{\partial f_0}{\partial\mathbf{c}} - \mathbf{P}\cdot\frac{\partial f}{\partial\mathbf{c}},\\
\mathbf{P} &= -\frac{1}{\rho}\nabla\cdot\mathbf{p} = -\frac{1}{\rho}\nabla p - \frac{1}{\rho}\nabla\cdot\boldsymbol{\pi},\\
\hat{\mathbf{P}} &= \mathbf{P} + \boldsymbol{\alpha}, \quad \mathbf{e} = \tfrac{1}{2}(\nabla\mathbf{v} + \widetilde{\nabla\mathbf{v}}),\\
\boldsymbol{\alpha} &= \tau_1\left\{\mathbf{c}\cdot\mathrm{D}^*\overset{\circ}{\mathbf{e}} + \mathbf{cc}:\nabla\overset{\circ}{\mathbf{e}} - \mathcal{C}^2\nabla\cdot\overset{\circ}{\mathbf{e}} - \mathbf{c}\cdot\overset{\circ}{\overline{\overset{\circ}{\mathbf{e}}\cdot\overset{\circ}{\mathbf{e}}}}\right.\\
&\quad \left. - \tfrac{2}{3}\nabla\cdot\mathbf{v}\,\mathbf{c}\cdot\overset{\circ}{\mathbf{e}}\right\} + \tfrac{1}{3}\tau_2(\mathbf{cc} - \tfrac{5}{2}\mathcal{C}^2\mathbf{1})\cdot\nabla\nabla\cdot\mathbf{v},\\
\mathrm{D}^*\boldsymbol{\Phi}_n &= \mathrm{D}\boldsymbol{\Phi}_n - n\boldsymbol{\Omega}\times\boldsymbol{\Phi}_n \qquad (n = 0, 1),\\
\mathrm{D}^*\boldsymbol{\Phi}_2 &= \mathrm{D}\boldsymbol{\Phi}_2 - \boldsymbol{\Omega}\times\boldsymbol{\Phi}_2 + \boldsymbol{\Phi}_2\times\boldsymbol{\Omega},\\
\mathrm{D} &= \frac{\partial}{\partial t} + \mathbf{v}\cdot\nabla, \quad \boldsymbol{\Omega} = \tfrac{1}{2}\nabla\times\mathbf{v}, \quad \mathcal{C}^2 = 2kT/m,\\
\tau &= \tau_2 \qquad \text{(energy transport)},\\
& \qquad\qquad\qquad\qquad\qquad (\tau_2 = \tfrac{3}{2}\tau_1)\\
\tau &= \tau_1 \qquad \text{(momentum transport)},
\end{aligned}\right\} \tag{8.6}$$

$\boldsymbol{\Phi}_n$ is a tensor of order n, it is assumed that there is no body force, and the BGK approximation $\bar{f} = f_0$ has been adopted.

We shall express the kinetic equation in terms of the relative distribution function,

$$\varphi \equiv (f - f_0)/f_0, \qquad \varphi = \varphi_1 + \varphi_2 + \cdots. \tag{8.7}$$

The kinetic equation (8.5) takes the form

$$\varphi = -\tau f_0^{-1}\mathbb{D}[f_0(1+\varphi)] + \tau\hat{\mathbf{P}}\cdot\frac{\partial\ln f_0}{\partial\mathbf{c}} - \tau f_0^{-1}\mathbf{P}\cdot\frac{\partial}{\partial\mathbf{c}}[f_0(1+\varphi)],$$

or since $\partial\ln f_0/\partial\mathbf{c} = -2\mathbf{c}/\mathcal{C}^2$,

$$\varphi = -\tau f_0^{-1}\mathbb{D}[f_0(1+\varphi)] - 2\tau\mathcal{C}^{-2}\mathbf{c}\cdot\boldsymbol{\alpha} - \tau f_0^{-1}\mathbf{P}\cdot\frac{\partial}{\partial\mathbf{c}}(f_0\varphi), \tag{8.8}$$

which is correct only up to $\mathrm{O}(\epsilon^2)$ terms.

To develop (8.8) into a Chapman–Enskog series (powers of ϵ), it is necessary to expand both φ and the streaming derivative $\mathbb{D}$. The reason for including $\mathbb{D}$ is explained in §7.5.2.* Then, as $\tau\mathbb{D}$ is $O(\epsilon)$, (8.8) yields the first- and second-order kinetic equations:

$$\varphi_1 = -\tau\mathbb{D}_0 \ln f_0, \tag{8.9}$$

$$\varphi_2 = -\tau f_0^{-1}\mathbb{D}_0(f_0\varphi_1) - \tau\mathbb{D}_1 \ln f_0 - 2\tau\mathcal{C}^{-2}\mathbf{c}\cdot\boldsymbol{\alpha} - \tau f_0^{-1}\mathbf{P}\cdot\frac{\partial}{\partial\mathbf{c}}(f_0\varphi_1). \tag{8.10}$$

The BGK kinetic equation lacks the terms containing $\boldsymbol{\alpha}$ and $\mathbf{P}$ on the right-hand side of (8.10). It also has a streaming derivative with $\mathbf{c}\cdot\nabla\mathbf{v}$ in place of $\mathbf{c}\cdot\mathbf{e}$ (see §5.7.3).

On evaluating the derivative in (8.9), we obtain the first-order (relative) distribution function (cf. (5.47) and (8.18) below):

$$\varphi_1 = -\tau_2(\nu^2 - \tfrac{5}{2})\mathbf{c}\cdot\nabla\ln T - 2\tau_1\boldsymbol{\nu}\cdot\overset{\circ}{\mathbf{e}}\cdot\boldsymbol{\nu}. \tag{8.11}$$

8.2 The second-order distribution function

8.2.1 *Zero-order streaming derivatives*

From (7.44),

$$\left.\begin{aligned} \mathbb{D}_0 &= \mathrm{D}_0^* + \mathbf{c}\cdot\nabla + \left(\frac{1}{\rho}\nabla p - \mathbf{c}\cdot\mathbf{e}\right)\cdot\frac{\partial}{\partial\mathbf{c}} \\ \mathbb{D}_1 &= \mathrm{D}_1^* + \frac{1}{\rho}\nabla\cdot\boldsymbol{\pi}_1\cdot\frac{\partial}{\partial\mathbf{c}} \end{aligned}\right\}, \tag{8.12}$$

where we omit the asterisk on D when the operand is a scalar. From (8.4) and (8.6),

$$\tau_2 \propto \tau_1 = \mu/p \propto T^s/p,$$

hence

$$\mathbb{D}_0 \ln\tau_2 = \mathbb{D}_0\ln\tau_1 = s\mathbb{D}_0\ln T - \mathbb{D}_0\ln p. \tag{8.13}$$

By (7.42) and $p = nkT$,

$$\mathrm{D}_0\ln n = \tfrac{3}{2}\mathrm{D}_0\ln T = \tfrac{3}{5}\mathrm{D}_0\ln p = -\nabla\cdot\mathbf{v}, \tag{8.14}$$

and

$$\mathrm{D}_1\ln n = 0, \quad \mathrm{D}_1 T = -(\rho c_v)^{-1}(\boldsymbol{\pi}_1 : \nabla\mathbf{v} - \nabla\cdot\mathbf{q}_1). \tag{8.15}$$

It follows from these equations that

$$\left.\begin{aligned} \mathbb{D}\tau_2 &= \tfrac{1}{3}\tau_2(5-2s)\nabla\cdot\mathbf{v} + \tau_2(s\mathbf{c}\cdot\nabla\ln T - \mathbf{c}\cdot\nabla\ln p) \\ \mathbb{D}\tau_1 &= \tfrac{1}{3}\tau_1(5-2s)\nabla\cdot\mathbf{v} + \tau_1(s\mathbf{c}\cdot\nabla\ln T - \mathbf{c}\cdot\nabla\ln p) \end{aligned}\right\}. \tag{8.16}$$

The Maxwellian distribution,

$$f_0 = n\left(\frac{m}{2\pi kT}\right)^{3/2} e^{-c^2/\mathcal{C}^2} \qquad (\mathcal{C}^2 = 2kT/m),$$

*If the *swept* derivative (see §5.2.3) were used here, this expansion would be unnecessary. Of course the final expression is the same.

with $p = nkT$, yields

$$\ln f_0 = \ln p - \tfrac{5}{2}\ln T - c^2/\mathcal{C}^2 + \text{const.}, \tag{8.17}$$

the derivative of which is (see (7.52))

$$\mathbb{D}_0 \ln f_0 = (\nu^2 - \tfrac{5}{2})\mathbf{c}\cdot\nabla T + \boldsymbol{\nu}\cdot\mathring{\mathbf{e}}\cdot\boldsymbol{\nu} \qquad (\boldsymbol{\nu} \equiv \mathbf{c}/\mathcal{C}). \tag{8.18}$$

From (8.13) and $\mathcal{C}^2 = 2p/\rho = 2RT$,

$$\begin{aligned}\mathbb{D}_0 \ln c^2 &= (\rho^{-1}\nabla p - \mathbf{c}\cdot\mathbf{e})\cdot 2\mathbf{c}/c^2,\\ \nu^2\mathbb{D}_0 \ln c^2 &= \mathbf{c}\cdot\nabla\ln p - 2\boldsymbol{\nu}\cdot\mathbf{e}\cdot\boldsymbol{\nu},\\ \nu^2\mathbb{D}_0 \ln \nu^2 &= \mathbf{c}\cdot\nabla\ln p - 2\boldsymbol{\nu}\cdot\mathbf{e}\cdot\boldsymbol{\nu} + \tfrac{2}{3}\nu^2\nabla\cdot\mathbf{v} - \nu^2\mathbf{c}\cdot\nabla\ln T.\end{aligned}$$

Hence, as $\mathring{\mathbf{e}} = \mathbf{e} - \frac{1}{3}\mathbf{1}\,\nabla\cdot\mathbf{v}$,

$$\mathbb{D}_0(\nu^2 - \tfrac{5}{2}) = \mathbf{c}\cdot\nabla p - \nu^2\mathbf{c}\cdot\nabla\ln T - 2\boldsymbol{\nu}\cdot\mathring{\mathbf{e}}\cdot\boldsymbol{\nu}. \tag{8.19}$$

Also

$$\begin{aligned}\mathbb{D}_0\boldsymbol{\nu} = \mathbb{D}_0(\mathbf{c}/\mathcal{C}) &= -\tfrac{1}{2}\mathcal{C}^{-3}\mathbf{c}\,\mathbb{D}_0\mathcal{C}^2 + \mathcal{C}^{-1}\mathbb{D}_0\mathbf{c}\\ &= -\tfrac{1}{2}\mathcal{C}^{-1}\mathbf{c}\,\mathbb{D}_0\ln T + \mathcal{C}^{-1}(\rho^{-1}\nabla p - \mathbf{c}\cdot\mathbf{e})\\ &= \tfrac{1}{2}\mathcal{C}\nabla\ln p + \tfrac{1}{2}\mathcal{C}^{-1}(\tfrac{2}{3}\mathbf{c}\nabla\cdot\mathbf{v} - \mathbf{c}\mathbf{c}\cdot\nabla\ln T) - \boldsymbol{\nu}\cdot\mathbf{e},\end{aligned}$$

i.e.

$$\mathbb{D}_0\boldsymbol{\nu} = \tfrac{1}{2}\mathcal{C}\nabla p - \tfrac{1}{2}\boldsymbol{\nu}\mathbf{c}\cdot\nabla\ln T - \boldsymbol{\nu}\cdot\mathring{\mathbf{e}}\,. \tag{8.20}$$

From (8.6),

$$\mathbb{D}_0\nabla\ln T = \mathrm{D}_0\nabla\ln T - \boldsymbol{\Omega}\times\nabla\ln T + \mathbf{c}\cdot\nabla\nabla\ln T,$$

where

$$\begin{aligned}\mathrm{D}_0\nabla T &= \mathrm{D}_0 T^{-1}\,\nabla T + T^{-1}\mathrm{D}_0\nabla T\\ &= T^{-1}\{\tfrac{2}{3}\nabla\cdot\mathbf{v}\nabla T + \mathrm{D}_0\nabla T\}.\end{aligned}$$

Hence

$$T\mathbb{D}_0\nabla\ln T = \tfrac{2}{3}\nabla\cdot\mathbf{v}\nabla T + \big(\mathrm{D}_0\nabla T - \boldsymbol{\Omega}\times\nabla T\big) + T\mathbf{c}\cdot\nabla\nabla\ln T,$$

or

$$T\mathbb{D}_0\nabla\ln T = \tfrac{2}{3}\nabla\cdot\mathbf{v}\nabla T + \mathrm{D}_0^*\nabla T + \mathbf{c}\cdot\nabla\nabla T - T^{-1}\mathbf{c}\cdot\nabla T\nabla T. \tag{8.21}$$

8.2.2 *First-order streaming derivatives*

From (8.12), (8.15), and (8.17) we have

$$\begin{aligned}\mathbb{D}_1\ln f_0 &= \mathbb{D}_1 n + (\nu^2 - \tfrac{3}{2})\mathbb{D}_1\ln T - 2\mathcal{C}^{-2}\mathbf{c}\cdot\mathbb{D}_1\mathbf{c}\\ &= \frac{1}{\rho c_v T}(\nu^2 - \tfrac{3}{2})\big(-\boldsymbol{\pi}_1 : \nabla\mathbf{v} - \nabla\cdot\mathbf{q}_1\big) - \frac{2}{\rho\mathcal{C}^2}\nabla\cdot\boldsymbol{\pi}_1\cdot\mathbf{c},\end{aligned}$$

or, since $\rho c_v T = 3p/2$, and $\rho\mathcal{C}^2 = p/2$,

$$\mathbb{D}_1\ln f_0 = -\frac{1}{p}\Big\{(\tfrac{2}{3}\nu^2 - 1)(\boldsymbol{\pi}_1 : \nabla\mathbf{v} + \nabla\cdot\mathbf{q}_1) + \nabla\cdot\boldsymbol{\pi}_1\cdot\mathbf{c}\Big\}. \tag{8.22}$$

Writing (8.2) and (8.3) as

$$\mathbf{q}_1 = -\tfrac{5}{2}R\tau_2 p\nabla T, \qquad \boldsymbol{\pi}_1 = -2\tau_1 p\,\overset{\circ}{\mathbf{e}},$$

we find that (cf. (8.13))

$$\begin{aligned}\nabla\cdot\mathbf{q}_1 &= -\tfrac{5}{2}R\{\nabla(\tau_2 p)\cdot\nabla T + \tau_2 p\nabla^2 T\}\\ &= -\tfrac{5}{2}R\tau_2 p\{s\nabla T\cdot\nabla\ln T + \nabla^2 T\},\end{aligned}$$

and similarly $\quad \nabla\cdot\boldsymbol{\pi}_1 = -2\tau_1 p\{s\nabla\ln T\cdot\overset{\circ}{\mathbf{e}} + \nabla\cdot\overset{\circ}{\mathbf{e}}\}.$

Therefore (8.22) becomes

$$\begin{aligned}\mathbb{D}_1\ln f_0 &= (\tfrac{2}{3}\nu^2 - 1)\big[2\tau_1\,\overset{\circ}{\mathbf{e}}:\nabla\mathbf{v} + \tfrac{5}{2}R\tau_2(s\nabla T\cdot\nabla\ln T + \nabla^2 T)\big]\\ &\quad +2\tau_1\big[s\nabla\ln T\cdot\overset{\circ}{\mathbf{e}}\cdot\mathbf{c} + \nabla\cdot\overset{\circ}{\mathbf{e}}\cdot\mathbf{c}\big].\end{aligned} \tag{8.23}$$

8.2.3 *The 'Burnett' terms*

Let

$$\varphi_2^{(B)} \equiv -\tau f_0^{-1}\mathbb{D}_0(f_0\varphi_1) - \tau\mathbb{D}_1\ln f_0, \tag{8.24}$$

then, excepting the change in $\mathbb{D}_0$ noted at the end of §8.1.2, these are the terms that would result from taking the BGK kinetic equation to second order. Burnett calculated similar terms for Boltzmann's kinetic equation, obtaining coefficients that depend on the molecular force law. But the issue here is not the values of the coefficients, but the type of derivatives that appear, so for want of a more precise description, we are calling them the 'Burnett' terms. By (8.9),

$$\varphi_2^{(B)} = \varphi_1^2 - \tau\mathbb{D}_0\varphi_1 - \tau\mathbb{D}_1\ln f_0. \tag{8.25}$$

We have expressions for φ_1 and $\mathbb{D}_1\ln f_0$ in (8.11) and (8.23). It remains to calculate $\mathbb{D}_0\varphi_1$.

From (8.11),

$$\begin{aligned}-\mathbb{D}_0\varphi_1 &= \mathbb{D}_0\tau_2\,(\nu^2 - \tfrac{5}{2})\mathbf{c}\cdot\nabla\ln T + 2\mathbb{D}_0\tau_1\,\boldsymbol{\nu}\cdot\overset{\circ}{\mathbf{e}}\cdot\boldsymbol{\nu}\\ &\quad +\tau_2\mathbb{D}_0\{(\nu^2 - \tfrac{5}{2})\mathbf{c}\}\cdot\nabla\ln T + 4\tau_1\mathbb{D}_0\boldsymbol{\nu}\cdot\overset{\circ}{\mathbf{e}}\cdot\boldsymbol{\nu}\\ &\quad +\tau_2(\nu^2 - \tfrac{5}{2})\mathbf{c}\cdot\mathbb{D}_0\nabla\ln T + 2\tau_1\boldsymbol{\nu}\cdot\mathbb{D}_0\overset{\circ}{\mathbf{e}}\cdot\boldsymbol{\nu},\end{aligned}$$

where we have used the symmetry of $\overset{\circ}{\mathbf{e}}$ to write $\mathbb{D}_0\boldsymbol{\nu}\cdot\overset{\circ}{\mathbf{e}}\cdot\boldsymbol{\nu} = \boldsymbol{\nu}\cdot\overset{\circ}{\mathbf{e}}\cdot\mathbb{D}_0\boldsymbol{\nu}$. Expressions for the various derivatives occurring in (8.25) have been derived in §§8.2.1 and 8.2.2. After some straightforward algebra, we arrive at

$$
\begin{aligned}
\varphi_2^{(B)} &= \tau\Big\{\tfrac{2}{3}\tau_2(3-s)(\nu^2-\tfrac{5}{2})\nabla\cdot\mathbf{v}\,\mathbf{c}\cdot\nabla\ln T \\
&\quad +\tau_2(\nu^2-\tfrac{5}{2})T^{-1}\mathbf{c}\cdot(\mathrm{D}^*\nabla T-\overset{\circ}{\mathbf{e}}\cdot\nabla T) \\
&\quad +2\tau_1\nabla\ln p\cdot\overset{\circ}{\mathbf{e}}\cdot\mathbf{c}-2\tau_1\nabla\ln p\cdot\mathbf{c}\boldsymbol{\nu}\boldsymbol{\nu}:\overset{\circ}{\mathbf{e}} \\
&\quad +2\big[(\tau_2+\tau_1)(\nu^2-\tfrac{7}{2})+\tau_1 s\big]\nabla\ln T\cdot\mathbf{c}\boldsymbol{\nu}\boldsymbol{\nu}:\overset{\circ}{\mathbf{e}} \\
&\quad -2\tau_1 s\nabla\ln T\cdot\overset{\circ}{\mathbf{e}}\cdot\mathbf{c}+2\tau_1(\mathbf{c}\boldsymbol{\nu}\boldsymbol{\nu}\,\vdots\,\nabla\overset{\circ}{\mathbf{e}}-\nabla\cdot\overset{\circ}{\mathbf{e}}\cdot\mathbf{c})\Big\} \\
&\quad +\tau\Big\{\tfrac{4}{3}\tau_1(3-s)\nabla\cdot\mathbf{v}\,\boldsymbol{\nu}\boldsymbol{\nu}:\overset{\circ}{\mathbf{e}}+2\tau_1\boldsymbol{\nu}\cdot(\mathrm{D}^*\overset{\circ}{\mathbf{e}}-2\overset{\circ}{\mathbf{e}}\cdot\overset{\circ}{\mathbf{e}})\cdot\boldsymbol{\nu} \\
&\quad +2R\tau_1\big[(\nu^2-\tfrac{5}{2})\boldsymbol{\nu}\cdot\nabla\nabla T\cdot\boldsymbol{\nu}-\tfrac{5}{4}(\tfrac{2}{3}\nu^2-1)\nabla^2 T\big] \\
&\quad +\tfrac{1}{2}\tau_2(\nu^2-\tfrac{5}{2})\mathcal{C}^2\nabla\ln T\cdot\nabla\ln p \\
&\quad -\tau_2(\nu^2-\tfrac{7}{2})\mathbf{c}\mathbf{c}:\nabla\ln p\nabla\ln T-\tfrac{5}{2}\tau_2 Rs(1-\tfrac{2}{3}\nu^2)\nabla T\cdot\nabla\ln T \\
&\quad +2R\tau_2\big[(\nu^2-\tfrac{5}{2})(\nu^2-\tfrac{7}{2}+s)-\nu^2\big]\nabla T\cdot\boldsymbol{\nu}\boldsymbol{\nu}\cdot\nabla\ln T \\
&\quad +4\tau_1\,\overset{\circ}{\mathbf{e}}:\boldsymbol{\nu}\boldsymbol{\nu}\boldsymbol{\nu}\boldsymbol{\nu}:\overset{\circ}{\mathbf{e}}-2\tau_1(\tfrac{2}{3}\nu^2-1)\,\overset{\circ}{\mathbf{e}}:\overset{\circ}{\mathbf{e}}\Big\},
\end{aligned}
\tag{8.26}
$$

where we have omitted the subscript zero from D^*, since the distinction will not be required in the rest of this chapter.

We have arranged the terms into two groups. The first group contributes only to the energy transport, and the second only to the momentum transport. Therefore the 'τ' outside the first bracket is τ_2 and that outside the second is τ_1, but for the present we shall omit the subscripts so that the cancellations that we shall later describe will be seen to be independent of these allocations.

8.2.4 *The pressure terms*

Next consider the last term in (8.10),

$$
\varphi_2^{(P)} \equiv -\tau f_0^{-1}\mathbf{P}\cdot\frac{\partial}{\partial\mathbf{c}}(f_0\varphi_1), \tag{8.27}
$$

where to sufficient accuracy $\mathbf{P}=-\nabla p/\rho$. By (8.11) and (8.17),

$$
\begin{aligned}
\varphi_2^{(P)} &= \tau f_0^{-1}\mathbf{P}\cdot\frac{\partial}{\partial\mathbf{c}}\{f_0\tau_2(\nu^2-\tfrac{5}{2})\mathbf{c}\cdot\nabla\ln T+f_0 2\tau_1\boldsymbol{\nu}\cdot\overset{\circ}{\mathbf{e}}\cdot\boldsymbol{\nu}\} \\
&= \tau\mathbf{P}\cdot\frac{\partial}{\partial\mathbf{c}}\{\tau_2(\nu^2-\tfrac{5}{2})\mathbf{c}\cdot\nabla\ln T+2\tau_1\boldsymbol{\nu}\cdot\overset{\circ}{\mathbf{e}}\cdot\boldsymbol{\nu}\} \\
&\quad -2\tau\mathcal{C}^{-2}\mathbf{c}\cdot\mathbf{P}\{\tau_2(\nu^2-\tfrac{5}{2})\mathbf{c}\cdot\nabla\ln T+2\tau_1\boldsymbol{\nu}\cdot\overset{\circ}{\mathbf{e}}\cdot\boldsymbol{\nu}\} \\
&= \tau\mathbf{P}\cdot\{2\tau_2\boldsymbol{\nu}\boldsymbol{\nu}\cdot\nabla\ln T+\tau_2(\nu^2-\tfrac{5}{2})\nabla\ln T+4\mathcal{C}^{-1}\tau_1\,\overset{\circ}{\mathbf{e}}\cdot\boldsymbol{\nu}\} \\
&\quad +\tau\,\nabla\ln p\{\tau_2(\nu^2-\tfrac{5}{2})\mathbf{c}\cdot\nabla\ln T+2\tau_1\boldsymbol{\nu}\boldsymbol{\nu}\cdot\overset{\circ}{\mathbf{e}}\},
\end{aligned}
$$

where in the final step we used

$$\frac{\partial \nu^2}{\partial \mathbf{c}} = \frac{1}{\mathcal{C}^2}\frac{\partial}{\partial \mathbf{c}}(\mathbf{c}\cdot\mathbf{c}) = \frac{2\mathbf{c}}{\mathcal{C}^2}, \qquad \frac{\partial \mathbf{c}}{\partial \mathbf{c}} = \mathbf{1}, \qquad \frac{\partial \boldsymbol{\nu}}{\partial \mathbf{c}} = \frac{1}{\mathcal{C}}\mathbf{1}.$$

With $\mathcal{C}^2 = 2p/\rho$, we can write the expression for $\varphi_2^{(P)}$ as

$$\begin{aligned}\varphi_2^{(P)} &= -\tau\{2\tau_1\nabla\ln p\cdot \overset{\circ}{\mathbf{e}}\cdot\mathbf{c} - 2\tau_1\nabla\ln p\cdot\mathbf{c}\boldsymbol{\nu}\boldsymbol{\nu}:\overset{\circ}{\mathbf{e}}\}\\ &\quad -\tau\{\tfrac{1}{2}\tau_2(\nu^2-\tfrac{5}{2})\mathcal{C}^2\nabla\ln T\cdot\nabla\ln p - \tau_2(\nu^2-\tfrac{7}{2})\mathbf{c}\mathbf{c}:\nabla\ln p\,\nabla\ln T\}.\end{aligned} \tag{8.28}$$

Notice that every term in (8.26) that depends on the pressure gradient is repeated in (8.28), but with the opposite sign. Thus there is no term containing ∇p in $(\varphi_2^{(B)} + \varphi_2^{(P)})$. The physical reason why this should be so will be explained later.

8.2.5 *The acceleration terms*

The term in (8.10) remaining to be discussed is

$$\varphi_2^{(\alpha)} \equiv -2\tau\mathbf{c}\cdot\boldsymbol{\alpha}/\mathcal{C}^2. \tag{8.29}$$

First we shall derive a more convenient expression for the term in $\boldsymbol{\alpha}$ containing $\nabla\nabla\cdot\mathbf{v}$.

By (8.14),

$$-\tfrac{2}{3}\mathbf{c}\cdot\nabla\nabla\cdot\mathbf{v} = \mathbf{c}\cdot\nabla D\ln T.$$

Now

$$\nabla \mathrm{D}\ln T = -T^{-2}\nabla T\mathrm{D}T + T^{-1}\nabla\mathrm{D}T,$$

so

$$\begin{aligned}\nabla\mathrm{D}T &= \mathrm{D}\nabla T + \nabla\mathbf{v}\cdot\nabla T = \mathrm{D}\nabla T - \boldsymbol{\Omega}\times\nabla T + \mathbf{e}\cdot\nabla T\\ &= \mathrm{D}^*\nabla T + \overset{\circ}{\mathbf{e}}\cdot\nabla T + \tfrac{1}{3}\nabla\cdot\mathbf{v}\,\nabla T.\end{aligned}$$

Hence

$$T\nabla\mathrm{D}\ln T = \mathrm{D}^*\nabla T + \overset{\circ}{\mathbf{e}}\cdot\nabla T + \nabla\cdot\mathbf{v}\nabla T,$$

so that

$$-\tfrac{2}{3}T\nabla\nabla\cdot\mathbf{v} = T\nabla\mathrm{D}\ln T = \mathrm{D}^*\nabla T + \overset{\circ}{\mathbf{e}}\cdot\nabla T + \nabla\cdot\mathbf{v}\,\nabla T. \tag{8.30}$$

It follows from (8.6), (8.29), and (8.30) that

$$\begin{aligned}\varphi_2^{(\alpha)} &= \tau_1\tau_2\Big\{-2\mathbf{c}\boldsymbol{\nu}\boldsymbol{\nu}\,\vdots\,\nabla\overset{\circ}{\mathbf{e}} + 2\nabla\cdot\overset{\circ}{\mathbf{e}}\cdot\mathbf{c}\Big\}\\ &\quad +\tau_2{}^2T^{-1}(\nu^2-\tfrac{5}{2})\mathbf{c}\cdot\big(\mathrm{D}^*\nabla T + \overset{\circ}{\mathbf{e}}\cdot\nabla T + \nabla\cdot\mathbf{v}\nabla T\big)\\ &\quad +2\tau_1{}^2\Big\{-\boldsymbol{\nu}\cdot\mathrm{D}^*\overset{\circ}{\mathbf{e}}\cdot\boldsymbol{\nu} + \boldsymbol{\nu}\cdot\overset{\circ}{\overline{\overset{\circ}{\mathbf{e}}\cdot\overset{\circ}{\mathbf{e}}}}\cdot\boldsymbol{\nu} + \tfrac{2}{3}\nabla\cdot\mathbf{v}\,\boldsymbol{\nu}\boldsymbol{\nu}:\overset{\circ}{\mathbf{e}}\Big\},\end{aligned} \tag{8.31}$$

where the terms contributing to energy transport have attracted $\tau = \tau_2$ and those contributing to momentum transport, $\tau = \tau_1$.

We note that $\varphi_2^{(B)}$ contains the following terms:

$$2\tau_1^2\boldsymbol{\nu}\cdot\mathrm{D}^*\overset{\circ}{\mathbf{e}}\cdot\boldsymbol{\nu} + 2\tau_1\tau_2\mathbf{c}\boldsymbol{\nu}\boldsymbol{\nu}\,\vdots\,\nabla\overset{\circ}{\mathbf{e}} + \tau_1^2\tfrac{4}{3}(3-s)\nabla\cdot\mathbf{v}\,\boldsymbol{\nu}\boldsymbol{\nu}:\overset{\circ}{\mathbf{e}},$$

similar to those occurring in $\varphi_2^{(\alpha)}$, so cancellations will occur when the components of φ_2 are added (see remark at the end of §5.10.1).

8.2.6 *Complete expression for* φ_2

From (8.10), (8.24), (8.27), and (8.29),

$$\varphi_2 = \varphi_2^{(B)} + \varphi_2^{(P)} + \varphi_2^{(\alpha)}.$$

Therefore, on adding (8.26), (8.28), and (8.31), we arrive at

$$\left.\begin{aligned}
\varphi_2 = {} & \tau_2\Big\{\tau_2(3 - \tfrac{2}{3}s)(\nu^2 - \tfrac{5}{2})\nabla\cdot\mathbf{v}\,\mathbf{c}\cdot\nabla\ln T \\
& +2\tau_2(\nu^2 - \tfrac{5}{2})T^{-1}\mathbf{c}\cdot\mathrm{D}^*\nabla T - 2\tau_1 s\nabla\ln T\cdot\overset{\circ}{\mathbf{e}}\cdot\mathbf{c} \\
& +2\big[(\tau_1 + \tau_2)(\nu^2 - \tfrac{7}{2}) + \tau_1 s\big]\nabla\ln T\cdot\mathbf{c}\boldsymbol{\nu}\boldsymbol{\nu}:\overset{\circ}{\mathbf{e}}\Big\} \\
& +\tau_1\Big\{\tfrac{4}{3}\tau_1(4 - s)\nabla\cdot\mathbf{v}\,\boldsymbol{\nu}\boldsymbol{\nu}:\overset{\circ}{\mathbf{e}} - 2\tau_1\boldsymbol{\nu}\cdot\overset{\circ}{\overline{\overset{\circ}{\mathbf{e}}\cdot\overset{\circ}{\mathbf{e}}}}\cdot\boldsymbol{\nu} \\
& +2R\tau_1\big[(\nu^2 - \tfrac{5}{2})\boldsymbol{\nu}\cdot\nabla\nabla T\cdot\boldsymbol{\nu} - \tfrac{5}{4}(\tfrac{2}{3}\nu^2 - 1)\nabla^2 T\big] \\
& -\tfrac{5}{2}R\tau_2 s(1 - \tfrac{2}{3}\nu^2)\nabla T\cdot\nabla\ln T \\
& +2R\tau_2\big[(\nu^2 - \tfrac{5}{2})(\nu^2 - \tfrac{7}{2} + s) - \nu^2\big]\nabla T\cdot\boldsymbol{\nu}\boldsymbol{\nu}\cdot\nabla T \\
& +4\tau_1\,\overset{\circ}{\mathbf{e}}:\boldsymbol{\nu}\boldsymbol{\nu}\boldsymbol{\nu}\boldsymbol{\nu}:\overset{\circ}{\mathbf{e}} - 2\tau_1(\tfrac{4}{3}\nu^2 - 1)\,\overset{\circ}{\mathbf{e}}:\overset{\circ}{\mathbf{e}}\Big\}.
\end{aligned}\right\} \quad (8.32)$$

By (7.45) and (7.88) this function is required to satisfy

$$0 = \int \varphi_2 f_0 \boldsymbol{\psi}\, d\mathbf{c} = \frac{4n}{\sqrt{\pi}}\int_\Omega\int_0^\infty \varphi_2\boldsymbol{\psi}\,\nu^2 e^{-\nu^2}\,d\nu\,d\Omega, \quad (8.33)$$

where $\boldsymbol{\psi} = (1, \mathbf{c}, \frac{1}{2}c^2)$. For the odd functions of $\mathbf{c}$ in φ_2, only the second component of $\boldsymbol{\psi}$ needs to be considered. We find that the factors $(\nu^2 - \frac{5}{2})$ and $(\nu^2 - \frac{7}{2})$ give zeros for the terms containing them. The terms in $\tau_1 s$ require both integrals to be evaluated (use the integrals in §§2.6.2 and 5.6.1). For the even functions of $\mathbf{c}$ in φ_2, the integral over Ω removes the terms containing deviators, the integral over ν dispatches all the terms containing the temperature, while the terms containing $\overset{\circ}{\mathbf{e}}$ require both the Ω and ν integrations in order to vanish. The conclusion is that φ_2 satisfies (8.33).

8.3 Second-order transport

8.3.1 *The second-order heat flux vector*

From

$$\mathbf{q}_2 = \tfrac{1}{2}m\int c^2\mathbf{c}f_2\,d\mathbf{c} = \tfrac{1}{2}m\int c^2\mathbf{c}\varphi_2 f_0\,d\mathbf{c},$$

(7.88), and $2\mathcal{C}^2 = p/\rho$, we find

$$\mathbf{q}_2 = p\mathcal{C}\frac{4}{\sqrt{\pi}}\int_\Omega\int_0^\infty \varphi_2\nu^5 e^{-\nu^2}\,d\nu\,\hat{\mathbf{c}}\,d\Omega. \tag{8.34}$$

It follows from the first term of (8.32), (8.34) and the integrals in §§2.6.2 and 5.6.1 that

$$\mathbf{q}_2 = Rp\tau_m^2\Big\{\theta_1\nabla\!\cdot\mathbf{v}\,\nabla T + \theta_2(\mathrm{D}\nabla T - \boldsymbol{\Omega}\times\nabla T) + 3\theta_5\nabla T\cdot\overset{\circ}{\mathbf{e}}\Big\}, \tag{8.35}$$

where $\qquad \theta_1 = \tfrac{15}{4}(\tfrac{9}{2} - s), \qquad \theta_2 = \tfrac{45}{4}, \qquad \theta_5 = \tfrac{35}{4} + s. \qquad (8.36)$

The missing coefficients θ_3 and θ_4 apply to Burnett terms that are cancelled in intermediate kinetic theory. This cancellation is an important result, to which we shall return in §8.5.

8.3.2 *The second-order viscous stress tensor*

From (7.88) and

$$\boldsymbol{\pi}_2 = m\int \mathbf{c}\mathbf{c}f_2\,d\mathbf{c} = m\int \mathbf{c}\mathbf{c}\varphi_2 f_0\,d\mathbf{c},$$

we get

$$\boldsymbol{\pi}_2 = 2p\frac{4}{\sqrt{\pi}}\int_\Omega\int_0^\infty \varphi_2\nu^4 e^{-\nu^2}\,d\nu\,\hat{\mathbf{c}}\hat{\mathbf{c}}\,d\Omega. \tag{8.37}$$

With (8.32) substituted into (8.37), of the terms contributing to $\boldsymbol{\pi}_2$ (those even in $\mathbf{c}$), only

$$\begin{aligned}\mathcal{I} &\equiv 8p\tau_1\frac{4}{\sqrt{\pi}}\int_\Omega\int_0^\infty \overset{\circ}{\mathbf{e}} : \hat{\mathbf{c}}\hat{\mathbf{c}}\hat{\mathbf{c}}\hat{\mathbf{c}}\hat{\mathbf{c}}\hat{\mathbf{c}} : \overset{\circ}{\mathbf{e}}\,\nu^8 e^{-\nu^2}\,d\nu\,d\Omega \\ &= 8p\tau_1\,\tfrac{105}{8}\int_\Omega \overset{\circ}{\mathbf{e}} : \hat{\mathbf{c}}\hat{\mathbf{c}}\hat{\mathbf{c}}\hat{\mathbf{c}}\hat{\mathbf{c}}\hat{\mathbf{c}} : \overset{\circ}{\mathbf{e}}\;d\Omega \end{aligned} \tag{8.38}$$

offers any difficulty.

Let

$$\mathbf{J}_4 = \textstyle\int \hat{\mathbf{c}}\hat{\mathbf{c}}\hat{\mathbf{c}}\hat{\mathbf{c}}\,d\Omega, \qquad \mathbf{J}_6 = \int \hat{\mathbf{c}}\hat{\mathbf{c}}\hat{\mathbf{c}}\hat{\mathbf{c}}\hat{\mathbf{c}}\hat{\mathbf{c}}\,d\Omega$$

$$\overset{\circ}{\mathbf{J}}_4 = \textstyle\int \overset{\circ}{\hat{\mathbf{c}}\hat{\mathbf{c}}}\overset{\circ}{\hat{\mathbf{c}}\hat{\mathbf{c}}}\,d\Omega, \qquad \overset{\circ}{\mathbf{J}}_6 = \int \overset{\circ}{\hat{\mathbf{c}}\hat{\mathbf{c}}}\overset{\circ}{\hat{\mathbf{c}}\hat{\mathbf{c}}}\overset{\circ}{\hat{\mathbf{c}}\hat{\mathbf{c}}}\,d\Omega,$$

then by (1.93),

$$\overset{\circ}{\mathbf{e}} \vdots \mathbf{J}_6 \vdots \overset{\circ}{\mathbf{e}} = \overset{\circ}{\mathbf{e}} : \int \overset{\circ}{\hat{\mathbf{c}}\hat{\mathbf{c}}}\;\hat{\mathbf{c}}\hat{\mathbf{c}}\;\overset{\circ}{\hat{\mathbf{c}}\hat{\mathbf{c}}}\,d\Omega : \overset{\circ}{\mathbf{e}} - \overset{\circ}{\mathbf{e}} : \overset{\circ}{\mathbf{J}}_6 : \overset{\circ}{\mathbf{e}} + \tfrac{1}{3}\mathbf{1}\;\overset{\circ}{\mathbf{e}} : \overset{\circ}{\mathbf{J}}_4 : \overset{\circ}{\mathbf{e}}$$

i.e.

$$\overset{\circ}{\mathbf{e}} : \mathbf{J}_6 : \overset{\circ}{\mathbf{e}} = \overset{\circ}{\mathbf{e}} : \overset{\circ}{\mathbf{J}}_6 : \overset{\circ}{\mathbf{e}} + \tfrac{1}{3}\mathbf{1}\;\overset{\circ}{\mathbf{e}} : \mathbf{J}_4 : \overset{\circ}{\mathbf{e}}\,. \tag{8.39}$$

Now $\overset{\circ}{\mathbf{e}} : \overset{\circ}{\mathbf{J}}_6 : \overset{\circ}{\mathbf{e}}$ is a deviator and hence is of the form $a\,\overset{\circ}{\overline{\overset{\circ}{\mathbf{e}}\cdot\overset{\circ}{\mathbf{e}}}}$, where a is a constant. To determine a we adopt the coordinates used in §2.6.2 and in Mathematical note 5.6 and take the **kkkkkk** component of (8.39). Now

$$\mathbf{kkkkkk} \vdots\vdots\, \mathbf{J}_6 = \frac{1}{4\pi}\int_0^{2\pi}\int_0^\pi \cos^6\theta\,\sin\theta\,d\theta\,d\phi = \tfrac{1}{7}$$

and $$\mathbf{kkkkkk} \vdots \vdots \tfrac{1}{3}\mathbf{1}\,\mathsf{J}_4 = \tfrac{1}{3} \times \tfrac{1}{5}.$$

Hence
$$a = \mathbf{kkkkkk} \vdots \vdots \,\overset{\circ}{\mathsf{J}}_6 = \tfrac{1}{7} - \tfrac{1}{15} = \tfrac{8}{105}.$$

The integral in (8.38) is therefore

$$\mathcal{I} = 8p\tau_1 \overline{\overset{\circ}{\mathbf{e}} \cdot \overset{\circ}{\mathbf{e}}}^{\circ} + \tfrac{14}{3}p\tau_1 \mathbf{1}\, \overset{\circ}{\mathbf{e}} : \overset{\circ}{\mathbf{e}}\,.$$

The last term in $\mathcal{I}$ is cancelled by the integral of the final term in (8.32).

From the result just obtained and the integrals in §§2.6.2 and 5.6.1 we find that the second-order viscosity tensor is given by

$$\boldsymbol{\pi}_2 = p\tau_1^2 \Big\{ \tilde{\omega}_1 \nabla\cdot\mathbf{v}\ \overset{\circ}{\mathbf{e}} + \tilde{\omega}_3 R \overline{\nabla\nabla T}^{\circ} + \tilde{\omega}_5 R T^{-1} \overline{\nabla T\, \nabla T}^{\circ} + \tilde{\omega}_6\, \overline{\overset{\circ}{\mathbf{e}} \cdot \overset{\circ}{\mathbf{e}}}^{\circ} \Big\} \tag{8.40}$$

where $$\tilde{\omega}_1 = \tfrac{4}{3}(\tfrac{9}{2} - s), \qquad \tilde{\omega}_3 = 3, \qquad \tilde{\omega}_5 = 3s, \qquad \tilde{\omega}_6 = 6. \tag{8.41}$$

The missing coefficients $\tilde{\omega}_2$ and $\tilde{\omega}_4$ occur in the Burnett formula for $\boldsymbol{\pi}_2$ (see §8.5.1).

8.3.3 *The reversibility of second-order terms*

The standard method of checking whether equations are irreversible is to determine if they are changed under the transformations $t \to -t$, and $\mathbf{v} \to -\mathbf{v}$. Thus of the two differential equations

$$\frac{\partial^2 u}{c^2 \partial t^2} = \nabla^2 u, \qquad \frac{1}{\kappa}\frac{\partial T}{\partial t} = \nabla^2 T,$$

the transformation alters only the second, which therefore describes an irreversible process. This method cannot be applied to constitutive equations like Fourier's law,
$$\mathbf{q} = -\kappa \nabla T,$$
which has no explicit time-dependence. In such cases, we need the concept of 'motioned-reversed' parity, η.

To find the microscopic or *ideal* parity $\eta(\xi)$ of any property ξ, we reverse the particle motions and use the kinetic definition of ξ. This change will transform ξ into either ξ or $-\xi$; if the first, we write $\eta(\xi) = 1$, and if the second, $\eta(\xi) = -1$. For example, if ξ is the heat flux $\mathbf{q}$, reversing particle motions changes the sign of $\mathbf{q}$, so $\eta(\mathbf{q}) = -1$. Temperature is a quadratic function of the particle speeds, therefore $\eta(\nabla T) = 1$. As κ is a constant, we see that $\eta(\kappa\nabla T)$ has positive parity. Therefore Fourier's law is a relation between properties having *opposite* parities, and this identifies it as being an irreversible process.

Applying the above principle to equations (8.35) and (8.40), we conclude that, unlike the first-order terms in $\mathbf{q}$ and $\boldsymbol{\pi}$, the second-order terms are reversible. For example, this means that while second-order terms may modify wave speeds, they cannot affect the damping of waves.

8.4 The physical principles in second-order transport

The physical bases of the terms in equations (8.35) and (8.40) for the second-order heat flux and second-order viscosity tensor are obscure. And the more complicated expressions resulting from Burnett's development of the Chapman–Enskog theory—to be discussed in the next section—are even more arcane. As remarked by Ferziger and Kaper (1970, p. 150), Burnett's equations are seldom used because they have no direct physical derivation and moreover there is uncertainty about the appropriate boundary conditions to employ.

The explanations we shall give below of the terms could be re-phrased into a derivation of the second-order transport equations, but this method suffers from the defect that there may be *other* terms not revealed by *ad hoc* physical arguments.

8.4.1 *The heat flux vector*

As expressed in (8.35), it is not obvious what physical significance can be attached to the three terms in $\mathbf{q}_2$. Another form, more convenient for this purpose, is obtained as follows. From

$$\mathbf{q}_1 = -\kappa \nabla T, \qquad \kappa = \tfrac{15}{4} R p \tau_1,$$

and (8.4), we find that $\mathbf{q}_2$ can be expressed in terms of $\mathbf{q}_1$ thus:

$$\mathbf{q}_2 = -(3 + \tfrac{2}{3}s)\tau_2 \nabla \cdot \mathbf{v}\, \mathbf{q}_1 - (\tfrac{14}{3} + \tfrac{8}{15}s)\tau_2 \overset{\circ}{\mathbf{e}} \cdot \mathbf{q}_1 - 2\tau_2(\mathrm{D}\mathbf{q}_1 - \Omega \times \mathbf{q}_1). \quad (8.42)$$

The terms in $\mathbf{q}_2$ are thus exhibited as modifications to the first-order heat flux vector resulting from dilatation, rate of fluid strain, and convection.

First we note that $(\mathrm{D}\mathbf{q}_1 - \boldsymbol{\Omega} \times \mathbf{q}_1)$ is the rate of change of $\mathbf{q}_1$ in the convected (and spinning) frame P_c. Hence, taken together, the term $\mathbf{q}_1$ plus the last term in (8.42) admit the following interpretation. To (partially) correct the present value $\mathbf{q}_1(t)$ for second-order effects, we replace it by the value $\mathbf{q}_1(t - 2\tau_2)$ that it had $2\tau_2$ seconds earlier, as measured in the frame P_c. As the vector $\mathbf{q}_1$ is swept along with the fluid, it takes a time $2\tau_2$ to adjust to ambient conditions. At a microscopic level of description, it is actually the convection of a set of mean-free-path vectors—or rather the ends of these vectors—that causes the effect.

Strictly the relaxation time is τ_2 seconds, with the factor '2' arising because the second-order terms depend on higher powers of the particle speed c than the first-order terms (see (8.32)). In other words, it is the effective mean-free-path that is larger, with the relaxation time unchanged. The same remark applies to the other second-order terms to be discussed shortly.

The second right-hand term in (8.42) has a similar origin to the last term, except in this case it is the *relative* displacement of the mean-free-path vectors by fluid strain that generates it. Adding the first-order heat flux to it, we have

$$\mathbf{q}^* = \left\{\kappa(\mathbf{1} - a\tau_2 \overset{\circ}{\nabla}\mathbf{v})\right\} \cdot \nabla T \qquad (a = (\tfrac{14}{3} + \tfrac{8}{15}s)), \tag{8.43}$$

in which form the effect appear as a modification of the thermal conductivity tensor $\kappa\mathbf{1}$.

8.4.2 *Sheared flow, orthogonal to a temperature gradient*

To appreciate how this change in conductivity arises, we shall consider the simplest case, namely planar flow, sheared orthogonal to a temperature gradient. Let $\hat{\mathbf{s}}$ and $\hat{\mathbf{v}}$ be unit vectors parallel to ∇T and $\mathbf{v}$, and denote rates of change along $\hat{\mathbf{s}}$ by a dash. Then

$$\nabla T = T'\hat{\mathbf{s}}, \quad \nabla\mathbf{v} = v'\hat{\mathbf{s}}\hat{\mathbf{v}}, \quad \overset{\circ}{\nabla}\mathbf{v} = \tfrac{1}{2}v'(\hat{\mathbf{s}}\hat{\mathbf{v}} + \hat{\mathbf{v}}\hat{\mathbf{s}}),$$

whence (8.43) becomes

$$\mathbf{q}^* = -\kappa(\hat{\mathbf{s}} - \tfrac{1}{2}a\tau_2 v'\hat{\mathbf{v}})T' = \mathbf{q}_1 + \mathbf{q}_2^*.$$

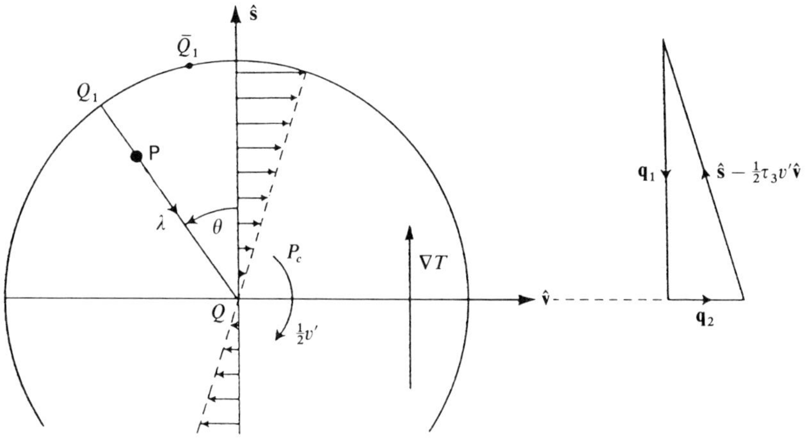

FIG. 8.1. The effect of fluid shear on heat flux

This heat flux may be deduced from elementary kinetic theory as follows. The convected fluid element P_c, shown in Fig. 8.1, rotates with an angular velocity $\frac{1}{2}v'\hat{\mathbf{s}} \times \hat{\mathbf{v}}$ about its centre Q. At a point Q_1, a vector distance $-\boldsymbol{\lambda} = \lambda(\hat{\mathbf{s}}\cos\theta - \hat{\mathbf{v}}\sin\theta)$ measured from Q, the relative fluid velocity divides into $\frac{1}{2}\lambda v'\cos\theta\,\hat{\mathbf{v}}$ due to spin and $\frac{1}{2}\lambda v'\cos\theta\,\hat{\mathbf{v}}$ due to the deviatoric rate of strain. Let λ be the mean-free-path for (second-order) energy transport, then molecules colliding at Q_1 and moving towards Q will, on average, deposit their excess energy at Q. The dominant velocity component of an average molecule P involved in this process is $-\mathcal{C}(\hat{\mathbf{s}}\cos\theta - \hat{\mathbf{v}}\sin\theta)$, where $\mathcal{C}$ is the most probable molecular speed. Owing to the fluid shear, in the convected frame P_c, P experiences an additional velocity component $\frac{1}{2}\lambda v' \cos\theta\, \hat{\mathbf{v}}$. Hence P approaches Q with the velocity $-\mathcal{C}(\hat{\mathbf{s}} - a\frac{1}{2}\tau_2 v'\hat{\mathbf{v}})\cos\theta + \mathcal{C}\hat{\mathbf{v}}\sin\theta$, where $\lambda = a\mathcal{C}\tau_2$ is the appropriate mean free path.

Particle P's excess energy is proportional to the displacement of Q_1 along $\hat{\mathbf{s}}$, and hence to $\cos\theta$, so weighting its velocity with this factor, and averaging over θ, we deduce that at Q the heat flux arrives from the direction $\hat{\mathbf{s}} - \frac{1}{2}a\tau_2 v'\hat{\mathbf{v}}$. An equivalent statement is that the sheared fluid behaves as if its original isotropic conductivity tensor, $\kappa\mathbf{1}$, were replaced by $\kappa\{\mathbf{1} - \frac{1}{2}a\tau_2 v'(\hat{\mathbf{s}}\hat{\mathbf{v}} + \hat{\mathbf{v}}\hat{\mathbf{s}})\}$, which confirms (8.43).

From the above description it is clear that the second term on the right of (8.42) results from small relative changes in the mean free path vector $\boldsymbol{\lambda}$, relative, that is, to a vector parallel to $\mathbf{q}_1$, which is itself subject to the similar process already discussed. As Q_1 and Q are convected with the fluid, it follows from an application of (1.94) that

$$\left(\frac{d\boldsymbol{\lambda}}{dt}\right)_{P_c} = \mathrm{D}\boldsymbol{\lambda} - \boldsymbol{\Omega}\times\boldsymbol{\lambda} = \overset{\circ}{\nabla\mathbf{v}}\cdot\boldsymbol{\lambda} + \tfrac{1}{3}\boldsymbol{\lambda}\nabla\cdot\mathbf{v}.$$

Hence, a time τ_2 earlier than the present instant, in P_c, $\boldsymbol{\lambda}$ had the value

$$\boldsymbol{\lambda}_\tau = (\mathbf{1} - a\tau_2\overset{\circ}{\nabla\mathbf{v}})\cdot\boldsymbol{\lambda} - \tfrac{1}{3}a\tau_2\boldsymbol{\lambda}\nabla\cdot\mathbf{v}.$$

The temperature gradient term in (8.11) can be written

$$\varphi_1^{(T)} = -(\nu^2 - \tfrac{5}{2})\boldsymbol{\lambda}\cdot\nabla\ln T,$$

and the second right-hand term in (8.42) is the consequence of replacing $\boldsymbol{\lambda}$ by $\boldsymbol{\lambda}_\tau$. The divergence term in (8.42) arises in a similar way.

8.4.3 *The viscous stress tensor*

With the help of $\boldsymbol{\pi}_1 = -2\mu\,\overset{\circ}{\mathbf{e}}$, $\mu = p\tau_1$, and the expression for $\mathbf{q}_1$, (8.40) can be written

$$\boldsymbol{\pi}_2 = -(3 - \tfrac{2}{3}s)\tau_1\nabla\cdot\mathbf{v}\,\boldsymbol{\pi}_1 - 3\tau_1\,\overline{\overset{\circ}{\overset{\circ}{\mathbf{e}}\cdot\boldsymbol{\pi}_1}} - \tfrac{4}{5}\tau_1\,\overline{\overset{\circ}{\nabla\mathbf{q}_1}}\,. \tag{8.44}$$

The first two terms on the right of (8.44) can be explained just as we have done for the corresponding terms in (8.42), i.e. they are generated by the relative displacement of the ends of mean-free-path vectors by fluid strain and dilatation.

To describe the physical basis for the last term in (8.44), we shall take the special case of flow in which there is no velocity shear, there is a temperature gradient along the axis $-OY$, and the resulting first-order heat flux along OY is sheared in a perpendicular direction, along OX. The existence of a $\nabla\mathbf{q}_1$ term in $\boldsymbol{\pi}_2$ can be deduced by the following reasoning. The gas may be divided into two superimposed pseudo-fluids, F_f containing the faster than average molecules and F_s containing the slower molecules. On average the fast molecules move along OY from the hot to the cold region, while the slow molecules move, again on average, along $-OY$. Transverse molecular motion along OX will transmit a shear force parallel to $-OY$ in

F_f and parallel to OY in F_s. Since the fast molecules have longer mean free paths than the slow ones, it follows that the shear force in F_f is larger than that in F_s, i.e. there is net force transmitted along $-OY$. By analogy with the case of fluid velocity shear, this force will be proportional to dq_1/dx, and to the average collision interval, τ_2. Thus this heuristic argument yields a force per unit area of $\hat{\mathbf{x}} \cdot \boldsymbol{\pi}_2 \propto -\tau_2 (dq_1/dx)\hat{\mathbf{y}}$, in agreement with (8.44).

8.5 Burnett's second-order transport equations

Although the terms we shall discuss in this section are usually associated with the name of Burnett, who in 1935 extended the Chapman–Enskog series to the $O(\epsilon^2)$ terms, other notable contributions to the theory were made by Maxwell (1879), Chapman and Cowling (1941), and Grad (1949, 1952).

To obtain the usual second-order expressions for the heat flux and the viscous stress tensor, we employ equations (8.26), (8.34) and (8.37), ignoring the pressure and acceleration terms of §§8.2.4 and 8.2.5. In fact this is not quite correct because, for the reason explained in §5.7.3, (8.26) has a streaming derivative with $\mathbf{c} \cdot \mathbf{e}$ in place of $\mathbf{c} \cdot \nabla \mathbf{v}$. We shall first return to the original, frame-dependent transport equations, which would follow from the standard BGK theory.

8.5.1 *The 'standard' second-order transport terms*

The second-order terms in $\mathbf{q}$ and $\boldsymbol{\pi}$ generally accepted are

$$\mathbf{q}_2 = Rp\tau_1^2 \Big\{ \theta_1 \nabla \cdot \mathbf{v}\, \nabla T + \theta_2 \overbrace{(\mathrm{D}\nabla T - \nabla \mathbf{v} \cdot \nabla T)}^{(i)} + \theta_3 T p^{-1} \overbrace{\nabla p \cdot \mathring{\mathbf{e}}}^{(ii)} + \theta_4 T \overbrace{\nabla \cdot \mathring{\mathbf{e}}}^{(iii)} + 3\theta_5 \nabla T \cdot \mathring{\mathbf{e}} \Big\}, \quad (8.45)$$

$$\text{with} \quad \theta_1 = \tfrac{15}{4}(\tfrac{7}{2} - s), \quad \theta_2 = \tfrac{45}{8}, \quad \theta_3 = -3, \quad \theta_4 = 3, \quad \theta_5 = \tfrac{35}{4} + s \quad (8.46)$$

$$\text{and} \quad \boldsymbol{\pi}_2 = p\tau_1^2 \Big\{ \tilde{\omega}_1 \nabla \cdot \mathbf{v}\, \mathring{\mathbf{e}} + \tilde{\omega}_2 \overbrace{(\mathrm{D}\mathring{\mathbf{e}} - 2\, \mathring{\overline{\nabla \mathbf{v} \cdot \mathring{\mathbf{e}}}})}^{(iv)} + \tilde{\omega}_3 R\, \mathring{\overline{\nabla\nabla T}} + \tilde{\omega}_4 (pT)^{-1} \overbrace{\mathring{\overline{\nabla p \nabla T}}}^{(v)} + \tilde{\omega}_5 R T^{-1} \mathring{\overline{\nabla T \nabla T}} + \tilde{\omega}_6\, \mathring{\overline{\mathring{\mathbf{e}} \cdot \mathring{\mathbf{e}}}} \Big\}, \quad (8.47)$$

$$\text{with } \tilde{\omega}_1 = \tfrac{4}{3}(\tfrac{7}{2} - s), \quad \tilde{\omega}_2 = 2, \quad \tilde{\omega}_3 = 3, \quad \tilde{\omega}_4 = 0, \quad \tilde{\omega}_5 = 3s, \quad \tilde{\omega}_6 = 8. \quad (8.48)$$

The five terms in $\mathbf{q}_2$ and $\boldsymbol{\pi}_2$ labelled (i) to (v) will be discussed below. Each owes their presence in second-order 'standard kinetic theory' (SKT) to that theory's neglect of fluid frame accelerations in defining diffusion. As we have shown in §8.3, in a theory that includes acceleration, the time derivatives in (i) and (iv) should appear in frame-indifferent forms, while terms (ii), (iii), and (v) should vanish.

8.5.2 *Time derivatives in SKT*

As $\nabla \mathbf{v} = \mathbf{e} - \boldsymbol{\Omega} \times \mathbf{1}$, where $\boldsymbol{\Omega} \equiv \frac{1}{2}\nabla \times \mathbf{v}$, the time derivative appearing in $\mathbf{q}_2$ is of the frame-*dependent* type $(\mathrm{D}\nabla T + \boldsymbol{\Omega} \times \nabla T)$. As shown in §§5.2.1 and 5.7.3, for frame indifference the combination $(\mathrm{D}\nabla T - \boldsymbol{\Omega} \times \nabla T)$ is required. In SKT the streaming derivative, namely

$$\mathbb{D}_{SK} = \mathrm{D} + \mathbf{c} \cdot \nabla - (\mathbf{P} + \mathbf{c} \cdot \nabla \mathbf{v}) \cdot \frac{\partial}{\partial \mathbf{c}}, \tag{8.49}$$

depends on the arbitrary choice of the laboratory reference frame. This same arbitrariness is reflected in $\mathbf{q}_2$. At the least, it is unsatisfactory to have a heat flux dependent on the observer's reference frame, and when this flux is dissipative—as is the case in magnetoplasma dynamics (see exercise 11.4)—we obtain the unphysical phenomenon of an observer-dependent dissipation rate. Similar remarks apply to the frame-dependent time derivative appearing in (8.47).

To remedy this defect of SKT, two changes are necessary to the streaming derivative defined in (8.49). In Mathematical note 7.5 we showed that in a fluid frame that both accelerates and spins with the fluid, between collisions the particle acceleration is $\dot{\mathbf{c}} = -\mathbf{c} \cdot \mathbf{e}$. Neither of the forces $\mathbf{P}$ and $\mathbf{F}$ that act on the fluid, and accelerate it according to $\mathrm{D}\mathbf{v} = \mathbf{P} + \mathbf{F}$, can contribute to $\dot{\mathbf{c}}$. Therefore the first change is to adopt the acceleration $-\mathbf{c} \cdot \mathbf{e}$ in place of $-\mathbf{c} \cdot \nabla \mathbf{v}$. The second is to replace D by the frame-indifferent form D^* given in (8.6). These changes alter $\mathbb{D}_{SK}$ into the operator $\mathbb{D}$.

8.5.3 *Pressure gradient terms in SKT*

If diffusion is calculated in a frame that moves with *both* the velocity and acceleration of the fluid element, the forces acting on this element cannot occur in the equation determining the diffusivity. This is true regardless of the order in Knudsen number of the process in question. As seen in such a fully convected frame, particle motion is independent of ∇p. It follows that term (ii) in (8.45) and term (iv) in (8.47) should not appear.

8.5.4 *Divergence of rate of strain tensor in SKT*

One conclusion from term (iii) in (8.45) is that a heat flux could be generated in a neutral gas by fluid shear *alone*; this is a strange result. We shall illustrate this by taking a special case.

Consider sheared flow, initially at uniform temperature. At the first moment $\mathbf{q}_1$ is zero, and (8.45) reduces to

$$\mathbf{q} = 3RpT\tau_1^2 \nabla \cdot \overset{\circ}{\mathbf{e}}\,.$$

Let $\hat{\mathbf{x}}$, $\hat{\mathbf{y}}$ be unit orthogonal vectors, and take $\nabla = \hat{\mathbf{y}} d/dy$, $\mathbf{v} = v\hat{\mathbf{x}}$. Then despite the absence of temperature gradients, SKT predicts an initial heat flux

$$\hat{\mathbf{x}} \cdot \mathbf{q} = -3RpT\tau_1^2 v'' \qquad (' = d/dy) \tag{8.50}$$

in the direction of fluid motion. Viscous heating will generate a temperature gradient in the $\hat{\mathbf{y}}$ direction, yet at the first instant, before this develops, according to SKT, there is a substantial flow of heat parallel to the fluid motion. Heat transfer is a thermodynamic process, that in the absence of electric fields or other driving forces, requires a temperature gradient to drive it. There are no other mechanisms in a neutral gas. Short of invoking convection (which is not allowed), the heat flux in (8.50) has no possible physical explanation, which is strong evidence that the standard theory is flawed.

8.5.5 *Viscomagnetic heat flux*

The above remarks apply only to a neutral gas, i.e. one that cannot be influenced by applied electromagnetic fields. However, there is a process by which it appears that sheared flow *alone* can produce a heat flux. It is termed 'viscomagnetic' heat flux, and it occurs in a rarefied polyatomic gas flowing along a channel across which a strong magnetic field B is applied. For an account of the physical origin of magnetic effects on paramagnetic molecules the reader can consult Beenakker and McCourt (1970). The phenomenon has been observed in a narrow rectangular channel (Eggermont *et al.* 1978), with the gas flow produced by a pressure difference Δp along the channel length. The resulting heat flux q, across the width of the channel, is found to satisfy an empirical law of the type

$$\frac{pq}{\Delta p} = \frac{2a(B/B_m)}{1 + (B/B_m)^2}, \tag{8.51}$$

the ratio achieving its maximum value of a at $B = B_m$. This maximum depends on the angle of the magnetic field in the plane orthogonal to the flow and on the nature of the gas. For example, when B is at right angles to the channel width and the gas is nitrogen, $a = 0.15\,\mathrm{W\,m^{-2}}$. With methane gas $a = 0.05\,\mathrm{W\,m^{-2}}$.

By the Navier–Stokes equations applied to steady channel flow, Δp is proportional to v'', allowing the interpretation that the transverse heat flux in (8.51) is a second-order phenomenon, similar in character to that predicted by (8.50). As this flux is orthogonal to the fluid motion and disappears in the absence of the magnetic field, it cannot be mistaken as support for (8.50). Nevertheless, the physical argument advanced against (8.50), should also dispose of 'viscomagnetic' heat flux.

Heat flux orthogonal to temperature gradients with crossed magnetic fields is a familiar phenomenon. It occurs in magnetoplasmas (see (9.96)) and also in polyatomic gases in magnetic fields (Hermans *et al.* 1970). An alternative interpretation of (8.51) is that it is the first-order, transverse heat flux resulting from a very small temperature gradient parallel to the fluid flow. If it is assumed that the channel flow (away from the boundaries) is a polytropic process, satisfying $p \propto \rho^s$, where s is the constant index of expansion, then isothermal conditions along the channel would require $s = 1$. It is not difficult to show that for this case (see exercise 8.5) the parameter a in (8.51) is given by

$$a = \frac{s-1}{2s}\frac{\kappa T}{\ell},$$

where ℓ is the length of the channel. In the experiments cited above, $\ell = 0.08$m, $T = 300$°K and for nitrogen $\kappa = 2.42 \times 10^2$ (in MKS units). Then the experimental value of a of 0.15 W m^{-2} can be accounted for by $s = 1.003$, which represents a very small departure from isothermal conditions parallel to the flow. It therefore seems unlikely that 'viscomagnetic' heat flux has been observed.

8.6 Ultrasonic sound waves

8.6.1 *Basic equations*

The simplest physical problem requiring a kinetic theory accurate to second order in Knudsen number is the propagation of very high frequency sound waves. Forced sound waves in a rarefied monatomic gas have been investigated experimentally and the effect of frequency on wave speed and attenuation determined (Greenspan 1956). This offers us a chance of comparing theory and experiment and to determine whether or not the intermediate kinetic theory of Chapter 5, identified as 'IKT' in the following, is an improvement over SKT.

The basic equations are:

Conservation

$$\mathrm{D}\rho + \rho\nabla\cdot\mathbf{v} = 0,$$

$$\rho\mathrm{D}\mathbf{v} + \nabla p = -\nabla\cdot\boldsymbol{\pi},$$

$$\rho T\mathrm{D}s + \nabla\cdot\mathbf{q} = -\boldsymbol{\pi} : \nabla\mathbf{v}.$$

Thermodynamics

$$p = R\rho T, \qquad p \propto \rho^\gamma \exp\{(\gamma-1)s/R\}.$$

Transport

$$\mathbf{q} = -\kappa\nabla T + \mu R\tau_1\left(\theta_2\mathrm{D}\nabla T + \theta_4 T\nabla\cdot\overset{\circ}{\nabla\mathbf{v}}\right),$$

$$\boldsymbol{\pi} = -2\mu\,\overset{\circ}{\nabla\mathbf{v}} + \mu\tau_1\left(\tilde{\omega}_2\mathrm{D}\overset{\circ}{\nabla\mathbf{v}} + \tilde{\omega}_3 R\overset{\circ}{\overline{\nabla\nabla T}}\right).$$

Coefficients

	θ_2	θ_4	$\tilde{\omega}_2$	$\tilde{\omega}_3$
Navier–Stokes	0	0	0	0
IKT	45/4	0	0	3
SKT	45/8	3	2	3

Only the linear terms have been retained in the transport equations and the coefficients are taken from (8.36), (8.41), (8.46), and (8.48).

8.6.2 *The dispersion relation*

We now express the dependent variables in the wave form

$$X(\mathbf{r}, t) = X_0 + \hat{X} \exp\{i(\mathbf{k} \cdot \mathbf{r} - \omega t)\},$$

and linearize the basic equations. It is convenient to adopt the variables

$$v_P = \frac{\omega}{k}, \quad \alpha = \frac{a}{v_P}, \quad \beta = \omega\tau_1, \quad a = \left(\frac{\gamma k T}{m}\right)^{1/2},$$

where v_P is the phase velocity, a is the low-frequency sound speed and β, being the ratio of the collision interval to a measure of the macroscopic time scale, is the Knudsen number. On eliminating the perturbation amplitudes, we obtain the consistency relation

$$1 - A\alpha^2 + B\alpha^4 - C\alpha^6 = i\beta\alpha^2(D\alpha^2 - 23/6\gamma), \tag{8.52}$$

where

$$\begin{aligned}
A &\equiv 1 + a_0\beta^2, & a_0 &= \frac{2}{3\gamma}\tilde{\omega}_2 + \frac{(\gamma-1)}{\gamma}\theta_2, \\
B &\equiv a_1\beta^2 + b_0\beta^4, & a_1 &= \frac{2(\gamma-1)}{3\gamma^2}\left(\tfrac{3}{2}\theta_2 + \tilde{\omega}_3 + \theta_4\right) - \frac{10}{3\gamma^2}, \\
C &\equiv b_1\beta^4, & b_1 &= \frac{4(\gamma-1)}{9\gamma^3}\tilde{\omega}_3\theta_4, \quad b_0 = \frac{2(\gamma-1)}{3\gamma^2}\theta_2\tilde{\omega}_2, \\
D &\equiv \frac{5}{2\gamma^2} + a_2\beta^2, & a_2 &= \frac{4(\gamma-1)}{3\gamma^2}\theta_2 + \frac{5}{3\gamma^2}\tilde{\omega}_2.
\end{aligned}$$

As we are considering only monatomic gases in the following, we shall set $\gamma = 5/3$. Let α_r and α_i denote the real and imaginary parts of α. Solving (8.52) for α_r and α_i, we find

$$\alpha_r = 1 - \mathcal{F}\beta^2 + \mathrm{O}(\beta^4) \quad \left(\mathcal{F} \equiv \tfrac{1}{2}\left(a_0 - a_1 + \tfrac{21}{100}\right)\right), \tag{8.53}$$

and

$$\alpha_i = 0.7\beta(1 - \mathcal{H}\beta^2) + \mathrm{O}(\beta^4) \quad \left(\mathcal{H} \equiv a_2 + \tfrac{6}{7}a_0 - \tfrac{13}{7}a_1 - \tfrac{223}{200}\right). \tag{8.54}$$

Using the coefficients given above, we obtain the following values for $\mathcal{F}$ and $\mathcal{H}$:

	Navier–Stokes	SKT	IKT
$\mathcal{F}$	0.71	1.08	1.37
$\mathcal{H}$	1.11	2.44	2.67

It is of interest to note that in the formal—and invalid—limit, $\beta \to \infty$, (8.52) shows that $\alpha_r \to 0$, i.e. wave speeds greatly exceeding that of low frequency sound waves are possible in second-order IKT theory. As we shall show in §8.7.3, this unphysical result plays a disruptive role in the calculation of shock-wave structures. Also we see that even terms in the Knudsen number expansion affect the wave speed but not their damping (see remarks in §8.3.3).

8.6.3 *Comparison with experiment*

Table 8.1 shows experimental values of β^{-1} and α_r taken from Greenspan's (1956) observations of neon, argon, krypton, and xenon over the range $0.14 < \beta < 0.34$. His results for helium showed too much scatter to be reliable. The values of $\mathcal{H}$ were deduced from the experimental points using (8.54), and the average values given for $\mathcal{H}$ are taken from the last three entries for each gas, since the theory applies to *small* β. The values for $\mathcal{F}$ were estimated from Greenspan's figures 5 and 6 by drawing straight lines through points in the range $0.1 \leq \beta < 0.3$.

Table 8.1 *Experimental values of $\mathcal{H}$ and $\mathcal{F}$ for four monatomic gases (from Greenspan 1956)*

Neon

β^{-1}	α_r	$\mathcal{H}$	$\mathcal{F}$
2.91	0.215	2.20	
3.87	0.151	2.47	
5.02	0.122	3.15	
6.52	0.100	2.91	
Average values		2.84	1.38

Argon

β^{-1}	α_r	$\mathcal{H}$	$\mathcal{F}$
3.13	0.174	2.18	
4.01	0.146	2.63	
5.27	0.119	2.89	
7.00	0.095	2.45	
Average values		2.66	1.31

Krypton

β^{-1}	α_r	$\mathcal{H}$	$\mathcal{F}$
2.97	0.180	2.08	
3.80	0.150	2.68	
5.00	0.126	2.50	
6.56	0.100	2.70	
Average values		2.63	1.24

Xenon

β^{-1}	α_r	$\mathcal{H}$	$\mathcal{F}$
3.12	0.176	2.10	
4.15	0.145	2.42	
5.45	0.116	2.88	
7.00	0.094	2.84	
Average values		2.71	1.37

We are justified in omitting the values of $\mathcal{H}$ at the largest values of β, because we can expect agreement with observation only for small β. This is also supported by the fact that every entry in the region of $\beta \approx 0.3$ is smaller than even the SKT prediction.

The values of $\mathcal{H}$ in Table 8.1 are quite sensitive to small errors in β^{-1} and α_r, especially at small β, while those at larger β are affected by the omitted $O(\beta^4)$ terms. Also because of uncertainties in measurements taken from graphs, the experimental values in the table are subject to appreciable errors. But the average values are less vulnerable. The overall average values for $\mathcal{F}$ and $\mathcal{H}$ are 1.33 and 2.71.

Since there is no essential distinction between the gases for the phenomenon in question, we may treat the values of $\mathcal{H}$ as providing a set of twelve observations to which we can apply Student's t-test for small samples. Doing this, we find that the 95 per cent confidence limits for the unknown mean of $\mathcal{H}$ give the range

$$2.59 \leq \mu_{\mathcal{H}} \leq 2.83; \qquad \text{(SKT 2.44, IKT 2.67)}.$$

Similarly

$$1.24 \leq \mu_{\mathcal{F}} \leq 1.41; \qquad \text{(SKT 1.08, IKT 1.37)}.$$

We conclude that the intermediate kinetic theory is well-supported by the experimental evidence from ultrasonic sound waves.

8.7 Shock-wave structure

8.7.1 *Thickness of shock waves*

Normal shock waves provide a second, straightforward test of second-order kinetic theory, the shock thickness being a good indicator of transport within the shock front. Figure 8.2 is a schematic representation of typical profiles of temperature and density through a shock wave.

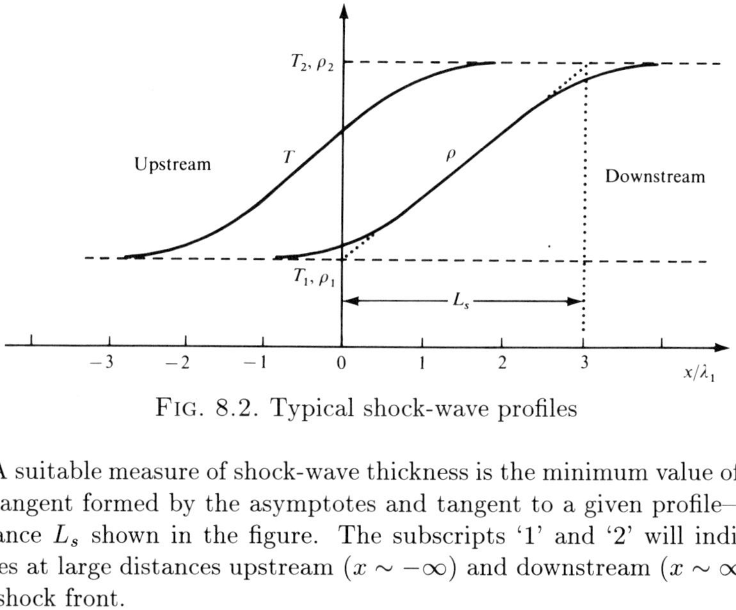

FIG. 8.2. Typical shock-wave profiles

A suitable measure of shock-wave thickness is the minimum value of the subtangent formed by the asymptotes and tangent to a given profile—the distance L_s shown in the figure. The subscripts '1' and '2' will indicate values at large distances upstream ($x \sim -\infty$) and downstream ($x \sim \infty$) of the shock front.

The upstream Mach number M_1 and a convenient measure of the upstream mean free path are defined by

$$M_1 \equiv \frac{v_1}{\sqrt{\frac{5}{3}RT_1}}, \qquad \lambda_1 \equiv \frac{\mu_1}{p_1}\sqrt{2RT_1}, \tag{8.55}$$

where $\mu_1 = p_1\tau_{11}$ is the coefficient of viscosity and we have restricted attention to monatomic gases ($\gamma = 5/3$).

A typical experimental value of the Knudsen number, $\epsilon = \lambda_1/L_s$, is ~ 0.3 for $3 < M_1 < 7$, but for these Mach numbers, the Navier–Stokes theory (first-order transport only) gives $\epsilon \sim 0.45$, i.e. it predicts shock-wave thicknesses some 50 per cent too thin. Adding the standard Burnett transport terms given in §8.5.1, improves agreement with experiment for small values of ϵ. But there is a critical Mach number (Sherman and Talbot 1960) above which the inclusion of the Burnett terms makes the situation dramatically worse. Above $M_1 = 1.5$ the solution curves develop upstream oscillations and at a Mach number just less than two these become so large that no physically meaningful solution can be found. A similar phenomenon has been found with Grad's version of second-order transport equations (see account in Chapman and Cowling, 1970). If the third-order terms $\mathbf{q}_3, \boldsymbol{\pi}_3$ (known as 'super Burnett' terms) are added, the situation is a little improved, with the critical Mach number increasing slightly (Elliot and Baganoff 1974).

8.7.2 *Shock-wave equations*

Starting with the steady-state forms of (1.41) to (1.43) and integrating them from the uniform upstream state to some position x in the shock structure, we obtain

$$\rho v = \rho_1 v_1,$$

$$p + \rho v^2 + \pi_{xx} = p_1 + \rho_1 v_1,$$

$$\rho v(c_v T + \tfrac{1}{2}v^2) + pv + v\pi_{xx} + q_x = \rho_1 v_1(c_v T_1 + \tfrac{1}{2}v_1^2) + p_1 v_1.$$

With $\mathbf{v} = v\hat{\mathbf{x}}$ and $\nabla = \hat{\mathbf{x}}d/dx$, we get $\boldsymbol{\Omega} = \tfrac{1}{2}\nabla \times \mathbf{v} = 0$, $\mathrm{D}\nabla T = vT''\hat{\mathbf{x}}$, $\nabla \cdot \mathbf{v} = v'$, and

$$\overset{\circ}{\mathbf{e}} = \tfrac{1}{3}v'(2\hat{\mathbf{x}}\hat{\mathbf{x}} - \hat{\mathbf{y}}\hat{\mathbf{y}} - \hat{\mathbf{z}}\hat{\mathbf{z}}),$$

where the dash denotes d/dx and $\hat{\mathbf{x}}$, $\hat{\mathbf{y}}$, $\hat{\mathbf{z}}$ is a triad of unit, orthogonal vectors.

It follows from (8.2), (8.35), (8.36), and $q_x \approx q_{1x} + q_{2x}$ that

$$q_x = -\tfrac{15}{4}R\mu T' + R(\mu^2/p)\{\Theta v'T' + \theta_2 vT''\} \qquad (\Theta \equiv \theta_1 + 2\theta_5).$$

Similarly,

$$\pi_{xx} = -\tfrac{4}{3}\mu v' + \tfrac{2}{3}(\mu^2/p)\{\Omega v'^2 + \tilde{\omega}_3 RT'' + \tilde{\omega}_5 RT^{-1}(T')^2\} \quad (\Omega \equiv \tilde{\omega}_1 + \tfrac{1}{3}\tilde{\omega}_6),$$

where the values of the coefficients are given in (8.36) and (8.41).

8.7.3 *Non-dimensional form of the equations*

The shock-wave equations may be written as

$$\pi_{xx} = R\rho_1 v_1(T_1/v_1 - T/v) + \rho_1 v_1(v_1 - v),$$

$$q_x = \tfrac{3}{2}R\rho_1 v_1(T_1 - T) + p_1(v_1 - v) + \tfrac{1}{2}\rho_1 v_1(v_1 - v)^2.$$

To non-dimensionalize these equations, we introduce the variables

$$w = v/v_1, \qquad \vartheta = T/T_1, \qquad X = x/\lambda_1, \tag{8.56}$$

and adopt the power law $\mu \propto T^s$, so that $\mu = \mu_1 \vartheta^s$. If dashes now denote d/dX, the above equations yield

$$-\tfrac{4}{3}\sqrt{\tfrac{5}{6}}M_1\vartheta^s w' + \tfrac{2}{3}\vartheta^{2s-1}w\Big\{\tfrac{5}{6}\Omega w'^2 M_1^2 + \tilde{\omega}_3\vartheta'' + \tilde{\omega}_5\vartheta^{-1}(\vartheta')^2\Big\}$$
$$= 1 - w^{-1}\vartheta + \tfrac{5}{3}M_1^2(1-w), \qquad (8.57)$$

and

$$-\sqrt{\tfrac{15}{8}}M_1^{-1}\vartheta^s\vartheta' + \tfrac{1}{3}w\vartheta^{2s-1}\Big\{\Theta w'\vartheta' + \theta_2 w\vartheta''\Big\}$$
$$= 1 - \vartheta + \tfrac{2}{3}(1-w) + \tfrac{5}{9}M_1^2(1-w)^2. \qquad (8.58)$$

At the downstream point the right-hand sides of (8.57) and (8.58) vanish, giving the relations

$$w_2 = \tfrac{1}{4}\big(1 + 3M_1^{-2}\big), \qquad \vartheta_2 = \tfrac{1}{16}M_1^{-2}(M_1^2+3)(5M_1^2-1). \qquad (8.59)$$

Equations (8.57) to (8.59) have been solved numerically for a wide range of Mach numbers. The standard method of numerical integration of this stiff set of differential equations, advancing upstream from a linearized downstream perturbation, runs into a difficulty with upstream oscillations similar to that described in 8.7.1 for the Burnett equations, although the critical Mach number is increased to about five (Reese 1991). The problem arises because the second-order transport terms, being *reversible*, can admit waves able to travel at speeds greater than the shock front (see §§8.3.3 and 8.6.2). These waves are unphysical. Their origin is mathematical rather than numerical, in the sense that they result from the *equations* adopted to describe the phenomenon and not from the numerical method of their solution. In the process of obtaining a numerical solution, small errors rapidly grow into 'waves', that above the critical Mach number, are insufficiently damped by the first-order terms. This disrupts the solution, but only in the region of the upstream singular point. As the shock wave thickness is determined by the density profile downstream of these spurious waves, their presence does not prevent us from using the theory to obtain shock-wave thicknesses for Mach numbers above the critical value. The origin of the problem is the absence of higher-order terms in the Knudsen number expansion. Such terms would both reduce the wave speed and also damp the oscillations.

8.7.4 *Other methods of determining shock-wave structure*

A good introduction to the shock-wave structure problem is found in the text by Vincenti and Kruger (1965). They describe the several attempts that have been made to calculate the shock front profiles. Mott–Smith's use of a bimodal distribution function, with parameters chosen to satisfy certain moments of the Boltzmann equation, is one of the more successful. Another, due to Leipmann et al. (1962), employs an integral form of the BGK kinetic equation (cf. (5.95)), allowing for a variable collision frequency. The Navier–Stokes equations are found to be in satisfactory agreement with experiment only for Mach numbers less than two.

Fiscko and Chapman (1988) have treated the problem by starting with the time-dependent equations and arbitrary initial conditions and allowing the system to relax to a steady state. This method also fails with the Burnett equations above a certain Mach number. To obtain solutions they found it necessary to augment these equations with some third-order terms, adjusted so as to stabilize the equations against high frequency waves.* As remarked at the end of §8.7.1, Elliot and Baganoff (1974) found that adding the formally correct third-order terms made little difference: additional damping from some source is required to suppress the oscillations. Fiscko and Chapman obtained continuous solutions over a wide range of Mach numbers, with shock-thicknesses a little thinner than found in experiments (see Fig. 8.3). For a review of this and related work, see Lumpkin and Chapman (1991).

Another approach, recently devised by Candel and Thivet (1992) for the equations in §8.7.2, avoids the difficulty encountered by numerical integration. They adopted a finite difference, global method, relaxing from assumed profiles to smooth solutions, without encountering a Mach number limit. This method appears not to excite the upstream waves, perhaps because of numerical damping.

The earlier failures with the Burnett equations resulted in some researchers abandoning the Chapman–Enskog series approach altogether and seeking direct numerical methods of solving Boltzmann's equation. The best known of these is Bird's (1978) direct simulation Monte Carlo method (see §7.12). This approach gives shock thicknesses in fair agreement with experiment (Pham-Van-Diep and Erwin 1989).

8.7.5 *Comparison of theory and experiment*

Most experimental data are expressed in terms of the 'Maxwellian' mean free path, which from (2.72) and (3.50) is

$$\lambda^{(M)} \equiv \frac{8}{5\sqrt{\pi}}\frac{\mu}{p}\sqrt{(2RT)}.$$

In Fig. 8.3 the inverse thickness ratio defined by

$$\frac{\lambda_1^{(M)}}{L_s} = \frac{(d\rho/dX)_{max}}{\rho_2 - \rho_1} \tag{8.60}$$

is plotted as a function of the upstream Mach number for argon gas. The experimental points, which are indicated by different symbols for each of seven research groups, have been taken from Alsmeyer's (1976) account. Also shown are the functions obtained from the Navier–Stokes theory, from the Burnett theory as modified by Lumpkin and Chapman (1991), and from

*Notice from (8.54) that waves become unstable if $\beta \geq \mathcal{H}^{-1/2}$, but as this value is well beyond the range of validity of the theory, the conclusion has no physical relevance.

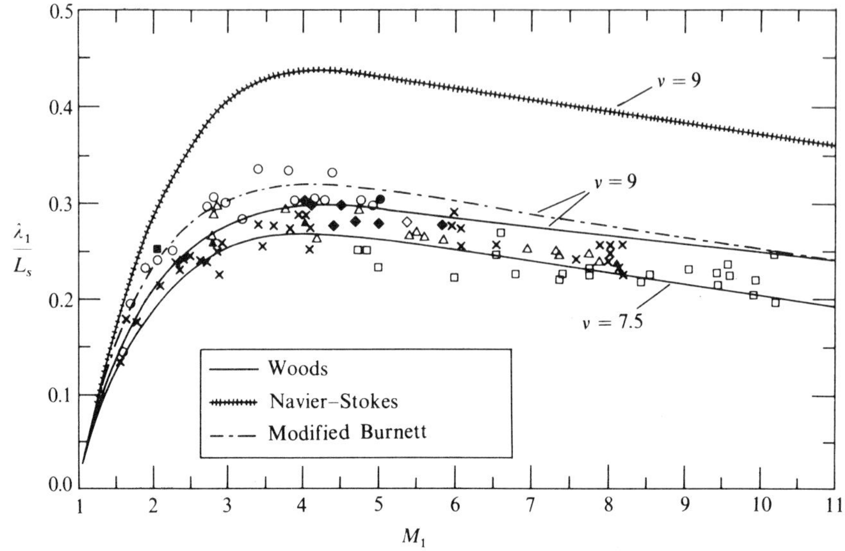

FIG. 8.3. Comparison of theory and experiment for shock-wave thicknesses in argon

the theory of this book. The latter was calculated by Reese (1991), and Candel and Thivet (1992), using the different methods described above; they give much the same results.

The good agreement between theory and experiment is further support for the modifications to the kinetic equation with which this chapter began.

8.8 Boundary conditions

Boltzmann's equation is an accurate description of the behaviour of a dilute gas, provided effects second-order in the Knudsen number (ϵ) can be disregarded. Boundary conditions for first-order transport theory are those usually employed in gas dynamics and offer no special difficulty. At the second-order, the Chapman–Enskog expansion produces several spurious transport terms, whose origin we have traced, not to the expansion itself, but to the neglect of acceleration and pressure effects in the *formulation* of Boltzmann's equation. We shall discuss the difficulty of finding appropriate boundary conditions for second-order transport theory shortly.

That unphysical second-order terms occur in the Chapman–Enskog expansion of Boltzmann's equation has been remarked by several authors, notably by Cercignani (1988), who advances a different resolution of the problem. He accepts Boltzmann's equation as being sound—at least within the limitations evident from its classical formulation—but finds an ambiguity in the Chapman–Enskog expansion that allows terms to appear that are either irrelevant, or worse, non-existent in reality. He argues that it is necessary to have 'strong' boundary conditions to discriminate against such spurious solutions. He also notes the limitation imposed by adopting a power series in ϵ; for example terms containing exponentials like $\exp(-1/\epsilon)$ (cf. 3.14) cannot be expanded in a power series in ϵ, and yet would be important in the transition regime $\epsilon \sim 1$.

Cercignani's approach is particularly important when conditions at solid walls need to be considered, large Knudsen numbers being unavoidable in boundary layer regions. There is no conflict between our amendment of the kinetic equation and Cercignani's modification of the Chapman–Enskog expansion—they relate to distinct difficulties in transport theory. Our theory is strictly confined to flows with small Knudsen numbers *throughout* and can be applied only to problems—like those in §§8.6 and 8.7—for which the boundary conditions do not impose rapid changes in the macroscopic variables. A similar observation applies to initial conditions.

Readers interested in mathematical questions about the existence and uniqueness of solutions to Boltzmann's equation, subject to given initial and boundary conditions, should consult Cercignani's text. He also presents several interesting examples of linearized, transition flows, and provides extensive lists of references.

A complete description of a problem in kinetic theory requires the correct initial and boundary conditions to be specified. Apart from simple cases like those mentioned above, this is a difficult subject, beyond the scope of an introductory text. Ferziger and Kaper (1972), Cercignani (1988) and Williams (1971) deal with this topic and list many references to the literature.

We have given two examples where second-order effects are significant for transport in neutral gases, namely ultrasonic waves and shock waves. In both cases the 'standard kinetic theory', i.e. that due to Boltzmann, Chapman, Enskog, Burnett and others, fails to give agreement with experiment. In Chapter 11 our treatment of transport will be extended to magnetoplasmas, where, for example, the failure of the received theory to predict losses from tokamaks is measured in 'orders of magnitude' rather than in 'percentages'.

It would be incorrect to conclude that Boltzmann's equation is 'wrong', for his main objective was to find a mechanical argument leading to the second law of thermodynamics and this he managed via his H-theorem. Only a few pages of his lengthy paper were concerned with transport, which for him was what we now term 'first-order' transport.

Exercises 8

8.1 A gas is flowing a long a cylinder with a steady velocity v_z. There is a radial temperature gradient. Show that there is heat flux parallel to the axis of the cylinder of magnitude

$$q_z = \tfrac{3}{2}(5+s)Rp\tau_1^2 v_z' T',$$

where the dashes denote radial derivatives.

8.2 Suppose that in exercise 8.1 the flow is around the axis in the azimuthal (θ) direction. Show that in this case there is an azimuthal energy flux

$$\tfrac{1}{2}Rp\tau_1^2 T'\left\{3\theta_5(v_\theta' - v_\theta/r) - \theta_2(v_\theta' + v_\theta/r)\right\},$$

where v_θ is the fluid velocity.

8.3 A cylinder of stationary fluid has a radial temperature gradient. Show that this generates a force on the fluid in the radial direction.

8.4 Show that in a uniform temperature standard kinetic theory predicts an energy flux of

$$\mathbf{Q}_2 = -\tfrac{3}{2}R\tau_1(\nabla\cdot\mathbf{p} - \nabla p)$$

and explain why $\mathbf{Q}_2$ could not be due a genuine *diffusion* process.

8.5 By assuming a polytropic process, $p \propto \rho^s$, for the flow of a polyatomic gas along a channel and using the expression for the first- order transverse heat flux given in §9.8.2, show that this heat flux is given by (cf. (8.51))

$$q_\wedge = -\kappa\frac{(B/B_m)}{1+(B/B_m)^2}\,\frac{s-1}{s}\,\frac{T}{p}\,\frac{\Delta p}{\ell},$$

where Δp is the pressure drop along a channel of length ℓ.

8.6 Find a physical explanation of the fact that ultrasonic waves travel faster than ordinary sound waves.

9

DYNAMICS OF CHARGED PARTICLES

9.1 The electromagnetic fields

9.1.1 *Introduction*

A plasma is a mixture of positive ions, electrons and neutral particles, electrically neutral over macroscopic volumes, and usually permeated by macroscopic electrical and magnetic fields. In addition to these 'smoothed' or averaged electromagnetic fields, which with laboratory plasmas are often imposed from outside the plasma volume, there are the microfields due to the individual particles. The trajectories of the charged particles are thus continuously modified by a range of electromagnetic forces, the average fields acting like body forces and the microfields like collisional forces. The microfields are responsible for the transmission of pressure and viscous forces, for the conduction of particle energy, and for the friction forces between diffusing components of the plasma. Some care is needed in dividing the continuum of electromagnetic forces into their macroscopic and microscopic components, but once this has been achieved, there is little formal distinction between the kinetic theory of neutral gases and the kinetic theory of magnetoplasmas. The most important practical distinction is that in strong magnetic fields, for cross-field transport, the terms second-order in the Knudsen number ϵ dominate the first-order terms. This occurs because the $O(\epsilon)$ terms are much more strongly suppressed by the magnetic fields than are the $O(\epsilon^2)$ terms.

Our principal aim in the remaining chapters of this book is to explain how to extend the intermediate kinetic theory of Chapter 5 to to magnetoplasmas and then to apply it to some important problems. It is necessary to use the intermediate theory, because only that theory is accurate to second order in the Knudsen number.

Much of the material to follow is based on the account of transport in magnetoplasmas presented in the author's text *Principles of Magnetoplasma Dynamics*, (Woods 1987). But there are some important changes in the kinetic equation from which the second-order terms are derived. The previous approach, given in Chapter 6 of the cited text, was based on a heuristic extension of the BGK kinetic equation, which lead to the 'correct' expression for the transverse heat flux vector. That the expression for the heat flux vector, $\mathbf{q}_2$, *was* correct followed from an earlier, unequivocal mean-free-path derivation of $\mathbf{q}_2$ (Woods 1983b), that we shall present later in §9.7.3.* Where the results of the new kinetic theory of magnetoplasmas furnished in this book differ from those of the author's previous account is described in §§11.11.2, 11.12, and the Appendix. Few of the applications to tokamak plasmas made in the earlier work need alteration.

9.1.2 *Maxwell's equations*

The four basic electromagnetic fields are the magnetic induction **B**, the magnetic field **H**, the electric field **E** and the electric displacement **D**. To these are added the current density **j** and the charge density Q. Maxwell's equations relating these variables are

$$\nabla \times \mathbf{E} = -\frac{\partial \mathbf{B}}{\partial t}, \tag{9.1}$$

$$\nabla \times \mathbf{H} = \mathbf{j} + \frac{\partial \mathbf{D}}{\partial t}, \tag{9.2}$$

and

$$\nabla \cdot \mathbf{j} = -\frac{\partial Q}{\partial t}. \tag{9.3}$$

With the assumption that at some past time the vector fields were zero, the divergences of (9.1) and (9.2) give

$$\nabla \cdot \mathbf{B} = 0, \qquad \nabla \cdot \mathbf{D} = Q. \tag{9.4}$$

To complete the set of equations, constitutive equations for **B** and **D** are required. In an isotropic medium these are

$$\mathbf{B} = \mu \mathbf{H}, \qquad \mathbf{D} = \epsilon \mathbf{E}, \tag{9.5}$$

where the permeability μ and the permittivity ϵ depend on the nature of the medium. In free space their values are $\mu_0 = 4\pi \times 10^{-7}\mathrm{H\,m^{-1}}$ and $\epsilon_0 = 8.8542 \times 10^{-12}\mathrm{F\,m^{-1}}$.

In a magnetoplasma it can be shown that the electric and magnetic polarizations defined by

$$\mathbf{P} \equiv \mathbf{D} - \epsilon_0 \mathbf{E}, \qquad \mathbf{M} \equiv \mu_0^{-1}\mathbf{B} - \mathbf{H},$$

are negligible. This allows us to eliminate **H** and **D** and write (9.2) and (9.5) as

$$\nabla \times \mathbf{B} = \mu_0 \mathbf{j} + \mu_0 \epsilon_0 \frac{\partial \mathbf{E}}{\partial t}, \tag{9.6}$$

and

$$\epsilon_0 \nabla \cdot \mathbf{E} = Q. \tag{9.7}$$

*At that time the author was unaware of how much of standard kinetic theory needed to be abandoned to give correct second-order expressions for the transport of particle energy and momentum.

9.1.3 Plasma neutrality

A plasma is a mixture of fluids, some of which consist of charged particles. We recall the notation of §1.12, in which each component fluid is identified by a subscript, i. We shall change this to 'α' in order to free 'i' to denote 'ions'. For a mixture of perfect gases, the density, momentum density, energy density, and pressure are defined by the sums over α:

$$\rho = \sum \rho_\alpha, \quad \rho\mathbf{v} = \sum \rho_\alpha \mathbf{v}_\alpha, \quad \rho u = \sum \rho_\alpha u_\alpha, \quad p = \sum p_\alpha. \tag{9.8}$$

The charge and current densities are given by

$$Q = \sum Q_\alpha n_\alpha, \quad \mathbf{j} = \sum Q_\alpha n_\alpha \mathbf{v}_\alpha, \tag{9.9}$$

where Q_α denotes the charge on the particles of the α-species. For example, in a plasma consisting of electrons ($Q_e = -e$) and one species of ion with a charge number Z ($Q_i = Ze$),

$$Q = e(Zn_i - n_e), \quad \mathbf{j} = e(Zn_i\mathbf{v}_i - n_e\mathbf{v}_e). \tag{9.10}$$

Most plasmas of interest are electrically neutral over sufficiently long distance and time scales. Precise meanings will be attached to these scales in the next section. If the subscript zero is adopted to indicate averaged or 'background' values, then (9.10) are reduced to

$$n_{e0} = Zn_{i0}, \quad \mathbf{j} = en_e(\mathbf{v}_i - \mathbf{v}_e). \tag{9.11}$$

9.2 Basic plasma parameters

9.2.1 The cyclotron frequency

Consider the motion of a particle P of mass m, charge Q and velocity $\mathbf{w}$ moving in (macroscopic) fields $\mathbf{E}$ and $\mathbf{B}$. The electromagnetic force acting on P is the *Lorentz* force, $Q(\mathbf{E} + \mathbf{w} \times \mathbf{B})$, and if in addition there is a body force $\mathbf{f}$ per unit mass, then the acceleration of P is

$$\dot{\mathbf{w}} = \frac{Q}{m}(\mathbf{E} + \mathbf{w} \times \mathbf{B}) + \mathbf{f}. \tag{9.12}$$

The following notation will be employed in the remainder of this book:

$$\mathbf{b} \equiv \mathbf{B}/|\mathbf{B}|, \quad \mathbf{A}_\perp \equiv \mathbf{b} \times \mathbf{A} \times \mathbf{b} = \mathbf{A} - \mathbf{b}\mathbf{b}\cdot\mathbf{A}, \quad \mathbf{A}_\| \equiv \mathbf{b}\mathbf{b}\cdot\mathbf{A}. \tag{9.13}$$

Thus $\mathbf{b}$ is unit vector parallel to the magnetic field, and $\mathbf{A}_\perp$, $\mathbf{A}_\|$ are termed the 'perpendicular' and 'parallel' components of a vector $\mathbf{A}$.

If $\mathbf{E}$ and $\mathbf{f}$ are zero,
$$\dot{\mathbf{w}} = \omega_c \mathbf{w} \times \mathbf{b}, \tag{9.14}$$

where
$$\omega_c \equiv QB/m. \tag{9.15}$$

Let $\mathbf{r}$ be the position vector of P, then $\dot{\mathbf{r}} = \mathbf{w}$, and if $\mathbf{B}$ is a uniform magnetic field*, (9.14) has the integral

---
*Strictly this is the magnetic *induction*, but because in plasmas the induction is proportional to the field, the distinction is ignored.

$$\dot{\mathbf{r}} = \omega_c \mathbf{a} \times \mathbf{b} \qquad (\mathbf{a} = \mathbf{r} - \mathbf{X}), \tag{9.16}$$

where $\mathbf{X}$ is the constant of integration.

It follows from this result that the motion of P projected on to a plane perpendicular to $\mathbf{b}$ is a circle about the point G, of position vector $\mathbf{X}$. The angular velocity ω_c is the *cyclotron frequency* and the point G($\mathbf{X}$) is termed the *guiding centre* for P. The distance $a = |\mathbf{a}|$ is called the *Larmor radius* of the particle; it is related to P's perpendicular speed by $a = w_\perp/\omega_c$. When a particle average is required for the Larmor radius, we replace $w_\perp$ by the thermal speed $\mathcal{C} = (2kT/m)^{1/2}$ and denote the radius by r_L:

$$a = w_\perp/\omega_c, \qquad r_L = \mathcal{C}/\omega_c. \tag{9.17}$$

9.2.2 *The plasma frequency*

For electrons (9.12) reads

$$m_e \dot{\mathbf{w}} = -e(\mathbf{E} + \mathbf{w} \times \mathbf{B}) + m_e \mathbf{f}_e. \tag{9.18}$$

When electron collisions can be ignored (and hence pressure gradients omitted; see §1.5.1) the particle average of this equation gives

$$m_e \frac{d\mathbf{v}_e}{dt} = -e(\mathbf{E} + \mathbf{v}_e \times \mathbf{B}) + m_e \mathbf{f}_e, \tag{9.19}$$

where $\mathbf{v}_e$ is the electron fluid velocity. Our present interest is in the case that $\mathbf{B}$ and $\mathbf{f}_e$ are zero. Thus from (9.19) and (1.46) written for the electrons,

$$m_e \frac{d\mathbf{v}_e}{dt} = -e\mathbf{E}, \qquad \frac{dn_e}{dt} = -n_e \nabla \cdot \mathbf{v}_e.$$

Let $n_e = n_{e0} + \hat{n}_e$, $\mathbf{v}_e = \hat{\mathbf{v}}_e$, where n_{e0} is the uniform background density and the circumflex indicates a small perturbation. Then to first order in these perturbations,

$$m_e \frac{\partial \hat{\mathbf{v}}_e}{\partial t} = -e\mathbf{E}, \qquad \frac{\partial \hat{n}_e}{\partial t} = -n_{e0} \nabla \cdot \hat{\mathbf{v}}_e. \tag{9.20}$$

By (9.7), (9.10), and (9.11),

$$\epsilon_0 \nabla \cdot \mathbf{E} = \hat{Q} = eZ\hat{n}_i - e\hat{n}_e, \qquad Zn_{i0} = n_{e0}. \tag{9.21}$$

The ion mass greatly exceeds the electron mass ($m_i \geq 1836\, m_e$), so with sufficiently high frequency waves it may be assumed that the ions remain stationary. In which case

$$\hat{n}_i = 0, \qquad \mathbf{j} = -en_e \hat{\mathbf{v}}_e. \tag{9.22}$$

By (9.20) to (9.22),

$$m_e \frac{\partial}{\partial t} \nabla \cdot \hat{\mathbf{v}}_e = -\frac{m_e}{n_{e0}} \frac{\partial^2 \hat{n}_e}{\partial t^2} = -e\nabla \cdot \hat{\mathbf{v}}_e = \frac{e^2}{\epsilon_0} \hat{n}_e,$$

i.e.

$$\frac{\partial^2 \hat{n}_e}{\partial t^2} + \omega_{pe}^2 \hat{n}_e = 0, \tag{9.23}$$

where

$$\omega_{pe} \equiv \left(\frac{n_{e0} e^2}{\epsilon_0 m_e} \right)^{\frac{1}{2}} \tag{9.24}$$

is called the *plasma frequency*. This is evidently the natural frequency at which the highly mobile electrons (relative to the heavy ions) seek to preserve charge neutrality.

9.2.3 The Debye length

In steady-state conditions (9.1) becomes $\nabla \times \mathbf{E} = 0$, permitting the introduction of a scalar potential ϕ via $\mathbf{E} = -\nabla \phi$. Consider the small perturbation $\hat{\phi}$ due to a point charge at the origin. On assuming the ion density to be unchanged by the charge, we obtain from (9.21)

$$\epsilon_0 \nabla^2 \hat{\phi} = -e\hat{n}_e. \tag{9.25}$$

The force per unit mass is $e\mathbf{E}/m_e = -\nabla(\phi/m_e)$, and therefore to apply Boltzmann's distribution law to the present case, we replace Φ in (2.96) by $e\phi/m_e$. Also, since it is the distribution function for the electrons that is relevant here, we set $T = T_e$. This gives

$$n_e = n_{e0} \exp(-e\phi/kT_e). \tag{9.26}$$

Hence for a small perturbation,

$$\hat{n}_e = -n_e e\hat{\phi}/kT_e.$$

Thus
$$\lambda_D^2 \nabla^2 \hat{\phi} = \hat{\phi}, \tag{9.27}$$

where
$$\lambda_D \equiv \left(\frac{\epsilon_0 kT_e}{n_{e0} e^2}\right)^{\frac{1}{2}} = \frac{(kT_e/m_e)^{\frac{1}{2}}}{\omega_{pe}}. \tag{9.28}$$

With spherical symmetry about the origin, the solution of (9.27) is

$$\hat{\phi} = \frac{Q_0}{r} \exp(-r/\lambda_D), \tag{9.29}$$

where Q_0 is the charge at the origin. This result shows that the movement of the electrons has masked the Coulomb potential Q_0/r, due to the fixed charge at the origin, almost completely in a distance λ_D. This shielding distance is known as the *Debye length*—it is the scale length below which average charge neutrality cannot be guaranteed.

9.3 Conservation equations

9.3.1 The one-fluid equations

A particle of the α-species, with mass m_α, charge Q_α and velocity $\mathbf{w}_\alpha$, is subject to a force per unit mass:

$$\mathbf{F}_\alpha = \frac{Q_\alpha}{m_\alpha}(\mathbf{E} + \mathbf{w}_\alpha \times \mathbf{B}). \tag{9.30}$$

On averaging this force over the α-particles at a macroscopic point, we obtain

$$\rho_\alpha \mathbf{F}_\alpha = n_\alpha Q_\alpha (\mathbf{E} + \mathbf{v}_\alpha \times \mathbf{B}), \tag{9.31}$$

where $\mathbf{F}_\alpha$ is the force acting on the α-fluid per unit mass.

For a 'one-fluid' treatment of the mixture, we require the sums

$$\sum \rho_\alpha \mathbf{F}_\alpha = Q\mathbf{E} + \mathbf{j} \times \mathbf{B}, \qquad \sum \rho_\alpha \mathbf{F}_\alpha \cdot \mathbf{v}_\alpha = \mathbf{j} \cdot \mathbf{E}.$$

Then, with average charge neutrality, the conservation equations in (1.46) to (1.48) become

$$\mathrm{D}\rho + \rho \nabla \cdot \mathbf{v} = 0, \tag{9.32}$$

$$\rho \mathrm{D}\mathbf{v} + \nabla \cdot \mathbf{p} = \mathbf{j} \times \mathbf{B}, \tag{9.33}$$

$$\rho \mathrm{D}u + \mathbf{p} : \nabla \mathbf{v} = -\nabla \cdot \mathbf{q} + \mathbf{j} \cdot (\mathbf{E} + \mathbf{v} \times \mathbf{B}). \tag{9.34}$$

9.3.2 *The two-fluid equations*

Some details are evidently lost in the one-fluid treatment. The principal of these is the distinction between the temperatures of the electrons and ions. When this is significant, it is necessary to retain separate equations for the fluids, and include terms for the interactions between them. With one ion fluid plus the electron fluid comprising the plasma, the equations are

$$\mathrm{D}_e \rho_e + \rho_e \nabla \cdot \mathbf{v}_e = 0, \tag{9.35}$$

$$\mathrm{D}_i \rho_i + \rho_i \nabla \cdot \mathbf{v}_i = 0, \tag{9.36}$$

$$\rho_e \mathrm{D}_e \mathbf{v}_e + \nabla \cdot \mathbf{p}_e = -e n_e (\mathbf{E} + \mathbf{v}_e \times \mathbf{B}) + \mathbf{R}_e, \tag{9.37}$$

$$\rho_i \mathrm{D}_i \mathbf{v}_i + \nabla \cdot \mathbf{p}_i = Z e n_e (\mathbf{E} + \mathbf{v}_i \times \mathbf{B}) + \mathbf{R}_i, \tag{9.38}$$

$$\tfrac{3}{2} n_e k \mathrm{D}_e T_e + p_e \nabla \cdot \mathbf{v}_e = -\nabla \cdot \mathbf{q}_e - \boldsymbol{\pi}_e \cdot \nabla \mathbf{v}_e + \psi_e + \varphi_e + \xi_e, \tag{9.39}$$

$$\tfrac{3}{2} n_i k \mathrm{D}_i T_i + p_i \nabla \cdot \mathbf{v}_i = -\nabla \cdot \mathbf{q}_i - \boldsymbol{\pi}_i \cdot \nabla \mathbf{v}_i + \psi_i + \varphi_i + \xi_i, \tag{9.40}$$

where

$$\mathrm{D}_e = \frac{\partial}{\partial t} + \mathbf{v}_e \cdot \nabla, \qquad \mathrm{D}_i = \frac{\partial}{\partial t} + \mathbf{v}_i \cdot \nabla, \tag{9.41}$$

and

$$p_e = k n_e T_e, \qquad p_i = k n_i T_i. \tag{9.42}$$

There are four interaction terms in the equations. Per unit volume of the electron gas, $\mathbf{R}_e$ is the force due to friction with the ion gas, ξ_e is the heating rate due to friction, φ_e is the heating rate due to the electron-ion temperature difference, and ψ_e is the heating rate due to radiation. The quantities $\mathbf{R}_i$, ξ_i, φ_i, and ψ_i have similar roles for the ion gas.

By Newton's third law and energy conservation during particle interactions,

$$\mathbf{R}_i = -\mathbf{R}_e, \qquad \varphi_i = -\varphi_e. \tag{9.43}$$

As the electron fluid moves with velocity $(\mathbf{v}_e - \mathbf{v}_i)$ relative to the ion fluid and exerts a force $\mathbf{R}_i$ on it, the friction force does work on unit volume at the rate $\mathbf{R}_i \cdot (\mathbf{v}_e - \mathbf{v}_i) = \mathbf{R}_e \cdot (\mathbf{v}_i - \mathbf{v}_e)$. It is dissipated into the two fluids at the rate $(\xi_e + \xi_e)$. The theory of isotropic, elastic collisions implies that this energy is distributed between the two fluids inversely as the particle masses, whence

$$\xi_e = \frac{m_i}{m_i + m_e}\mathbf{R}_e \cdot (\mathbf{v}_i - \mathbf{v}_e), \qquad \xi_i = \frac{m_e}{m_i + m_e}\mathbf{R}_e \cdot (\mathbf{v}_i - \mathbf{v}_e).$$

As $m_e \ll m_i$, we can adopt the approximations

$$\xi_e = \mathbf{R}_e \cdot (\mathbf{v}_i - \mathbf{v}_e), \qquad \xi_i = 0. \tag{9.44}$$

To complete the set of two-fluid equations, we require constitutive equations for $\mathbf{q}_e$, $\mathbf{q}_i$, $\boldsymbol{\pi}_e$, $\boldsymbol{\pi}_i$, $\mathbf{R}_e$, φ_e, and expressions for the rate at which radiation is emitted or absorbed. In the following we shall omit the radiation term.

9.4 Generalized Ohm's law

Here the general expressions for diffusion, given in §§4.2 and 4.5, will be applied to the two-fluid plasma for the case of zero magnetic field. The effects of magnetic fields will be addressed in §10.6.

9.4.1 *Electron-ion diffusion*

The Coulomb force law index is $\nu = 2$, so by (4.37), $1 - s = -3/2$. Also, since m_e is some three orders of magnitude smaller than m_i, the thermal diffusion is almost entirely due to the electron temperature gradient (see exercise 4.8). Thus (4.24) gives

$$\mathbf{v}_e - \mathbf{v}_i = -\frac{n^2}{n_e n_i} D_{ei}\mathbf{d}_{ei} - D_{ei}\alpha_T \nabla \ln T_e, \tag{9.45}$$

where from (4.8), (4.9), (4.38), and (9.31),

$$\begin{aligned} \mathbf{d}_{ei} &= \frac{\rho_i \rho_e}{p\rho}\left\{\frac{e}{m_e}(\mathbf{E} + \mathbf{v}_e \times \mathbf{B}) + \frac{1}{n_e m_e}\nabla p_e\right\} \\ D_{ei} &= \frac{\rho \tau_{ei} kT}{n m_i m_e} \qquad (n = n_i + n_e), \end{aligned} \tag{9.46}$$

and $\qquad \alpha_T = \frac{3}{2}n/n_i, \qquad \rho = \rho_i + \rho_e \approx n_i m_i.$

By (9.11) we can write (9.45) in the form of a generalized Ohm's law:

$$\mathbf{j} = \sigma(\mathbf{E} + \mathbf{v}_e \times \mathbf{B} + \frac{1}{en_e}\nabla p_e) + \beta \nabla T_e, \tag{9.47}$$

where
$$\sigma = \frac{e^2 n_e \tau_{ei}}{m_e}, \qquad \beta = \frac{3k}{2e}\left(\frac{e^2 n_e \tau_{ei}}{m_e}\right). \tag{9.48}$$

A good approximation to the friction force can be deduced from (9.37) by ignoring the electron viscosity and inertia terms:

$$\mathbf{R}_e = en_e(\mathbf{E} + \mathbf{v}_e \times \mathbf{B} + \frac{1}{en_e}\nabla p_e). \tag{9.49}$$

Then from (9.47),
$$\mathbf{R}_e = en_e(\eta\mathbf{j} - \delta\nabla T_e), \tag{9.50}$$
where
$$\eta = 1/\sigma, \qquad \delta = \beta/\sigma = \eta\beta. \tag{9.51}$$

9.4.2 *The collision interval*

We require an expression for τ_{ei}. The result in exercise 4.6 shows that the mutual diffusivity is principally determined by the electron temperature, which accounts for the appearance of T_e in the following.

From the last equation in §6.8.1 and $M = m_i m_e/(m_i + m_e) \approx m_e$, we have
$$D_{ei} = 3kT_e/(16 m_e n \Omega_D),$$
then from (9.46),
$$\tau_{ei}^{-1} = \tfrac{16}{3} n_i \Omega_D. \tag{9.52}$$

The value of Ω_D is obtained by setting $Z_1 = Z, Z_2 = 1$ in (6.76) and (6.68). Then we find
$$\tau_{ei} = \tau_e, \tag{9.53}$$
where
$$\tau_e \equiv \frac{3m_e^2}{n_i Z^2 \ln\Lambda}\frac{\epsilon_0^2}{e^4}\left(\frac{2\pi kT_e}{m_e}\right)^{\frac{3}{2}}. \tag{9.54}$$

The formula for Λ follows from (6.73), (6.75), and the average value $3kT_e/m_e$ for g^2, given in (2.44):
$$\Lambda = \lambda_D\left(\frac{12\pi\epsilon_0^2 kT_e}{Ze^2}\right) = 12\pi\frac{n_e}{Z}\left(\frac{\epsilon_0 kT_e}{e^2 n_e}\right)^{\frac{3}{2}}. \tag{9.55}$$

The theory is readily modified to allow for the presence of several types of ion in the plasma. If the species 's' has number density n_s and charge number Z_s, in place of the frequency $\tau_e^{-1} \propto n_i Z^2$, we write $\tau_e^{-1} \propto n_s Z_s^2$ and sum over s. Ignoring the weak dependence of $\ln\Lambda$ on Z, this is equivalent to replacing $n_i Z^2 = n_e Z$ in (9.54) by $n_e Z^*$, where
$$n_e Z^* = \textstyle\sum_s n_s Z_s^2, \qquad n_e = \sum_s n_s Z_s. \tag{9.56}$$
The average Z^* is termed 'Z-effective' and usually written as Z_{eff}.

A convenient practical unit for the temperature is the electron volt. We shall denote practical units by a tilde. Thus if $\tilde{T}_e$ is in electron volts and $\tilde{n}_e$ in particles per cubic centimetre,
$$kT_e = e\tilde{T}_e, \qquad n_e = 10^6\tilde{n}_e.$$
Then using the table of numerical values given in Appendix A1, we find
$$\tau_e = \frac{3.44\times 10^5}{\ln\Lambda}\frac{\tilde{T}_e^{3/2}}{Z^*\tilde{n}_e}\ \text{secs.} \tag{9.57}$$

Equation (9.53) gives only the *first* approximation $[\tau_{ei}]_1$ to τ_{ei}, and as remarked at the end of §6.8.4, it is not accurate. We shall return to this point shortly.

9.4.3 *The Coulomb logarithm*

The factor $\ln\Lambda$ is termed the 'Coulomb logarithm'. Above a temperature of $T_e = 4.2 \times 10^5$ K, quantum mechanical effects increase the effective impact parameter and hence reduce Λ. According to Spitzer (1962) this may be allowed for by replacing the value of Λ in (9.52) by

$$\Lambda^* = \left(\frac{4.2 \times 10^5}{T_e}\right)^{\frac{1}{2}} \Lambda = \left(\frac{36.2}{\tilde{T}_e}\right)^{\frac{1}{2}} \Lambda, \tag{9.58}$$

when $T_e > 4.2 \times 10^5$ K.

Values for $\ln\Lambda$ (for electron–electron collisions and electron–ion collisions at small Z) follow from (9.55) and (9.58):

$$\ln\Lambda = \begin{cases} 23.46 + 1.5\ln\tilde{T}_e - 0.5\ln\tilde{n}_e & (\tilde{T}_e < 36.2\text{eV}) \\ 25.25 + \ln\tilde{T}_e \quad - 0.5\ln\tilde{n}_e & (\tilde{T}_e > 36.2\text{eV}) \end{cases}. \tag{9.59}$$

9.4.4 *Electrical conductivity for zero magnetic field*

The results quoted below are obtained from an algebraically complicated analysis, not appropriate to this text (see Spitzer (1962), Shkarofsky et al. (1966), and Ferziger and Kaper (1972)).

Consider first the case of single ionization, i.e. $Z = 1$. The collision interval τ_e defined in (9.54) is adopted as a standard, in terms of which related relaxation times are expressed. The Chapman–Enskog approximations, explained in §§6.8.3 and 7.10.1, are found to be

$$[\tau_{ei}]_1 = \tau_e, \quad [\tau_{ei}]_2 = 1.931\tau_e, \quad \ldots, \quad [\tau_{ei}]_\infty = 1.98\cdot\cdot\tau_e.$$

It follows from (9.48) that

$$\sigma = 1.98\frac{e^2 n_e \tau_e}{m_e} \qquad (Z = 1). \tag{9.60}$$

The Lorentzian plasma is a special case of theoretical interest. This is a model plasma, in which there are no electron–electron collisions and the ions have no temperature. It approximates to a real plasma with very large Z-values. Its conductivity is (Lorentz 1923)

$$\sigma_L = \frac{32}{3\pi}\frac{e^2 n_e \tau_e}{m_e} \qquad (Z \gg 1). \tag{9.61}$$

We shall derive this result in §10.3.2 from the Fokker–Planck kinetic equation.

Spitzer and Härm (1953) showed that the general expression for the conductivity could be expressed as

$$\sigma = \gamma_E \sigma_L, \tag{9.62}$$

where the ratio γ_E depends on Z^*. They obtained the values set out in Table 9.1.

Table 9.1 *Ratio of Spitzer to Lorentzian conductivity*

Z^*	1	2	4	16	∞
γ_E	0.582	0.683	0.785	0.923	1.000

An approximate analytical formula, agreeing with the table to within 0.5 per cent, is

$$\sigma = \left(0.295 + \frac{0.39}{0.85 + Z^*}\right)^{-1} \frac{e^2 n_e \tau_e}{m_e}. \tag{9.63}$$

9.4.5 *Thermoelectric coefficient*

The modifications to τ_{ei} described above apply only to its role in determining the conductivity. The thermoelectric coefficient β, defined in (9.47), has the form given in (9.48) as a first approximation. It is found that the ratio δ defined in (9.51) varies between 0.71 at $Z^* = 1$, to 1.52 at $Z^* = \infty$, whereas (9.48) gives the fixed value $\delta = 1.5$. We shall not pursue this further, but in §10.10 we shall provide a complete list of the constitutive equations required to complete the two-fluid description of a magnetoplasma described in §9.3.2.

Collision intervals like τ_e and corresponding expressions for the ion fluid are not the only microscopic times of importance in magnetoplasmas. Variations in the magnetic field result in the trapping of particles, a process that can introduce times much shorter than τ_e into the kinetic equation. An introduction to this topic will occupy the following pages.

9.5 Guiding centre motion

9.5.1 *Particle acceleration in a convected frame*

In §3.4.2 we showed that, measured in a convected frame P_c, a neutral particle P experiences an acceleration given by

$$\dot{\mathbf{c}} = -\mathbf{c}\cdot\mathbf{e} + \mathcal{F} \qquad (\mathcal{F} = \dot{\mathbf{w}} - \overline{\dot{\mathbf{w}}}).$$

Here, the agitation acceleration $\mathcal{F}$ is due only to the impulsive collisional forces. With charged particles, in the presence of a macroscopic magnetic field, the Lorentz force adds a term to $\mathcal{F}$, since by (9.12) $\dot{\mathbf{w}}$ includes the term

$$\dot{\mathbf{w}}_L = \frac{Q}{m}(\mathbf{E} + (\mathbf{v}+\mathbf{c})\times\mathbf{B}) = \overline{\dot{\mathbf{w}}}_L + \frac{Q}{m}\mathbf{c}\times\mathbf{B}.$$

Thus

$$\dot{\mathbf{c}} = \frac{Q}{m}\mathbf{c}\times\mathbf{B} - \mathbf{c}\cdot\mathbf{e} + \mathcal{F}_s, \quad \left(\mathbf{e} \equiv \tfrac{1}{2}(\nabla\mathbf{v} + \widetilde{\nabla\mathbf{v}})\right), \tag{9.64}$$

where $\mathcal{F}_s$ is due only to scattering collisions.

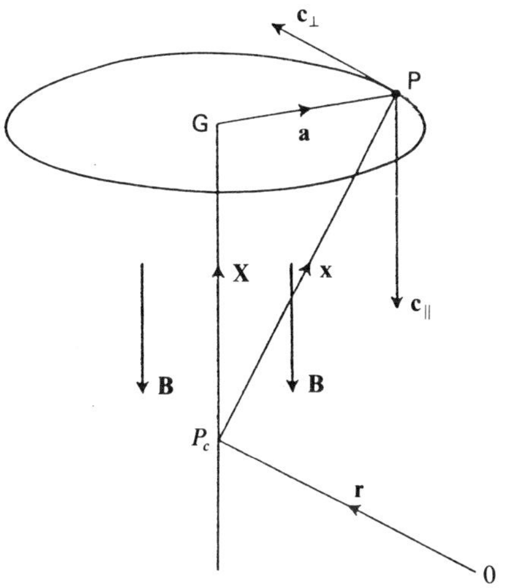

FIG. 9.1. Guiding centre motion in a uniform field

As $\mathcal{F}_s$ is proportional to the collision frequency, τ^{-1} and

$$\frac{Q}{m}\mathbf{c}\times\mathbf{B}=\omega_c\mathbf{c}\times\mathbf{b} \qquad \big(\omega_c=QB/m, \quad \mathbf{b}=\mathbf{B}/|\mathbf{B}|\big), \tag{9.65}$$

the time scales for the right-hand terms in (9.64) are ω_c^{-1}, T, and τ, where T is the macroscopic time scale for changes in the fluid velocity. The Knudsen number is $\epsilon=\tau/\mathsf{T}$, and as we shall require $\epsilon\ll 1$, it follows that $\omega_c\tau\gg 1$ is sufficient to allow (9.64) to be approximated by

$$\dot{\mathbf{c}}=\frac{Q}{m}\mathbf{c}\times\mathbf{B} \qquad (\omega_c\tau\gg 1). \tag{9.66}$$

9.5.2 *Guiding centre drifts*

Let $\mathbf{x}$ denote the position vector of P relative to P_c, then $\dot{\mathbf{x}}=\mathbf{c}$ and (9.66) is

$$\ddot{\mathbf{x}}=\omega_c\dot{\mathbf{x}}\times\mathbf{b}. \tag{9.67}$$

Now assume that $\mathbf{B}$ is steady and uniform. The solution of (9.67) is

$$\mathbf{c}=\dot{\mathbf{x}}=\omega_c(\mathbf{x}-\mathbf{X})\times\mathbf{b}+\mathbf{c}_\parallel \qquad (\mathbf{c}_\parallel\equiv\mathbf{bb}\cdot\mathbf{c}), \tag{9.68}$$

where $\mathbf{X}_\parallel$ is arbitrary. Chose $\mathbf{X}_\parallel$ so that

$$\mathbf{a}=\mathbf{x}-\mathbf{X} \tag{9.69}$$

is always perpendicular to $\mathbf{b}$, then $\dot{\mathbf{X}}_\parallel=c_\parallel\mathbf{b}$, and the derivative of (9.68), i.e.

$$\ddot{\mathbf{x}}=\omega_c(\dot{\mathbf{x}}-\dot{\mathbf{X}})\times\mathbf{b}+\dot{\mathbf{c}}_\parallel$$

is the same as (9.66) if $\dot{\mathbf{X}}=\dot{\mathbf{X}}_\parallel$ and $\dot{\mathbf{c}}_\parallel=c_\parallel\mathbf{b}=0$. Hence $\ddot{\mathbf{X}}=0$. It now follows from (9.66) to (9.69) that

$$\ddot{\mathbf{a}}=-\omega_c\mathbf{a}, \qquad \mathbf{a}=\omega_c^{-1}\mathbf{b}\times\mathbf{c}, \qquad a=c_\perp/\omega_c, \tag{9.70}$$

which describes the motion of a vector rotating about an end point G at $\mathbf{X}$ (relative to P_c), with angular velocity ω_c. The point G is the guiding centre, introduced in §9.2.1.

The situation is depicted in Fig. 9.1. The point G follows P along the magnetic field lines, but relative to the fluid particle centred on P_c, has no motion perpendicular to the field. Of course P_c moves perpendicular to the field lines, for (9.37) and (9.38), expressed in general form, yield

$$\mathbf{v}_\perp = \frac{\mathbf{E}}{B} \times \mathbf{b} + \frac{m}{QB}\left(\mathbf{f} - \frac{1}{\rho}\nabla \cdot \mathsf{p} + \frac{1}{\rho}\mathbf{R} - \mathrm{D}\mathbf{v}\right) \times \mathbf{b}, \tag{9.71}$$

where for generality we have added an external force $\mathbf{f}$. Let

$$\mathbf{u} = \dot{\mathbf{x}} + \dot{\mathbf{r}} = \dot{\mathbf{X}} + \mathbf{v} \tag{9.72}$$

denote the guiding centre velocity in the laboratory frame, then we have established that in a steady, uniform magnetic field,

$$\mathbf{u}_\perp = \mathbf{v}_\perp, \quad \mathbf{u}_\parallel = \mathbf{v}_\parallel + c_\parallel \mathbf{b}, \quad \mathbf{u} = \mathbf{v} + \mathbf{c}_\parallel. \tag{9.73}$$

From (9.71) and (9.73) we have

$$\mathbf{u}_\perp = \frac{\mathbf{E} \times \mathbf{b}}{B} + \frac{m}{QB}\left(\mathbf{f} \times \mathbf{b} - \frac{1}{\rho}\nabla \cdot \mathsf{p} \times \mathbf{b} + \frac{1}{\rho}\mathbf{R} \times \mathbf{b} - \dot{\mathbf{u}} \times \mathbf{b}\right), \tag{9.74}$$

where
$$\dot{\mathbf{u}} = \frac{\partial \mathbf{u}}{\partial t} + \mathbf{u} \cdot \nabla \mathbf{u},$$

is the rate of change of $\mathbf{u}$ in the guiding centre frame.

Perpendicular motions of guiding centres are called 'drifts', and these are named according to the forces that cause them. Thus the drifts in (9.71) are 'electric', 'external force', 'pressure gradient', and so on. In general a force $\mathbf{F}'$ causes a drift $\mathbf{u}'_\perp$, and conversely a drift $\mathbf{u}'_\perp$ requires the presence of a force $\mathbf{F}'$, where

$$\mathbf{F}' = \frac{QB}{m}\mathbf{b} \times \mathbf{u}'_\perp, \qquad \mathbf{u}'_\perp = \frac{m}{QB}\mathbf{F}' \times \mathbf{b}. \tag{9.75}$$

9.5.3 *Drifts due to variations in the magnetic field*

There are also drifts caused by magnetic field inhomogeneities that depend on the particle's peculiar velocity. As their contribution to transport is less important than that due to the second-order terms to be discussed later in this chapter, we shall not derive the drift formula here. By using the method of variation of parameters, it can be shown (Woods 1987) that the additional drift is

$$\delta\mathbf{u}_\perp = \frac{1}{QnB}\{\rho(\tfrac{1}{2}c_\perp^2 + c_\parallel^2) - (p_\perp + p_\parallel)\}\mathbf{b} \times \nabla \ln B + \frac{\mu_0}{QnB^2}(\rho c_\parallel^2 - p_\parallel)\mathbf{j}_\perp. \tag{9.76}$$

It follows from equations (1.26) that $\langle \delta\mathbf{u}_\perp \rangle = 0$. In place of (9.73) we now have

$$\mathbf{u} = \mathbf{v} + \delta\mathbf{u}_\perp + c_\parallel \mathbf{b}, \tag{9.77}$$

for the relation between the guiding centre velocity $\mathbf{u}$ and the fluid velocity $\mathbf{v}$. The mean guiding centre velocity, $\langle \mathbf{u} \rangle$, is identical with the fluid velocity, provided of course that $\omega_c \tau \gg 0$.

9.6 Magnetic mirrors

9.6.1 *Constants of the motion of gyrating particles*

Because of their gyratory motion, each charged particle behaves like a dipole, with a magnetic moment $\mathbf{M}$ given by

$$\mathbf{M} = -m\mathbf{b}, \qquad M = \frac{mc_\perp^2}{2B}, \qquad \langle M \rangle = \frac{p_\perp}{nB}, \tag{9.78}$$

where we have included an expression for the average value of the moment for a given species of particle. Provided the magnetic field changes slowly enough, the particles tend to move so as to enclose a constant magnetic flux within their Larmor orbits, i.e. $\pi r_L^2 B \propto c_\perp^2 / B$ is approximately constant. We can show this as follows.

From (9.66), in a frame convected with the fluid,

$$m\dot{\mathbf{c}} = m\dot{\mathbf{c}}_\perp + m\dot{\mathbf{c}}_\parallel = Q\mathbf{c}_\perp \times \mathbf{B}, \tag{9.79}$$

where in general $\mathbf{B}$ is a function of $\mathbf{r}$ and t. In the fluid frame only convective changes due to the peculiar velocity remain, hence

$$\dot{\mathbf{c}}_\parallel = c_\parallel \dot{\mathbf{b}} + \dot{c}_\parallel \mathbf{b}, \qquad \dot{\mathbf{b}} = (\mathbf{c}_\parallel + \mathbf{c}_\perp) \cdot \nabla \mathbf{b}.$$

Therefore, since $\overline{\mathbf{c}_\perp} = 0$, the scalar product of (9.79) with $\mathbf{c}_\perp$, followed by averaging over a gyration, yields

$$\frac{d}{dt}\left(\tfrac{1}{2} m c_\perp^2\right) = -m c_\parallel \overline{\mathbf{c}_\perp \cdot \nabla \mathbf{b} \cdot \mathbf{c}_\perp}. \tag{9.80}$$

Let $\hat{\mathbf{r}}$, $\hat{\boldsymbol{\theta}}$ denote unit polar vectors orthogonal to $\mathbf{b}$, then $\mathbf{c}_\perp = c_\perp \hat{\boldsymbol{\theta}}$. From the gyro-average of

$$\mathbf{1} \cdot \mathsf{A} = (\hat{\mathbf{r}}\hat{\mathbf{r}} + \hat{\boldsymbol{\theta}}\hat{\boldsymbol{\theta}} + \mathbf{b}\mathbf{b}) \cdot \mathsf{A},$$

where A is any second order tensor, it follows that

$$\overline{\hat{\mathbf{r}}\hat{\mathbf{r}}} \cdot \mathsf{A} = \overline{\hat{\boldsymbol{\theta}}\hat{\boldsymbol{\theta}}} \cdot \mathsf{A} = \tfrac{1}{2}(\mathsf{A} - \mathbf{b}\mathbf{b} \cdot \mathsf{A}) = \tfrac{1}{2}(\mathbf{1} - \mathbf{b}\mathbf{b}) \cdot \mathsf{A},$$

i.e.

$$\overline{\hat{\mathbf{r}}\hat{\mathbf{r}}} = \overline{\hat{\boldsymbol{\theta}}\hat{\boldsymbol{\theta}}} = \tfrac{1}{2}(\mathbf{1} - \mathbf{b}\mathbf{b}). \tag{9.81}$$

Therefore, since $\nabla \mathbf{b} \cdot \mathbf{b}$ is zero,

$$\overline{\mathbf{c}_\perp \cdot \nabla \mathbf{b} \cdot \mathbf{c}_\perp} = \tfrac{1}{2} c_\perp^2 (\mathbf{1} - \mathbf{b}\mathbf{b}) : \nabla \mathbf{b} = \tfrac{1}{2} c_\perp^2 \nabla \cdot \mathbf{b}.$$

Now

$$\nabla \cdot \mathbf{b} = \nabla \cdot (\mathbf{B}/B) = -\mathbf{b} \cdot \nabla \ln B, \tag{9.82}$$

whence (9.80) yields

$$\frac{d}{dt}\left(\tfrac{1}{2} m c_\perp^2\right) = \tfrac{1}{2} m c_\perp^2 \, \mathbf{c}_\parallel \cdot \nabla \ln B = \tfrac{1}{2} m c_\perp^2 \frac{d}{dt}(\ln B),$$

or

$$\frac{dM}{dt} = 0, \qquad M = \frac{mc_\perp^2}{2B}. \tag{9.83}$$

It also follows directly from (9.79) that $d(\frac{1}{2}mc^2)/dt = 0$. Thus there are two 'constants' of the particle motion, namely

$$M = \frac{mc_\perp^2}{2B} = \text{const.}, \qquad \mathsf{E} = \tfrac{1}{2}mc^2 = \text{const.} \tag{9.84}$$

In general neither of these quantities are *exact* constants. They require variations of B across the Larmor radius $\mathbf{a}$ to be small, i.e.

$$\epsilon_L \equiv |\mathbf{a} \cdot \nabla \ln B| \ll 1. \tag{9.85}$$

And of course collisions must be relatively rare, which in this context means that $\omega_c \tau \gg 1$.

9.6.2 *Magnetically trapped particles*

An immediate consequence of (9.84) is that as a particle P approaches a region of magnetic field of increasing strength, the increase required in $c_\perp$ to balance the increase in B can be found only at the expense of $c_\parallel$. Thus P's guiding centre G has a reducing value of $c_\parallel$ and eventually it may stop, reverse its motion and move away from the strong magnetic field region. For this reason the term 'mirror' is an apt description of the region where B has its maximum value. Two mirrors, as depicted in Fig. 9.2, make a 'magnetic bottle' and some of the particles within the bottle will be *trapped*, with their guiding centres oscillating between the mirrors.

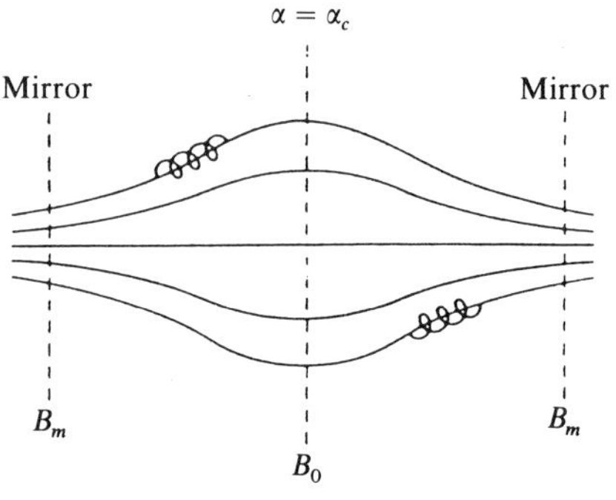

FIG. 9.2. Magnetic bottle

By (9.84) $c_\parallel$ and B are related by

$$\mathsf{E} = \tfrac{1}{2}mc_\parallel^2 + MB = \mathsf{E}(\sin^2\vartheta + MB/\mathsf{E}), \tag{9.86}$$

where
$$\cos^2\vartheta = \frac{c_\perp^2}{c_\perp^2 + c_\parallel^2}.$$

Here ϑ is the pitch angle of P's trajectory, measured from a plane perpendicular to the magnetic field. Thus as P approaches the mirror field, B may increase sufficiently to reduce its value of $c_\parallel$ to zero, at which point P is reflected from the mirror.

Let B_m be the maximum field strength of a mirror, then particles with $\mathsf{E} < MB_m$ will be reflected, and hence in a symmetrical field as shown in the figure, they will be trapped. It follows that those particles in any region where the field strength is B, having pitch angles satisfying

$$\alpha < \vartheta < \pi/2 \qquad (\alpha \equiv \sin^{-1}(1 - B/B_m)^{\frac{1}{2}}), \tag{9.87}$$

will escape through the magnetic 'throat' at the mirror. The boundary between *captured* $(0 < \vartheta < \alpha)$ and *passing* $(\alpha < \vartheta < \pi/2)$ particles is determined by the pitch angle α.

9.6.3 *The fraction of trapped particles*

Consider the magnetic bottle of Fig. 9.2. Let α_c be the critical pitch angle at the point of minimum magnetic field B_0, then by (9.87) the bottle holds all particles for which

$$0 < \vartheta < \alpha_c, \qquad \cos^2 \alpha_c = R_m^{-1} \qquad (R_m \equiv B_m/B_0), \tag{9.88}$$

where R_m is termed the 'mirror ratio'. Collisions will steadily scatter these particles into one of the two loss cones, $\alpha_c < \vartheta < \pi/2$, $-\pi/2 < \vartheta < -\alpha_c$, allowing the bottle to leak on the collision time scale τ.

A simple expression for the fraction of trapped particles can be deduced by assuming that they are part of an equilibrium distribution. With a velocity-space, spherical coordinate system (c, θ, φ), oriented so that $c_\parallel$ lies along the axis from which θ is measured, the element of solid angle is $\sin\theta \, d\theta \, d\varphi$, and the trapped particles lie in $\pi/2 - \alpha_c < \theta < \pi/2 + \alpha_c$. Then by (2.37) the fraction of trapped particles is

$$f_T = \frac{1}{4\pi} \int_0^{2\pi} d\varphi \int_{\pi/2-\alpha_c}^{\pi/2+\alpha_c} \sin\theta \, d\theta \int_0^\infty \frac{4}{\sqrt{\pi}} \nu^2 e^{-\nu^2} d\nu,$$

i.e.

$$f_T = \sin\alpha_c = (1 - R_m^{-1})^{\frac{1}{2}}. \tag{9.89}$$

That particles can be trapped by magnetic fields has important consequences for energy confinement in torodial magnetic fields, as will be explained in §11.8.

9.7 Heat flux in strong magnetic fields

Our objective here is to make a start on the problem of transport in strong magnetic fields by extending the elementary kinetic theory of Chapter 3 to a magnetoplasma. The free-path model is conceptually easy to understand, although uncertainty about its completeness later requires us to adopt a more formal treatment based on a kinetic equation. This we shall do in the next chapter. But our preliminary account of heat flux using elementary methods will provide a clear physical description of the origin of the most important term in this flux.

9.7.1 *Elementary kinetic theory*

The elementary kinetic theory described in §§3.4 to 3.7 was based on equation (3.40). Using (3.19) to replace τ_1 by τ_2 in the temperature-dependent term of this equation, writing $\mathcal{C}\boldsymbol{\nu} = \mathbf{c}$ and omitting the error term, we have

$$\mathbf{c} = (\mathcal{C}\boldsymbol{\nu} - t\overset{\circ}{\mathbf{e}})e^{-t/\tau_1} - \tfrac{1}{3}\tau_2(\nu^2 - \tfrac{5}{2})(\mathbf{c}\mathrm{D}\ln T + \mathbf{cc}\cdot\nabla\ln T)e^{-3t/2\tau_2}. \quad (9.90)$$

We shall alter the description given in §3.5.1 by subtracting instead of adding collision times, which is merely a change in the time origin. Thus (9.90) gives the peculiar velocity of a bunch of molecules $\mathcal{B}$ arriving at a 'target' $Q_c(\mathbf{r}, t)$ having left a 'source' $P_c(\mathbf{r}-\mathbf{c}\tau, t-\tau)$ a time τ earlier, where τ is either τ_1 for momentum transport or τ_2 for energy transport. Both Q_c and P_c are convected with the fluid. To enter $\mathcal{B}$ the molecules experience a collision at P_c and then have no further collisions until they pass Q_c. As explained in §5.11, this synchronous model is easily avoided, but as it leads to no errors and facilitates comprehension of the mechanism of transport, we shall adopt it.

In the free-path model the molecules are assumed to belong to an equilibrium distribution at P_c, so that when they arrive at Q_c without experiencing collisions *en route*, they will be in disequilibrium relative to the local conditions. An essential feature of the model is that the peculiar velocity of the molecules at Q_c depends on the macroscopic conditions prevailing earlier at P_c. We can think of P_c as being a source that 'labels' the molecules in $\mathcal{B}$ with its macroscopic variables, T and $\mathbf{v}$.

To extend elementary kinetic theory to a magnetoplasma, we need to allow for the fact that the straight trajectories between collisions in a neutral gas are replaced by circular orbits about the field lines. By 'strong' magnetic fields we mean fields for which there are a great number of gyrations per collision interval, i.e. the parameter

$$\varpi \equiv \omega_c\tau \quad (9.91)$$

is very much larger than unity. In this case the projections of particle orbits on to a plane normal to $\mathbf{b}$ are tight, repeating circles. Therefore the simple notion of a physical source, lying a mean free path back along the trajectory, and that labels the particle with its temperature and fluid velocity, needs to be revised. It is evidently necessary to relate particle speeds at Q_c to macroscopic conditions at some point P_s lying *within* the circular path. As we shall see in §9.8.1 the location of this point is determined by the magnitude of ϖ.

9.7.2 *Guiding centres as labelling sources*

Consider the ion P, shown in Fig. 9.3, that has just passed through $Q_c(\mathbf{r}, t)$ with the peculiar velocity $\mathbf{c} = c_\perp\hat{\mathbf{c}}_\perp + c_\parallel\mathbf{b}$, where $\hat{\mathbf{c}}_\perp$ is unit vector. According to (9.70), relative to Q_c, its guiding centre $\mathsf{G}(\mathbf{X}, t)$ is at

$$\mathbf{X} = \mathbf{r} - \mathbf{a}, \qquad \mathbf{a} = \omega_c^{-1}\mathbf{b}\times\mathbf{c}, \qquad |\mathbf{a}| = c_\perp/\omega_c. \quad (9.92)$$

Typically P may gyrate about G millions of times per collision interval. The distribution function from which its peculiar velocity is chosen, i.e. the one that labels P, has macroscopic variables obtained by averaging over the orbits. When ϖ is very large, the ambient temperatures at all points on the orbits have equal weight in this averaging process, which means that it will yield the temperature at the guiding centre, G, subject to the usual collision-interval time delay. In this case the velocity distribution from which P's *present* peculiar velocity is chosen, is based on the temperature and fluid velocity that G *had* one collision interval earlier. In the following we shall restrict attention to the temperature of G.

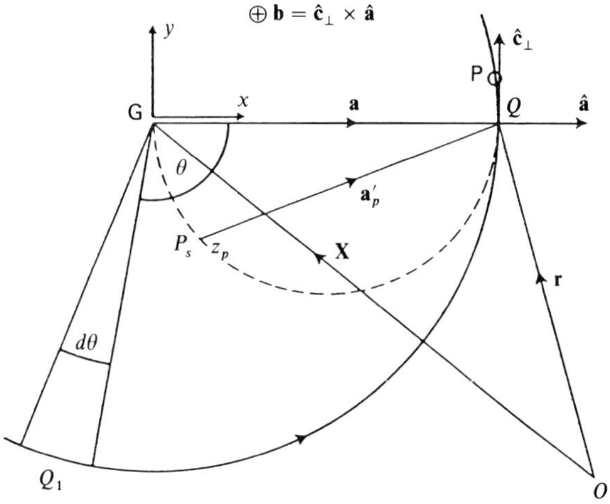

FIG. 9.3. Calculating the heat flux

Let $\mathbf{u}$ and $\mathbf{v}$ denote the velocities of G and Q_c in the laboratory frame, then since at time t, Q_c is a vector distance $\mathbf{a}$ from G, at time $t - \tau$ this distance was $\mathbf{a} - \tau(\mathbf{v} - \mathbf{u})$. Hence if G was at $\mathsf{G}_s = \mathsf{G}(\mathbf{X}_s, t - \tau)$, and $\mathbf{v}_G$ denotes the fluid velocity at G,

$$\mathbf{X}_s = \mathbf{r} - \mathbf{a} + \tau(\mathbf{v} - \mathbf{u}) = \mathbf{r} - \mathbf{a} - \tau(\mathbf{u} - \mathbf{v}_G) + \tau(\mathbf{v} - \mathbf{v}_G). \tag{9.93}$$

Now

$$\mathbf{v} = \mathbf{v}_G + \mathbf{a} \cdot \nabla \mathbf{v} + \mathrm{O}(a^2),$$

whence by (1.91),

$$\mathbf{v} - \mathbf{v}_G = \mathbf{a} \cdot \overset{\circ}{\mathbf{e}} - \mathbf{a} \times \boldsymbol{\Omega} + \tfrac{1}{3}\mathbf{a}\nabla \cdot \mathbf{v} + \mathrm{O}(a^2). \tag{9.94}$$

The vectors $\mathbf{q}$ and $\boldsymbol{\pi}$ are defined in frames rotating with the fluid element (see §5.7) and therefore $\boldsymbol{\Omega}$ cannot contribute to them. The Knudsen number (see (3.30)) is $\epsilon = \tau|\nabla \mathbf{v}|$, so the term $\tau\mathbf{a}\nabla \cdot \mathbf{v}$ is $\mathrm{O}(\epsilon)$ smaller than $\mathbf{a}$ in (9.93) and being parallel to it, can be omitted. Then by (9.73) we arrive at an effective value for $\mathbf{X}_s$, namely

$$\mathbf{X}_s^* = \mathbf{r} - \tau\mathbf{c}_\parallel - \mathbf{a}\cdot(\mathbf{1} - \tau\overset{\circ}{\mathbf{e}}) + \mathrm{O}(a^2).$$

By (9.92) this can be written

$$\mathbf{X}_s^* = \mathbf{r} - \tau\mathbf{c}\cdot\left\{\mathbf{bb} + \varpi^{-1}\mathbf{b}\times(\mathbf{1} - \tau\overset{\circ}{\mathbf{e}})\right\} + \mathrm{O}(a^2), \tag{9.95}$$

which gives the location of the source G_s.

9.7.3 *The heat flux vector in a strong magnetic field*

The arguments yielding (9.90) can now be applied to the magnetoplasma, except that instead of a source Q_c at $\mathbf{r} - \tau\mathbf{c}$, we now have a source G_s at $\mathbf{X}_s^*$. By (9.95) this is equivalent to the replacement

$$\mathbf{c} \to \mathbf{c}\cdot\left\{\mathbf{bb} + \varpi_2^{-1}\mathbf{b}\times(\mathbf{1} - \tau_2\overset{\circ}{\mathbf{e}})\right\} \qquad (\varpi_2 \equiv \omega_c\tau_2).$$

Therefore $\tau_2\mathbf{c}\cdot\nabla T$ in (9.90) is replaced by

$$\tau_2\mathbf{c}\cdot\left\{\mathbf{bb} + \varpi_2^{-1}\mathbf{b}\times(\mathbf{1} - \tau_2\overset{\circ}{\mathbf{e}})\right\}\cdot\nabla T,$$

which is equivalent to substituting

$$\left\{\mathbf{bb} + \varpi_2^{-1}\mathbf{b}\times\mathbf{1} - \varpi_2^{-1}\tau_2\mathbf{b}\times\overset{\circ}{\mathbf{e}}\right\}\cdot\nabla T$$

for ∇T in Fourier's law, $\mathbf{q} = -\kappa\nabla T$. It follows that in very strong magnetic fields the heat flux vector is given by

$$\mathbf{q} = -\kappa\nabla_\parallel T + \kappa\varpi_2^{-1}\mathbf{b}\times\nabla T - \kappa\varpi_2^{-1}\tau_2\mathbf{b}\times\overset{\circ}{\mathbf{e}}\cdot\nabla T.$$

By (3.19), (3.60), and (9.15), $\kappa = 5kp\tau_2/2m$ and $\kappa\varpi_2^{-1} = 5kp/2QB$, whence

$$\mathbf{q} = -\kappa\nabla_\parallel T + \frac{5kp}{2QB}\mathbf{b}\times\nabla T - \frac{5kp}{2QB}\tau_2\mathbf{b}\times\overset{\circ}{\nabla\mathbf{v}}\cdot\nabla T. \tag{9.96}$$

In order, the right-hand terms in this equation are the 'parallel', 'transverse' and 'second-order' heat fluxes.* Their coefficients have magnitudes $\mathrm{O}(\epsilon)$, $\mathrm{O}(\epsilon/\varpi)$ and $\mathrm{O}(\epsilon^2/\varpi)$, and with typical laboratory values of $\epsilon \sim 0.1$, $\varpi \sim 10^6$, they have relative magnitudes of 1, 10^{-6} and 10^{-8}. The parallel heat flux is dominant, so geometries in which $\nabla_\parallel T$ remains zero under steady conditions are of special interest. In these cases, except for a negligible term of $\mathrm{O}(\epsilon/\varpi^2)$ to be derived below, only the small, second-order term (Woods 1983b), can produce the familiar phenomenon of heat flux *down* the temperature gradient. But it can also drive heat *up* the temperature gradient, at least for short time intervals. We shall return to this interesting case in §9.9.2.

*Strictly, since no collisions are involved, the transverse flux is *energy* rather than heat (see §2.1), but as we shall see in §9.8.2, we are dealing here with a limiting case of the collisional transfer of energy, i.e. heat.

9.8 Heat flux for all strengths of magnetic field

We shall now generalize (9.96) to give an expression for $\mathbf{q}$ over the range $0 \le \varpi \le \infty$. Only the $\mathrm{O}(\epsilon)$ terms will be considered here, which allows us to set the gradient of the fluid velocity equal to zero.

9.8.1 *The location of the labelling source*

Our objective is to locate that point P_s, the temperature at which determines the distribution of peculiar velocities at Q_c. Regardless of the value of ϖ, the term $-\tau\mathbf{c}\cdot\mathbf{bb}$ remains on the left-hand side of (9.95), and this produces the parallel heat flux in (9.96) unchanged. We can therefore set $\mathbf{c}_\parallel$ equal to zero for the moment and concentrate on the other $\mathrm{O}(\epsilon)$ terms. Referring to Fig. 9.3, we have in effect dealt with the case $\varpi = \infty$ by setting the source P_s at the guiding centre G. We shall show below that as ϖ is reduced, P_s moves along the semi-circular arc, shown dotted in the figure, ending at Q_c when ϖ is zero.

By moderating the particle velocities, interactions with the electromagnetic microfields maintain the temperature of the gyrating particles at a value equal to a weighted average taken over the whole orbit. The weighting factor must give that part of the orbit most recently traversed the greatest importance by allowing for an exponential decline in the relevance of interactions as they recede into the past. The appropriate factor is $\tau^{-1}\exp(t/\tau)$, $-\infty < t \le 0$ (cf. (3.17)), or equivalently

$$W(\theta)\,d\theta = \exp(-\theta/\varpi)\,d\theta/\varpi \qquad (0 \le \theta < \infty),$$

where $\theta = c_\perp t/a = \omega_c t$ is the angle subtended at G by the remaining orbit $Q_c'Q_c$ (Fig. 9.3) and $W(\theta)$ has been normalized.

In an Argand plane, $z = x + iy$, with origin at G and real axis along $\mathsf{G}Q_c$, the point Q_c' on the orbit is at $a(\cos\theta - i\sin\theta)$. On weighting this with the probability $W(\theta)$, we deduce that P_s is at

$$z_P = x_P + iy_P = a\int_0^\infty \exp(-i\theta)\exp(-\theta/\varpi)\,d\theta/\varpi = a\frac{1 - i\varpi}{1+\varpi^2}.$$

It is easily verified that the locus of P_s, as ϖ varies from zero to infinity, is the semicircle shown in the figure. The point

$$a_P' = a - z_P = \varpi(\varpi + i)/(1+\varpi^2)$$

corresponds to the vector

$$\mathbf{a}_P' = \frac{a\varpi}{1+\varpi^2}(\varpi\mathbf{b}\times\hat{\mathbf{c}} + \hat{\mathbf{c}}_\perp) = \tau\mathbf{c}\cdot\left\{-\frac{\varpi}{1+\varpi^2}\mathbf{1}\times\mathbf{b} + \frac{1}{1+\varpi^2}(\mathbf{1} - \mathbf{bb})\right\},$$

where $\mathbf{b}\times\hat{\mathbf{c}}$ and $\hat{\mathbf{c}}_\perp$ are the unit vectors parallel to OX, OY in the Argand plane.

9.8.2 *The first-order heat flux vector*

Now restoring the parallel movement of P_s due to $\mathbf{c}_\parallel$, which gives the temperature source a vector displacement $\tau\mathbf{c}\cdot\mathbf{bb}$ relative to Q_c, and adding this to $\mathbf{a}'_P$, we obtain the total vectorial displacement

$$\boldsymbol{\lambda} = \tau\mathbf{c}\cdot\mathsf{k}, \tag{9.97}$$

where
$$\mathsf{k} \equiv \mathbf{bb} - \frac{\varpi}{1+\varpi^2}\mathbf{b}\times\mathbf{1} + \frac{1}{1+\varpi^2}(\mathbf{1}-\mathbf{bb}). \tag{9.98}$$

We can view $\boldsymbol{\lambda}$ as being a generalized mean free path.

The replacement used in §9.7.3 to deduce (9.96) is now generalized to $\mathbf{c}\to\mathbf{c}\cdot\mathsf{k}$, which is equivalent to substituting $\mathsf{k}\cdot\nabla T$ for ∇T in Fourier's law. Our expression for the first-order heat flux is therefore

$$\mathbf{q}_1 = -\kappa\mathsf{k}\cdot\nabla T, \tag{9.99}$$

so the thermal conductivity now has the tensor form

$$\mathsf{K} = \kappa\mathsf{k} = \kappa\left(\mathbf{bb} - \frac{\varpi}{1+\varpi^2}\mathbf{b}\times\mathbf{1} + \frac{1}{1+\varpi^2}(\mathbf{1}-\mathbf{bb})\right). \tag{9.100}$$

It has been assumed above that τ is a constant, whereas—before averaging—it is a function of the peculiar speed c (see §10.3.2). When there is no magnetic field, the two averages τ_1 and τ_2 prove sufficient. Otherwise a different average for τ is necessary for each field strength. Alternatively, if just one average is adopted, k becomes a rather more complicated function of ϖ than that in (9.98). Braginskii (1965) has obtained approximations to these functions that we shall describe later (§10.10).

9.8.3 *Strong magnetic fields*

From (9.99) and (9.100),

$$\mathbf{q}_1 = -\mathsf{K}\cdot\nabla T = \mathbf{q}_\parallel + \mathbf{q}_\wedge + \mathbf{q}_\perp, \tag{9.101}$$

where
$$\mathbf{q}_\parallel = -\kappa_\parallel\nabla_\parallel T, \quad \mathbf{q}_\wedge = -\kappa_\wedge\mathbf{b}\times\nabla T, \quad \mathbf{q}_\perp = -\kappa_\perp\nabla_\perp T, \tag{9.102}$$

in which
$$\kappa_\parallel = \kappa, \quad \kappa_\wedge = -\frac{\kappa\varpi}{1+\varpi^2}, \quad \kappa_\perp = \frac{\kappa}{1+\varpi^2}, \qquad (\varpi\equiv\omega_c\tau_2). \tag{9.103}$$

We shall discuss $\mathbf{q}_\wedge$ for the strong field case shortly. From the practical standpoint of confining energy in a plasma by strong a magnetic field, the first-order term of interest is

$$\mathbf{q}_\perp \simeq -\frac{\kappa}{\varpi^2}\nabla_\perp T = -\rho\chi\nabla_\perp(c_v T), \tag{9.104}$$

where
$$\chi_\perp = \frac{\kappa_\perp}{\rho c_v} = \frac{5}{6}\frac{C^2}{\omega_c^2}\frac{1}{\tau_2} = \frac{5}{6}\frac{r_L^2}{\tau_2} \tag{9.105}$$

is the perpendicular thermal diffusivity. Comparing this with (3.62), we see that in the perpendicular direction very strong fields reduce the effective mean free path to the Larmor radius, r_L. Another way of expressing this result is to identify r_L/τ_2 as being the *speed* at which energy is transmitted across a distance r_L, and then identifying diffusivity as 'speed of energy transfer' times 'distance transferred'.

9.9 Physical mechanisms for heat flux

9.9.1 *The transverse heat flux*

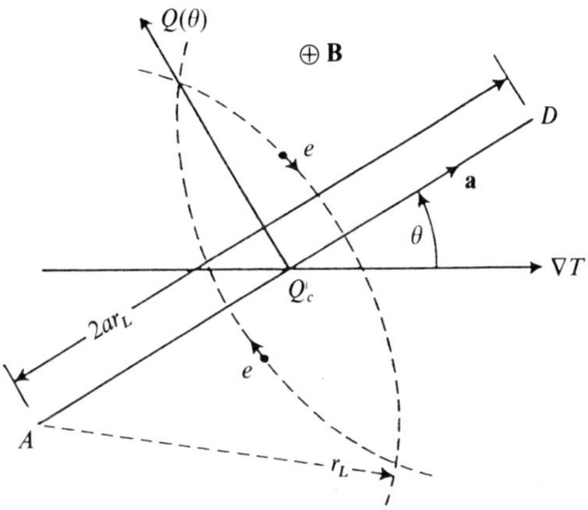

FIG. 9.4. Origin of the transverse heat flux

In strong magnetic fields it follows from (9.102) and (9.103) that

$$\mathbf{q}_\wedge = \frac{5kp}{2_Q B}\mathbf{b} \times \nabla T. \tag{9.106}$$

A physical picture of the origin of $\mathbf{q}_\wedge$ can be obtained by taking the particles in pairs, each member of which passes through Q_c in opposite directions. In Fig. 9.4 (drawn for the electrons) the guiding centres at A and D are placed a distance $2hr_L$ apart, where h is a constant of order unity. Let $\hat{\mathbf{a}}$ be unit vector along AD and θ the angle between $\hat{\mathbf{a}}$ and ∇T, then there will be an energy difference of amount

$$2hr_L \cos\theta |\nabla(c_v T)| = 3hC\frac{k}{|_Q|B}\cos\theta\, |\nabla T|$$

per unit mass between the two groups of particles crossing AD in opposite directions and with A and D as their guiding centres. Since C is the average speed normal to AD and nm is the particle mass per unit volume, it follows that the net energy flux normal to AD is $\xi(\theta)\hat{\mathbf{a}} \times \mathbf{b}$, where

$$\xi(\theta) = \pm 3hC^2\frac{nmk}{|_Q|B}\cos\theta|\nabla T| = 6h\frac{kp}{_Q B}\cos\theta|\nabla T|, \tag{9.107}$$

the sign choice being + for ions and − for electrons.

The components of energy flux normal and parallel to ∇T are $\cos\theta\,\xi(\theta)$ and $-\sin\theta\,\xi(\theta)$. To include all the contributions from all the particles gyrating about guiding centres a distance hr_L from the point Q_c at which the heat flux is being estimated, we average these components over $-\pi/2 < \theta < \pi/2$ the result is zero flux along ∇T and a flux $3hkp|\nabla T|/(QB)$ parallel to $\mathbf{b}\times\nabla T$, and with $h = \frac{5}{6}$, we arrive at the formula in (9.106). Of course a complete calculation along these lines would require our averaging over a Maxwellian distribution as well. Our objective is merely to give an explanation why it is possible for a temperature gradient to generate a heat flux normal to itself.

9.9.2 *The second-order heat flux in sheared flow*

Next consider the third term in (9.96), viz.

$$\mathbf{q}^* = -\frac{5kp}{2QB}\tau_2\mathbf{b}\times \overset{\circ}{\nabla\mathbf{v}}\cdot\nabla T. \tag{9.108}$$

The importance of this flux is that unlike the transverse term $\mathbf{q}_\wedge$, of which it is a modification due to fluid strain, it has a component down the temperature gradient. To illustrate its physical origin, we shall apply it to a simple flow problem.

Let $\hat{\mathbf{s}}$ be unit vector parallel to ∇T, then $\nabla T = \hat{\mathbf{s}}T'$, where the dash denotes the spatial rate of change in a direction along $\hat{\mathbf{s}}$. Equation (9.108) gives

$$\hat{\mathbf{s}}\cdot\mathbf{q}^* = -\kappa^* T', \tag{9.109}$$

where

$$\kappa^* = -\frac{5kp}{2QB}\tau_2 H \qquad \left(H \equiv \mathbf{b}\times\hat{\mathbf{s}}\cdot \overset{\circ}{\nabla\mathbf{v}}\cdot\hat{\mathbf{s}}\right).$$

The simplest example of the theory is provided by a transverse planar flow, sheared in a direction parallel to the temperature gradient, assumed to be orthogonal to $\mathbf{B}$. We find that $H = \frac{1}{2}v'$ and

$$\hat{\mathbf{s}}\cdot\mathbf{q}^* = \frac{5kp}{4QB}\tau_2 v'T'. \tag{9.110}$$

Notice that if v'/Q is positive, the heat flows *up* the temperature gradient and the flow is thermally unstable.

9.9.3 *Physical origin of second-order heat flux*

The sheared flow just described is illustrated in Fig. 9.5. The x-axis lies in stationary fluid and the guiding centres D and A are a distance $r_L\cos\theta$ on either side of the axis, moving parallel to it with speeds $\pm v'r_L\cos\theta$, where $v' = dv/dy$. We may treat the whole of the x-axis as being the fluid particle Q_c at which the heat flow parallel to the temperature gradient along OY is to be calculated. All the guiding centres, with particle orbits crossing the

x-axis, can be paired in the way indicated in the figure. Of course the orbits shown are actually the projections of the three-dimensional particle orbits on to the XY-plane. But this is not important since we are interested only in heat transport in planes orthogonal to the magnetic field.

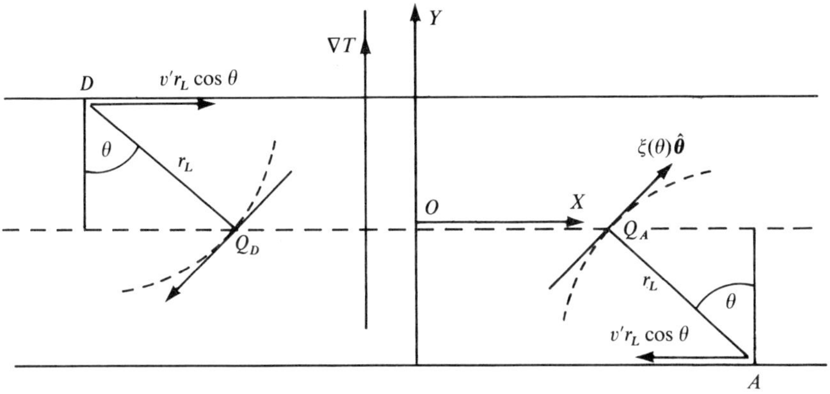

FIG. 9.5. Second-order heat flux in sheared flow

The separation of the paired orbits does not alter the fact that the sum of the energies transferred across the x-axis at an angle θ to it, at points Q_D and Q_A, is the flux

$$\xi(\theta) = \frac{5kp}{QB} \cos\theta |\nabla T| \tag{9.111}$$

calculated in (9.107).

The plasma at Q_A is moving along the x-axis towards the orgin with the speed $v' r_L \cos\theta$. This has components $v' r_L \cos\theta \sin\theta$ parallel to AQ_A and $v' r_L \cos^2\theta$ along the tangent to the particle orbit about A. The latter velocity is swamped in magnitude by the peculiar velocity (which lies in the same direction), and its contribution to the flux vector $\xi(\theta)\hat{\boldsymbol{\theta}}$ is negligible. On the other hand, the velocity along AQ_A *deflects* $\xi(\theta)\hat{\boldsymbol{\theta}}$ in a manner similar to that described in §8.4.2 for sheared flow in a neutral gas. Of course, in the latter case the primary or first-order heat flux is parallel to ∇T and the fluid shear creates a secondary heat flux at right angles to it (see Fig. 8.1). In the magnetoplasma case directions are interchanged—the primary flux is *normal* to ∇T and the fluid shear induces a heat flux *parallel* to it.

In §8.4.2 we showed that the angle of deflection of the primary heat flux is equal to the ratio of the normal convection speed remaining after removal of the vorticity, namely $\frac{1}{2}\lambda v'$, to the speed of energy transfer in the primary flux, λ/τ_2. This gives the angle $\frac{1}{2}\tau_2 v'$ shown in Fig. 8.1. We shall use the same principle for the magnetoplasma.

The convection speed normal to the primary flux $\xi(\theta)\hat{\boldsymbol{\theta}}$ is $v' r_L \cos\theta \sin\theta$ and in §9.8.3 we showed that the speed of energy transfer of the primary flux across a Larmor orbit is r_L/τ_2. With two distinct Larmor orbits, this rate is doubled. It follows that the required deflection angle is $2\tau_2 v' \cos\theta \sin\theta$. Applying this to the primary flux, we conclude that the fluid shear generates a heat flux $2\tau_2 v' \cos\theta \sin\theta\, \xi(\theta)$ along AQ_A. Resolved along OY, this gives

$$2\tau_2 v' \cos\theta \sin^2\theta\, \xi(\theta) = 2\tau_2 v' \cos^2\theta \sin^2\theta (5kp/QB)|\nabla T|$$

by (9.111). It remains to average this flux over $-\frac{1}{2}\pi < \theta \leq \frac{1}{2}\pi$, to obtain the final result, namely the expression given in (9.110), and to complete our physical explanation of the second-order heat flux vector.

Exercises 9 9.1 Let $\lambda_{90°}$ be the average distance a charged particle travels before experiencing a 90° deflection in a single encounter with a similar particle, and let n_D be the number of particles in a sphere of radius λ_D (known as a 'Debye' sphere). Use expressions in §6.8.4 to show that $\lambda_{90°} = 108\, n_D \lambda_D / Z^4$ and $\Lambda = 9n_D/Z$. Deduce that the average collision frequency for momentum transfer is more than 100 times larger than the 90° collision frequency.

9.2 A two-fluid system of plasma equations has six conservation laws—three for each component. Explain what happens to the 'information' lost when these laws are reduced to the three required in the one-fluid description.

9.3 Show that in a one-fluid description of a magnetoplasma, the entropy production rate is σ, where

$$T\sigma = -\boldsymbol{\pi} : \nabla \mathbf{v} - \mathbf{q} \cdot \nabla \ln T + \mathbf{j} \cdot (\mathbf{E} + \mathbf{v} \times \mathbf{B}) \geq 0$$

(cf. (1.51) and (9.34)). Infer that each of the linear constitutive equations for $\mathbf{q}$ and $\mathbf{j}$ must depend on *both* ∇T and $(\mathbf{E} + \mathbf{v} \times \mathbf{B})$.

9.4 The neutral gas in a partially-ionized plasma can be regarded as being stationary. Write down a constitutive equation for the friction force acting on the plasma (see (4.5)). What other constitutive equations are made necessary by the presence of the neutrals?

9.5 A cylindrical plasma is confined by an axial magnetic field, along which an electric current is flowing. The radial gradient of the electron fluid velocity is v'_e. Show that a suitable measure of the Knudsen number is $\epsilon = \tau_2 v'_e$. Ignoring the magnetic field due to the current, show that the ratio of the $\mathrm{O}(\epsilon^2)$ and $\mathrm{O}(\epsilon)$ radial heat fluxes in the electron gas is $\frac{1}{2}\varpi_e \epsilon = eB{\tau_E}^2 v'_e/2m_e$.

9.6 Show that, with some unimportant approximations, Ohm's law can be written

$$\mathbf{j} = \sigma\left\{\mathbf{E} + \mathbf{v} \times \mathbf{B} + \frac{1}{en_e}(\nabla p_e - \mathbf{j} \times \mathbf{B})\right\} + \beta \nabla T_e,$$

and explain the discrepancy between this equation and that referred to in exercise 9.3.

9.7 Use $\dot{\mathbf{w}} = \mathrm{D}\mathbf{v} + \dot{\mathbf{c}}$ to obtain an expression for particle acceleration in the laboratory frame, and provide a physical explanation of the presence of the pressure gradient in the formula.

9.8 Suppose there is a density gradient along the OX axis, with uniform $\mathbf{B}$ along OZ. In a frame with stationary guiding centres, at any elemental volume $P(0, y_0)$, the electrons passing through P in the OX direction have come from the denser gas in $y > y_0$, whereas those passing in the opposite direction have come from the less dense gas in $y < y_0$. The two groups of electrons have the same orbital speed. Comment on the conclusion, that because of the imbalance in particle numbers, there is a fluid velocity along OX, *relative to the guiding centres*.

10

KINETIC THEORY FOR MAGNETOPLASMAS

10.1 The Fokker–Planck equation

One of the basic assumptions of Boltzmann's equation is that only binary collisions contribute to the collision integral, whereas in a plasma each particle is simultaneously interacting with the vast number of particles lying within a Debye distance. Most of these interactions result in very small deflections (see final paragraph of §6.3.3). From this description it would *appear* unlikely that Boltzmann's equation has much relevance in plasma physics. In the first five sections of this chapter we shall introduce and apply a kinetic equation—known as the Fokker–Planck equation—which is usually considered to be more appropriate for plasmas, although in fact it can be deduced from Boltzmann's equation by restricting the scattering angles to small values.

10.1.1 *The Markovian hypothesis*

Consider the particles comprising a fully-ionized plasma. Choose one of them to be a 'test' particle and suppose that the species to which it belongs has a velocity distribution $f(\mathbf{r}, \mathbf{w}, t)$. The long-range Coulomb forces between the test particle p and the 'field' particles within a Debye distance will cause p to experience a multiplicity of 'distant' collisions, which as noted at the end of §6.3.3, will be far more numerous than close collisions. Consequently, almost all the changes in direction and speed experienced by p will be small. Let p have an initial velocity $\mathbf{w}$, then after a small time interval Δt, distance collisions will impose on p a random walk motion resulting in a small cumulative change $\Delta\mathbf{w}$ satisfying $|\Delta\mathbf{w}| \ll |\mathbf{w}|$. We are ignoring here the non-random, or streaming collisions (see §5.8.2).

Let $P(\mathbf{w}|\Delta\mathbf{w})$ denote the probability density that the test particle p experiences the change $\Delta\mathbf{w}$ in the time Δt, then

$$f(\mathbf{r}, \mathbf{w} - \Delta\mathbf{w},\ t - \Delta t)\, P(\mathbf{w} - \Delta\mathbf{w}|\Delta\mathbf{w}) d(\Delta\mathbf{w})$$

is the number of particles like p that are deflected from $(\mathbf{w} - \Delta\mathbf{w})$ into the element $(\mathbf{w},\ \mathbf{w} + d\mathbf{w})$ owing to interactions occurring in $(t - \Delta t,\ t)$. These particles contribute to the number $f(\mathbf{r},\ \mathbf{w},\ t)\, d\mathbf{w}$, and on the assumption that the process is *Markovian*, that is, that no earlier time intervals contribute to this number, we obtain the Chapman–Kolmogorov equation:

$$f(\mathbf{r},\ \mathbf{w},\ t) = \int f(\mathbf{r},\ \mathbf{w} - \Delta\mathbf{w},\ t - \Delta t)\, P(\mathbf{w} - \Delta\mathbf{w}|\Delta\mathbf{w})\, d(\Delta\mathbf{w}), \qquad (10.1)$$

the integration being over all possible changes in the velocity vector. The time interval Δt must be short enough for $\Delta\mathbf{w}$ to remain small compared with $\mathbf{w}$, but long compared with the transit time of p over the correlation length for microfield fluctuations. For electrons this lower limit is about ω_{pe}^{-1} (see §9.2.2).

The Markovian hypothesis on which (10.1) is based, requires that the scattering process be without 'memory' beyond the short interaction time Δt, yet in the derivation of the BGK kinetic equation given in §5.4.1 the distribution function f has a memory extending over the much longer 90° deflection ($\Delta\mathbf{w} \sim \mathbf{w}$) time τ. And the intermediate kinetic theory involves an even longer memory time (see §5.9.2). But when a *single* collision interval gives sufficient accuracy, the introduction of the streaming derivative and division by τ (§5.4.2) yields a Markovian kinetic equation, namely one that allows future f to be predicted from a given present state, without the history of this state being required. In this case 'history' is represented by time derivatives. The problem with memory emerges only when we need to take the effects of *two* successive collisions into account. As shown in §5.9, this is essential for the correct accounting of accelerations in kinetic theory, and it affects only the terms of second and higher order in the Knudsen number. What follows is therefore limited to first-order transport.

10.1.2 *Diffusion in velocity space*

Expanding the integral in (10.1) in a Taylor series to first order in Δt and to second order in $\Delta\mathbf{w}$, we obtain the approximate form

$$f(\mathbf{r},\mathbf{w},t) = \int \Big\{ (f - \Delta t \mathbb{D} f) P(\mathbf{w}|\Delta\mathbf{w}) - \Delta\mathbf{w}\cdot\frac{\partial}{\partial\mathbf{w}}[fP(\mathbf{w}|\Delta\mathbf{w})] + \tfrac{1}{2}\Delta\mathbf{w}\Delta\mathbf{w} : \frac{\partial^2}{\partial\mathbf{w}\,\partial\mathbf{w}}[fP(\mathbf{w}|\Delta\mathbf{w}]\Big\}\, d(\Delta\mathbf{w}),$$

where $\mathbb{D}f$ is the phase-space streaming derivative defined in (7.2), and the terms in the integrand are to be evaluated at $\mathbf{r}$, $\mathbf{w}$, t. As the probability of a transition of some kind occurring is unity,

$$\int P(\mathbf{w}|\Delta\mathbf{w})\, d(\Delta\mathbf{w}) - 1.$$

Hence the leading term in the integral expression for f cancels with the left-hand side. The remaining terms can arranged as the kinetic equation:

$$\left.\begin{aligned} \mathbb{D}f &= \mathbb{C}_{FP}, \\ \text{where}\qquad \mathbb{C}_{FP} &= -\frac{\partial}{\partial\mathbf{w}}\cdot(\mathbf{A}f) + \frac{1}{2}\frac{\partial^2}{\partial\mathbf{w}\,\partial\mathbf{w}}:(\mathbf{B}f), \end{aligned}\right\} \tag{10.2}$$

in which

$$\left.\begin{aligned} \mathbf{A} &= \langle \Delta\mathbf{w} \rangle \equiv \frac{1}{\Delta t}\int \Delta\mathbf{w}\, P(\mathbf{w}|\Delta\mathbf{w})\, d(\Delta\mathbf{w}), \\ \text{and} \qquad \mathbf{B} &= \langle \Delta\mathbf{w}\Delta\mathbf{w} \rangle \equiv \frac{1}{\Delta t}\int \Delta\mathbf{w}\Delta\mathbf{w}\, P(\mathbf{w}|\Delta\mathbf{w})\, d(\Delta\mathbf{w}). \end{aligned}\right\} \tag{10.3}$$

The relation (10.2) is known as the 'Fokker–Planck' kinetic equation. Notice that the collision term may be expressed as the divergence in velocity space of a 'flow' vector $\mathbf{Q}$:

$$\mathbb{C}_{FP} = -\frac{\partial}{\partial \mathbf{w}} \cdot \mathbf{Q}, \qquad \mathbf{Q} \equiv \mathbf{A}f - \frac{1}{2}\frac{\partial}{\partial \mathbf{w}} \cdot (\mathbf{B}f). \tag{10.4}$$

This vector describes the continuous flow of phase points due to the accumulative effect of many small-angle collisions. Despite being limited to binary collisions, Boltzmann's collision integral also yields this expression when restricted to small deflections (see exercises 7.7 and 10.8). For reasons that will become clear shortly, the averages $\langle \Delta\mathbf{w} \rangle$ and $\langle \Delta\mathbf{w}\, \Delta\mathbf{w} \rangle$ are termed the 'friction' and 'diffusion' coefficients. To apply (10.2) to a plasma, we must calculate these averages for the case of Coulomb collisions.

10.1.3 *The friction and diffusion coefficients for a plasma*

In general there will be several types of field particles or scatterers influencing the test particle p, including the species to which p itself belongs. We shall first consider just one type of scatterer, and denote its distribution function by $f_s(\mathbf{r}, \mathbf{w}_s, t)$. The probability that a single scatterer deflects p into the solid angle $d\Omega = \sin\chi\, d\chi\, d\epsilon$ (see Fig. 6.2) is $\alpha(g,\chi)\, d\Omega$, where $g = |\mathbf{w} - \mathbf{w}_s|$ is the relative speed between the interacting particles and $\alpha(g,\chi)$ is the Rutherford scattering cross-section defined in (6.20). The corresponding scattering rate is $g\alpha\, d\Omega$, i.e. of a group of N incident particles in a time dt, $Ng\alpha\, d\Omega\, dt$ will appear in $d\Omega$. The assumption of small scattering angles allows us to superimpose the contributions of all the scatterers lying in the appropriate element $d\mathbf{w}_s$ of velocity space.

The total probability that p is scattered into $d\Omega$ per second per unit volume is $f_s\, d\mathbf{w}_s g\alpha(g,\chi)\, d\Omega$. Hence the averages in (10.3) are

$$\left.\begin{aligned} \langle \Delta\mathbf{w} \rangle &= \iint \Delta\mathbf{w}\, g\alpha(g,\chi)\, d\Omega\, f_s\, d\mathbf{w}_s &&= \int [\Delta\mathbf{w}]_\Omega\, f_s\, d\mathbf{w}_s, \\ \langle \Delta\mathbf{w}\Delta\mathbf{w} \rangle &= \iint \Delta\mathbf{w}\Delta\mathbf{w} g\alpha(g,\chi)\, d\Omega f_s\, d\mathbf{w}_s &&= \int [\Delta\mathbf{w}\Delta\mathbf{w}]_\Omega\, f_s\, d\mathbf{w}_s, \end{aligned}\right\} \tag{10.5}$$

$$\left.\begin{aligned} \text{where} \qquad [\Delta\mathbf{w}]_\Omega &\equiv \int_0^{2\pi}\int_\chi \Delta\mathbf{w}\, g\alpha(g,\chi)\sin\chi\, d\chi\, d\epsilon, \\ [\Delta\mathbf{w}\Delta\mathbf{w}]_\Omega &\equiv \int_0^{2\pi}\int_\chi \Delta\mathbf{w}\Delta\mathbf{w}\, g\alpha(g,\chi)\sin\chi\, d\chi\, d\epsilon. \end{aligned}\right\} \tag{10.6}$$

When there are several types of scatterer, say $s = 1, 2, \ldots$, the integrals in (10.5) are required for each s and must be summed to give the required averages. In the next section we shall express these averages as the velocity space gradients of two 'superpotential' functions, first introduced into plasma theory by Rosenbluth *et al.* (1957).

10.2 The superpotentials

10.2.1 *Small scattering angles*

Let the test particle p have mass m and velocities $\mathbf{w}$, $\mathbf{w}'$ before and after an elastic collision with a scatterer of mass m_s and velocities $\mathbf{w}_s$, $\mathbf{w}'_s$ before and after the collision. With $\mathbf{g} = \mathbf{w} - \mathbf{w}_s$, $\mathbf{g}' = \mathbf{w}' - \mathbf{w}'_s$, denoting the relative velocities and

$$\mathbf{G} = (m\mathbf{w} + m_s\mathbf{w}_s)/(m+m_s), \quad \mathbf{G}' = (m\mathbf{w}' + m_s\mathbf{w}'_s)/(m+m_s)$$

denoting the centre of mass velocities, we have (see §6.4.1)

$$\mathbf{G} = \mathbf{G}', \quad g = g', \quad \mathbf{w} = \mathbf{G} + M\mathbf{g}/m \quad (M \equiv mm_s/(m+m_s)). \tag{10.7}$$

As $\mathbf{g}$ is unchanged in magnitude by the collision, $|\Delta\mathbf{g}| = |\mathbf{g}' - \mathbf{g}| = 2g\sin\frac{1}{2}\chi$. Resolving the vector $\Delta\mathbf{g}$ into a component Δg_1 parallel to $\mathbf{g}$ and components Δg_2, Δg_3 perpendicular to $\mathbf{g}$, (see Fig. 10.1), we have

$$\Delta\mathbf{g} = 2g\sin\tfrac{1}{2}\chi\left(-\sin\tfrac{1}{2}\chi,\ \cos\tfrac{1}{2}\chi\cos\epsilon,\ \cos\tfrac{1}{2}\chi\sin\epsilon\right). \tag{10.8}$$

By (10.7), $\mathbf{w}' - \mathbf{w} = (M/m)(\mathbf{g}' - \mathbf{g})$, i.e. $\Delta\mathbf{w} = M\Delta\mathbf{g}/m$.

The first average in (10.6) is calculated using (6.20) and (10.8). The

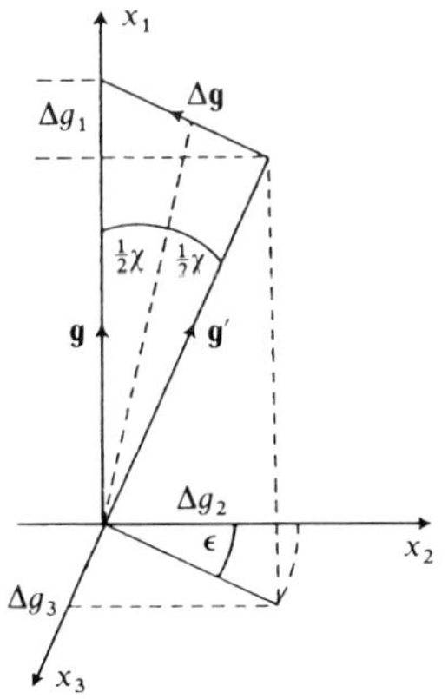

FIG. 10.1. Scattering in velocity space

integrals containing $\cos\epsilon$ and $\sin\epsilon$ vanish, leaving only the component parallel to the unit vector $\hat{\mathbf{g}} = \mathbf{g}/g$. Thus with χ lying in the range $\chi_{min} \le \chi \le \chi_{max}$, we get

$$[\Delta\mathbf{w}]_\Omega = -4\pi\frac{M}{m}(b_0 g)^2\left[\ln\sin\tfrac{1}{2}\chi\right]_{\chi_{min}}^{\chi_{max}}\hat{\mathbf{g}}. \tag{10.9}$$

As χ is assumed to be small, $\frac{1}{2}\pi \gg \chi_{max} \gg \chi_{min}$. As already explained in §6.8.4, a reasonable assumption for χ_{min} is that it corresponds to an impact parameter b equal to the Debye length λ_D. Then by the equation following (6.19), $\cot\frac{1}{2}\chi_{min} \approx 2/\chi = \lambda_D/b_0$. Therefore

$$\left[\ln\sin\tfrac{1}{2}\chi\right]_{\chi_{min}}^{\chi_{max}} \approx (1-\varepsilon)\ln\Lambda \qquad (\varepsilon \equiv -\ln\tfrac{1}{2}\chi_{max}/\ln\Lambda), \tag{10.10}$$

where $\Lambda \equiv \lambda_D/\bar{b}_0$ and as in §9.4.2, the weak dependence of the logarithm on particle velocity has been removed by replacing b_0 by the average $\bar{b}_0$.

10.2.2 *The first superpotential*

Let us assume that it is possible to chose χ_{max} to make $\varepsilon \ll 1$ and yet retain the small deflection condition $\chi_{max} \ll \frac{1}{2}\pi$. Then (6.17), (10.9), and (10.10) give

$$[\Delta\mathbf{w}]_\Omega = -\frac{(ZZ_s e^2)^2\ln\Lambda}{4\pi\epsilon_0^2 Mm}\hat{\mathbf{g}}, \tag{10.11}$$

where Z is p's ionization number. Finally by integrating over the distribution of field particles, we obtain

$$\langle\Delta\mathbf{w}\rangle = -\Gamma\frac{m}{M}\int f_s(\mathbf{w}_s)\frac{\mathbf{w}-\mathbf{w}_s}{|\mathbf{w}-\mathbf{w}_s|^3}\,d\mathbf{w}_s, \tag{10.12}$$

where

$$\Gamma \equiv \frac{(ZZ_s e^2)^2\ln\Lambda}{4\pi\epsilon_0^2 m^2}. \tag{10.13}$$

As

$$\mathbf{g}\cdot\frac{\partial\mathbf{g}}{\partial\mathbf{w}} = g\frac{\partial g}{\partial\mathbf{w}} = \mathbf{g},$$

we can write

$$[\Delta\mathbf{w}]_\Omega = \frac{m}{M}\Gamma\frac{\partial}{\partial\mathbf{w}}\left(\frac{1}{g}\right), \tag{10.14}$$

and therefore

$$\langle\Delta\mathbf{w}\rangle = \Gamma\frac{\partial\mathcal{H}}{\partial\mathbf{w}}, \tag{10.15}$$

where

$$\mathcal{H} \equiv \frac{m}{M}\int\frac{f_s(\mathbf{w}_s)}{|\mathbf{w}-\mathbf{w}_s|}\,d\mathbf{w}_s. \tag{10.16}$$

This scalar $\mathcal{H}$ is the first of the superpotentials.

10.2.3 *The second superpotential*

A similar method is applicable to the average $\langle\Delta\mathbf{w}\Delta\mathbf{w}\rangle$. We separately calculate the components $[\Delta w_i\,\Delta w_j]_\Omega$, $i, j = 1, 2, 3$. The integration over

ϵ eliminates the terms in which i and j are not equal. It is readily found that

$$[\Delta w_2\,\Delta w_2]_\Omega = [\Delta w_3\,\Delta w_3]_\Omega = \frac{\Gamma}{g} = \ln\Lambda[\Delta w_1\,\Delta w_1]_\Omega.$$

In many applications of plasma theory the Coulomb logarithm is much larger than unity, allowing us to neglect $[\Delta w_1\,\Delta w_1]_\Omega$ compared with the other components. And since the neglected component is the coefficient of the unit tensor $\hat{\mathbf{g}}\hat{\mathbf{g}}$, we arrive at

$$[\Delta\mathbf{w}\Delta\mathbf{w}]_\Omega = \frac{\Gamma}{g}(\mathbf{1} - \hat{\mathbf{g}}\hat{\mathbf{g}}). \tag{10.17}$$

By
$$\frac{\partial^2 g}{\partial\mathbf{w}\,\partial\mathbf{w}} = \frac{\partial}{\partial\mathbf{w}}\left(\frac{\mathbf{g}}{g}\right) = \frac{1}{g}(\mathbf{1} - \hat{\mathbf{g}}\hat{\mathbf{g}}),$$

(10.17) can be expressed
$$[\Delta\mathbf{w}\Delta\mathbf{w}]_\Omega = \Gamma\frac{\partial^2 g}{\partial\mathbf{w}\,\partial\mathbf{w}}. \tag{10.18}$$

We now introduce the second superpotential,

$$\mathcal{G} \equiv \int f_s(\mathbf{w}_s)\,|\mathbf{w} - \mathbf{w}_s|\,d\mathbf{w}_s, \tag{10.19}$$

then the average of (10.18) over the field particles gives

$$\langle\Delta\mathbf{w}\Delta\mathbf{w}\rangle = \Gamma\frac{\partial^2\mathcal{G}}{\partial\mathbf{w}\partial\mathbf{w}}. \tag{10.20}$$

By (10.2), (10.3), (10.16), and (10.20),

$$\frac{1}{\Gamma}\mathbb{C}_{FP} = -\frac{\partial}{\partial\mathbf{w}}\cdot\left(f\frac{\partial\mathcal{H}}{\partial\mathbf{w}}\right) + \frac{1}{2}\frac{\partial^2}{\partial\mathbf{w}\,\partial\mathbf{w}} : \left(f\frac{\partial^2\mathcal{G}}{\partial\mathbf{w}\,\partial\mathbf{w}}\right). \tag{10.21}$$

In general we should write $\mathcal{H} = \sum_s \mathcal{H}_s$, $\mathcal{G} = \sum_s \mathcal{G}_s$ and have a separate kinetic equation for each species. As $\mathcal{H}$ and $\mathcal{G}$ are integrals, we now have a set of rather complicated, coupled, integro-differential equations that can be solved accurately only by computation.

10.2.4 *Two useful relations*

Finally we note two useful relations involving $\mathcal{H}$ and $\mathcal{G}$. The first is

$$\frac{m}{M}\nabla_w^2\mathcal{G} = 2\mathcal{H} \qquad (\nabla_w = \partial/\partial\mathbf{w}), \tag{10.22}$$

which follows on integrating $\nabla_w\cdot\nabla_w g = 2/g$ over the field particles. The second is the Poisson equation

$$\nabla_w^2\mathcal{H} = -4\pi\frac{m}{M}f_s(\mathbf{w}_s) \tag{10.23}$$

implied by (10.12) and (10.15).

10.3 The Lorentzian plasma

10.3.1 *Collision operator*

In §9.4.4 we defined a Lorentzian plasma as being a hypothetical, fully-ionized gas, with ions of infinite mass (and therefore zero temperature) and no electron–electron collisions. Since their temperature is zero, the ions have negligible peculiar velocities. We shall take the laboratory frame to be fixed in the ion fluid. Therefore $f_i(\mathbf{w}_i) = n_i\delta(0)$, where n_i is the ion number density and $\delta(\mathbf{w}_i)$ is the delta function. Also $M = m_e m_i/(m_e + m_i) \to m_e$ as $m_i \to \infty$. In this case, with $f_s = f_i$ in (10.16) and (10.19), we get $\mathcal{H} = n_i/w$, $\mathcal{G} = n_i w$ where w is the electron speed. Therefore (10.15) and (10.20) yield

$$\langle \Delta\mathbf{w} \rangle = -\frac{n_i\Gamma}{w^2}\hat{\mathbf{w}}, \qquad \langle \Delta\mathbf{w}\Delta\mathbf{w} \rangle = \frac{n_i\Gamma}{w}\mathbf{1}_\perp, \tag{10.24}$$

where $\hat{\mathbf{w}}$ is unit vector parallel to $\mathbf{w}$ and $\mathbf{1}_\perp \equiv \mathbf{1} - \hat{\mathbf{w}}\hat{\mathbf{w}}$. The Fokker–Planck collision operator in (10.21) can be written

$$\mathbb{C}_{ei} = n_i\Gamma\nabla_w \cdot \left\{ \frac{\hat{\mathbf{w}}}{w^2} f + \tfrac{1}{2}\nabla_w \cdot \left(\frac{1}{w}\mathbf{1}_\perp f \right) \right\}, \tag{10.25}$$

where $\nabla_w = \partial/\partial\mathbf{w}$, and the subscript '$ei$' indicates that only electron–ion collisions are involved.

Since
$$\hat{\mathbf{w}} \cdot \mathbf{1}_\perp = 0, \quad \nabla_w w = \hat{\mathbf{w}}, \quad \nabla_w \hat{\mathbf{w}} = \frac{1}{w}\mathbf{1}_\perp, \quad \nabla_w \cdot \hat{\mathbf{w}} = \frac{2}{w},$$
and
$$\nabla_w \cdot \mathbf{1}_\perp = -(\nabla_w \cdot \hat{\mathbf{w}})\hat{\mathbf{w}} - \hat{\mathbf{w}} \cdot \nabla_w \hat{\mathbf{w}} = -\frac{2}{w}\hat{\mathbf{w}},$$
we obtain
$$\nabla_w \cdot \left(\frac{1}{w}\mathbf{1}_\perp f \right) = -\frac{2}{w}\hat{\mathbf{w}} f + \frac{1}{w}\mathbf{1}_\perp \cdot \nabla_w f;$$
whence
$$\mathbb{C}_{ei} = \tfrac{1}{2} n_i\Gamma\nabla_w \cdot \left(\frac{1}{w}\mathbf{1}_\perp \cdot \nabla_w f \right) = \tfrac{1}{2} n_i\Gamma\nabla_w \cdot \left\{ \frac{1}{w}(\nabla_w f - \hat{\mathbf{w}}\hat{\mathbf{w}} \cdot \nabla_w f) \right\}. \tag{10.26}$$

Mathematical note 10.3 *The distribution function in tensor form*

We shall assume that the distribution function can be expanded in the form

$$f = f_0 + \hat{\mathbf{w}} \cdot \mathbf{f}_1 + \hat{\mathbf{w}}\hat{\mathbf{w}} : \mathbf{f}_2 + \ldots, \tag{10.27}$$

where the tensors $f_0, \mathbf{f}_1, \mathbf{f}_2, \ldots$ have the order indicated by their subscripts, and depend on $\mathbf{r}$, t and $w = |\mathbf{w}|$. Also, f_0 is the local Maxwellian distribution. It is sufficient for $\mathbf{f}_2$ to be a deviator, since (cf. (1.91))

$$\hat{\mathbf{w}} \cdot \mathbf{f}_2 \cdot \hat{\mathbf{w}} = \hat{\mathbf{w}} \cdot (\overset{\circ}{\mathbf{f}}_2 - \mathbf{f}_2^v \times \mathbf{1} + \tfrac{1}{3}\overset{\times}{f}_2 \, \mathbf{1}) \cdot \hat{\mathbf{w}} = \hat{\mathbf{w}} \cdot \overset{\circ}{\mathbf{f}}_2 \cdot \hat{\mathbf{w}} + \tfrac{1}{3}\overset{\times}{f}_2,$$

the last term of which, being independent of $\hat{\mathbf{w}}$, may be assumed to be included in f_0. For an account of the completeness of (10.27) and its equivalence to an expansion in spherical harmonics, the reader can consult Shkarofsky *et al.* (1966).

The electron fluid velocity is

$$\mathbf{v}_e = \frac{1}{n_e}\int \mathbf{w} f \, d\mathbf{w} = \frac{4\pi}{n_e}\int_\Omega\int_0^\infty \mathbf{w} f w^2 \, dw \, d\boldsymbol{\Omega}.$$

Hence by (2.40) and (10.27),

$$\mathbf{v}_e = \frac{4\pi}{3n_e}\int_0^\infty w\mathbf{f}_1 \, dw. \tag{10.28}$$

Denoting $\partial f_0/\partial w$, $\partial \mathbf{f}_1/\partial w, \ldots$ by f_0', $\mathbf{f}_1', \ldots$ and using (10.27), we obtain

$$\nabla_w f_0 = \nabla_w w \frac{\partial f_0}{\partial w} = \hat{\mathbf{w}} f_0', \qquad \nabla_w(\hat{\mathbf{w}} \cdot \mathbf{f}_1) = \frac{1}{w}\mathbf{1}_\perp \cdot \mathbf{f}_1 + \hat{\mathbf{w}}\hat{\mathbf{w}} \cdot \mathbf{f}_1'.$$

Whence

$$\frac{\partial f}{\partial \mathbf{w}} = \nabla_w f = \hat{\mathbf{w}} f_0' + \hat{\mathbf{w}}\hat{\mathbf{w}} \cdot \mathbf{f}_1' + \frac{1}{w}\mathbf{1}_\perp \cdot \mathbf{f}_1, \tag{10.29}$$

$$\mathbf{1}_\perp \cdot \nabla_w f = \frac{1}{w}\mathbf{1}_\perp \cdot \mathbf{f}_1,$$

and

$$\nabla_w \cdot \left(\frac{1}{w}\mathbf{1}_\perp \cdot \nabla_w f\right) = \frac{1}{w^2}\nabla_w \cdot (\mathbf{1}_\perp \cdot \mathbf{f}_1) = \frac{1}{w^2}\left(-\frac{2}{w}\hat{\mathbf{w}} \cdot \mathbf{f}_1 + \mathbf{1}_\perp : \hat{\mathbf{w}}\mathbf{f}_1'\right),$$

i.e.

$$\nabla_w \cdot \left(\frac{1}{w}\mathbf{1}_\perp \cdot \nabla_w f\right) = -\frac{2}{w^3}\hat{\mathbf{w}} \cdot \mathbf{f}_1. \tag{10.30}$$

10.3.2 *The Lorentzian conductivity*

In the absence of a magnetic field, the kinetic equation for the electron gas is

$$\frac{\partial f}{\partial t} + \mathbf{w} \cdot \frac{\partial f}{\partial \mathbf{r}} - e\mathbf{E} \cdot \frac{\partial f}{\partial \mathbf{w}} = \mathbb{C} = \mathbb{C}_{ei} + \mathbb{C}_{ee}, \tag{10.31}$$

where by (10.26) and (10.30),

$$\mathbb{C}_{ei} = -\nu_{ei}\hat{\mathbf{w}} \cdot \mathbf{f}_1, \tag{10.32}$$

ν_{ei} being the collision frequency,

$$\nu_{ei} = \frac{n_i\Gamma_e}{w^3} = \frac{n_i}{w^3}\left(\frac{Z^2e^4 \ln \Lambda}{4\pi\epsilon_0^2 m_e^2}\right), \tag{10.33}$$

obtained by setting $Z = -1$ in (10.13). In a Lorentzian gas $\mathbb{C}_{ee}$ is zero. It follows from (10.29), (10.31), and (10.32) that in steady uniform conditions

$$e\mathbf{E} \cdot \left(\hat{\mathbf{w}} f_0' + \hat{\mathbf{w}}\hat{\mathbf{w}} \cdot \mathbf{f}_1' + \frac{1}{w}\mathbf{1}_\perp \cdot \mathbf{f}_1\right) = \nu_{ei}\hat{\mathbf{w}} \cdot \mathbf{f}_1.$$

Multiplying this equation by $\hat{\mathbf{w}}$, integrating over $\hat{\mathbf{w}}$ and evaluating the integrals with the aid of (5.57) and (5.58), we get

$$e\mathbf{E} f_0' = \nu_{ei}\mathbf{f}_1. \tag{10.34}$$

The current density is proportional to the electric field, i.e.

$$\mathbf{j} = -en_e\mathbf{v}_e = \sigma_L \mathbf{E}, \tag{10.35}$$

where σ_L is the Lorentzian conductivity. Hence by (10.28) and (10.34),

$$\sigma_L = -\tfrac{4}{3}\pi e^2 \int_0^\infty \frac{w^3}{\nu_{ei}} f_0' \, dw = \tfrac{4}{3}\pi e^2 \int_0^\infty f_0 \frac{\partial}{\partial w}\left(\frac{w^3}{\nu_{ei}}\right) dw,$$

on integrating by parts. Thus

$$\sigma_L = \left(\frac{4\pi\epsilon_0 m_e}{Ze}\right)^2 \frac{2}{n_i \ln\Lambda} \int_0^\infty w^5 f_0 \, dw.$$

With the approximation $\mathbf{w} - \mathbf{v}_e \approx \mathbf{w}$ in the Maxwellian distribution

$$f_0 = n_e \left(\frac{m_e}{2\pi kT_e}\right)^{\frac{3}{2}} \exp\{-m_e(\mathbf{w} - \mathbf{v}_e)^2/2kT_e\},$$

and (2.41), we arrive at (Lorentz 1923)

$$\sigma_L = \frac{2m_e}{\ln\Lambda}\left(\frac{n_e}{Z^2 n_i}\right)\left(\frac{4\pi\epsilon_0}{e}\right)^2 \left(\frac{2kT_e}{\pi m_e}\right)^{\frac{3}{2}}. \tag{10.36}$$

In a neutral plasma $n_e = Zn_i$, so with the definition

$$\tau_e^{-1} \equiv \frac{n_e Z\, e^4}{3m_e^2\, \epsilon_0^2}\left(\frac{m_e}{2\pi kT_e}\right)^{\frac{3}{2}} \ln\Lambda, \tag{10.37}$$

we find that

$$\sigma_L = \frac{32}{3\pi}\frac{e^2 n_e \tau_e}{m_e}. \tag{10.38}$$

In §9.4.4 we gave a table relating this conductivity to that found by Spitzer and Härm for a fully ionized plasma.

10.4 The friction and diffusion coefficients

10.4.1 *The first superpotential*

Returning to the general theory of §10.2, we shall now assume that the field particles have a Maxwellian distribution,

$$f_s \, dw_s = \frac{4}{\sqrt{\pi}}\frac{n_s}{\mathcal{C}^3} \exp(-w_s^2/\mathcal{C}_s^2)\, w_s^2 \, dw_s \qquad \left(\mathcal{C}_s \equiv (2kT_s/m_s)^{\frac{1}{2}}\right). \tag{10.39}$$

The function $\mathcal{H}$ in (10.16) is analogous to the gravitational potential due to a symmetrically distributed mass of density mf_s/M, centred on the origin $w_s = 0$. Using a theorem dating back to Newton, we shall first calculate the increment $d\mathcal{H}$ due to the 'mass' in a spherical shell Σ of radius w_s and thickness dw_s. Within Σ, $d\mathcal{H}$ is zero and outside Σ $d\mathcal{H}$ is the same as that due to a concentrated 'mass' at the origin. Hence

$$d\mathcal{H} = \frac{m}{M}\left\{\frac{f_s(w_s)}{w} - \frac{f_s(w_s)}{w_s}\right\} 4\pi w_s^2 \, dw_s \qquad \text{if} \quad w \geq w_s,$$

and $d\mathcal{H} = 0$ if $w \leq w_s$. Therefore by (10.39) and $y \equiv w_s/\mathcal{C}_s$, $x \equiv w/\mathcal{C}_s$,

$$\mathcal{H} = \frac{mn_s}{M\mathcal{C}_s}\frac{4}{\sqrt{\pi}}\int_0^x \left(\frac{1}{x} - \frac{1}{y}\right) \exp(-y^2)\, y^2\, dy + \text{const.}$$

Integrating by parts and introducing the error function $\Phi(x)$,

$$\mathcal{H}(x) = \frac{mn_s}{M\mathcal{C}_s}\left(\frac{\Phi(x)}{x} - \frac{2}{\sqrt{\pi}}\right) + \text{const.}, \tag{10.40}$$

where

$$\Phi(x) \equiv \frac{2}{\sqrt{\pi}}\int_0^x \exp(-y^2)\, dy. \tag{10.41}$$

To evaluate the constant of integration, we note that at $w = 0$, where $x = 0$, (10.16) and (10.39) yield

$$\mathcal{H}(0) = \frac{mn_s}{M\mathcal{C}_s}\frac{4}{\sqrt{\pi}}\int_0^\infty \exp(-y^2) y\, dy = \frac{mn_s}{M\mathcal{C}_s}\frac{2}{\sqrt{\pi}}. \tag{10.42}$$

For x small,

$$\Phi(x) = \frac{2}{\sqrt{\pi}}\int_0^\infty (1 - y^2 + \ldots)\, dy = \frac{2}{\sqrt{\pi}} x(1 - \tfrac{1}{3}x^2 + \ldots),$$

so

$$\frac{\Phi(x)}{x} = \frac{2}{\sqrt{\pi}}(1 - \tfrac{1}{3}x^2 + \ldots). \tag{10.43}$$

It follows from (10.40) and (10.42) that

$$\mathcal{H} = \frac{mn_s}{M\mathcal{C}_s}\frac{\Phi(x)}{x}. \tag{10.44}$$

10.4.2 *The friction coefficient*

By (10.15) the friction coefficient is

$$\langle \Delta \mathbf{w} \rangle = \Gamma \frac{\partial \mathcal{H}}{\partial \mathbf{w}} = \frac{\Gamma}{\mathcal{C}_s}\frac{\partial w}{\partial \mathbf{w}}\frac{\partial \mathcal{H}}{\partial x} = \frac{m\Gamma n_s}{M\mathcal{C}_s^2}\hat{\mathbf{w}}\frac{d}{dx}\left(\frac{\Phi(x)}{x}\right).$$

Therefore

$$\hat{\mathbf{w}} \cdot \langle \Delta \mathbf{w} \rangle \equiv \langle \Delta w_\parallel \rangle = -\frac{A_D}{\mathcal{C}_s^2}\left(1 + \frac{m}{m_s}\right) G\left(\frac{w}{\mathcal{C}_s}\right), \tag{10.45}$$

where by (10.13) the 'diffusion constant' is defined by

$$A_D \equiv 2\Gamma n_s = \frac{(ZZ_s)^2 e^4 n_s \ln \Lambda}{2\pi\epsilon_o^2 m^2}, \tag{10.46}$$

and $G(x)$ is the function

$$G(x) = -\frac{1}{2}\frac{d}{dx}\left(\frac{\Phi(x)}{x}\right) = \frac{\Phi(x) - x\Phi'(x)}{2x^2}. \tag{10.47}$$

10.4.3 *The second superpotential*

The second superpotential can be deduced from (10.22) and (10.44). As the functions are spherically symmetric, we have

$$\frac{1}{w^2}\frac{\partial}{\partial w}\left(w^2\frac{\partial \mathcal{G}}{\partial w}\right) = \frac{n_s}{\mathcal{C}_s}\frac{\Phi(x)}{x},$$

whence

$$\frac{\partial \mathcal{G}}{\partial x} = \frac{n_s \mathcal{C}_s}{x^2}\int_0^x x'\Phi(x')\,dx',$$

the constant of integration being zero in order to keep $\partial\mathcal{G}/\partial x$ finite at $x = 0$. Integration by parts yields

$$\begin{aligned}\frac{1}{x^2}\int_0^x x'\Phi(x')\,dx' &= \frac{1}{x^2}\left\{(x^2 - \tfrac{1}{2})\Phi(x) + \frac{1}{\sqrt{\pi}}\,x\exp(-x^2)\right\}\\ &= \Phi(x) - G(x);\end{aligned}$$

also

$$\Phi' - G' = \Phi' + \frac{2}{x}G + \frac{1}{2x}\Phi'' = \frac{2}{x}G(x).$$

Hence

$$\frac{\partial \mathcal{G}}{\partial x} = n_s\mathcal{C}_s\{\Phi(x) - G(x)\}, \qquad \frac{\partial^2 \mathcal{G}}{\partial x^2} = 2n_s\mathcal{C}_s\frac{G(x)}{x}. \tag{10.48}$$

Since

$$\frac{\partial \mathcal{G}}{\partial \mathbf{w}} = \frac{\partial w}{\partial \mathbf{w}}\frac{1}{\mathcal{C}_s}\frac{\partial \mathcal{G}}{\partial x} = \frac{\hat{\mathbf{w}}}{\mathcal{C}_s}\frac{\partial \mathcal{G}}{\partial x},$$

then

$$\frac{\partial^2 \mathcal{G}}{\partial \mathbf{w}\,\partial \mathbf{w}} = \frac{1}{\mathcal{C}_s w}(\mathbf{1} - \hat{\mathbf{w}}\hat{\mathbf{w}})\frac{\partial \mathcal{G}}{\partial x} + \frac{1}{\mathcal{C}_s}\hat{\mathbf{w}}\hat{\mathbf{w}}\frac{\partial^2 \mathcal{G}}{\partial x^2}. \tag{10.49}$$

10.4.4 *The diffusion coefficient*

The diffusion coefficient now follows from (10.20), (10.46), (10.48), and (10.49):

$$\langle \Delta\mathbf{w}\Delta\mathbf{w}\rangle = \frac{A_D}{2w}\left\{\Phi\left(\frac{w}{\mathcal{C}_s}\right) - G\left(\frac{w}{\mathcal{C}_s}\right)\right\}(\mathbf{1} - \hat{\mathbf{w}}\hat{\mathbf{w}}) + \frac{A_D}{w}G\left(\frac{w}{\mathcal{C}_s}\right)\hat{\mathbf{w}}\hat{\mathbf{w}}. \tag{10.50}$$

Its parallel and perpendicular components are

$$\hat{\mathbf{w}}\hat{\mathbf{w}} : \langle \Delta\mathbf{w}\Delta\mathbf{w}\rangle \equiv \langle(\Delta w_\parallel)^2\rangle = \frac{A_D}{w}G\left(\frac{w}{\mathcal{C}_s}\right), \tag{10.51}$$

and

$$(\mathbf{1} - \hat{\mathbf{w}}\hat{\mathbf{w}}) : \Delta\mathbf{w}\Delta\mathbf{w}\rangle \equiv \langle(\Delta w_\perp)^2\rangle = \frac{A_D}{w}\left\{\Phi\left(\frac{w}{\mathcal{C}_s}\right) - G\left(\frac{w}{\mathcal{C}_s}\right)\right\}. \tag{10.52}$$

From (10.43) and (10.47) we find the limits

$$\left.\begin{array}{llll} x \to 0: & \dfrac{G(x)}{x} \to \dfrac{2}{3\sqrt{\pi}}, & \dfrac{1}{x}\{\Phi(x) - G(x)\} \to \dfrac{4}{3\sqrt{\pi}}, \\ x \to \infty: & \dfrac{G(x)}{x} \to \dfrac{1}{2x^2}, & \dfrac{1}{x}\{\Phi(x) - G(x)\} \to \dfrac{1}{x}. \end{array}\right\} \quad (10.53)$$

It follows from (10.45), (10.51), (10.52), and (10.53) that

$$x = 0: \qquad \langle \Delta w_\parallel \rangle = 0 \qquad \langle (\Delta w_\parallel)^2 \rangle = \tfrac{1}{2}\langle (\Delta w_\perp)^2 \rangle = \frac{2}{3\sqrt{\pi}}\frac{A_D}{\mathcal{C}_s}; \quad (10.54)$$

$$x \to \infty: \quad \left\{\begin{array}{rcl} \langle \Delta w_\parallel \rangle & = & -\dfrac{A_D}{2}\left(1 + \dfrac{m}{m_s}\right)\dfrac{1}{w^2}, \\ \langle (\Delta w_\parallel)^2 \rangle & = & \dfrac{A_D \mathcal{C}_s^2}{2w^2}, \quad \langle (\Delta w_\perp)^2 \rangle = \dfrac{A_D}{w}. \end{array}\right\} \quad (10.55)$$

For intermediate values of x, Table 10.1 is adequate.

Table 10.1 *Values of $G(x)$ and $\Phi(x) - G(x)$*

x	0	0.2	0.4	0.6	0.8	1.0	1.2	1.4	1.6
$G(x)$	0	0.073	0.137	0.183	0.208	0.214	0.205	0.186	0.163
$\chi(x)$	0	0.149	0.292	0.421	0.534	0.629	0.708	0.766	0.813
x	1.8	2.0	2.5	3.0	3.5	4.0	5.0	6.0	8.0
$G(x)$	0.140	0.119	0.080	0.056	0.041	0.031	0.020	0.014	0.008
$\chi(x)$	0.849	0.876	0.920	0.944	0.959	0.969	0.980	0.986	0.992

$\chi(x) \equiv \Phi(x) - G(x)$

Equations (10.54) confirm what is physically evident, namely that stationary test particles experience no friction and that their diffusion is isotropic. On the other hand, (10.55) show that quite fast test particles in the main diffuse transversely to their original direction. Also notice that heavy test particles ($m \gg m_s$) tend to be dominated by friction. The theory for $\langle (\Delta w_\parallel)^2 \rangle$ given above is not valid if $x^2 > \ln \Lambda$ (Spitzer 1962).

10.5 Relaxation times

A 'relaxation time' is the time it takes for collisions to effect a substantial change in a given initial velocity or energy distribution. Several relaxation times can be deduced from (10.51) and (10.52). Those of importance in plasma physics are listed below.

10.5.1 *Slowing-down time*

$$\tau_s \equiv -\frac{w}{\langle \Delta w_\parallel \rangle} = \frac{w\mathcal{C}_s^2}{(1 + m/m_s)A_D G(w/\mathcal{C}_s)}. \quad (10.56)$$

Consider, for example, the slowing-down time of cold ions drifting through a background of thermal electrons. In this case the electrons are the scatterers and the ions are the test particles, so that $m_s = m_e$, $m = m_i$ and

$\mathcal{C}_s = \mathcal{C}_e = (2kT_e/m_e)^{\frac{1}{2}}$. In a frame fixed in the electron fluid, the ions have a drift velocity $w = w_i$. Provided $w_i \ll \mathcal{C}_e$, equations (9.54), (10.45), (10.46), (10.53), (10.56), and $m_e \ll m_i$ yield

$$\tau_s = 3(2\pi)^{3/2} \frac{\epsilon_0^2 m_e^{1/2}}{e^4 \ln \Lambda} \frac{(kT_e)^{3/2}}{Zn_e} \frac{m_i}{Zm_e} = \tau_e \frac{m_i}{Zm_e}, \tag{10.57}$$

which can be used to obtain an estimate for the electrical conductivity (see exercise 10.2).

10.5.2 *Deflection time*

$$\tau_D \equiv \frac{w^2}{\langle(\Delta w_\perp)^2\rangle} = \frac{w^3}{A_D\{\Phi(w/\mathcal{C}_s) - G(w/\mathcal{C}_s)\}}; \tag{10.58}$$

this is approximately the time it takes grazing collisions to deflect a test particle through 90° (see remarks in §1.1.1). For example, the deflection time for electrons moving with the r.m.s. speed $w = (3kT_e/m_e)^{\frac{1}{2}}$, and being scattered by ions at a similar temperature ($x \sim (m_i/m_e)^{\frac{1}{2}} \sim \infty$) is

$$\tau_D = 2\pi 3^{3/2} \frac{\epsilon_0^2 m_e^{1/2}}{e^4 \ln \Lambda} \frac{(kT_e)^{3/2}}{Zn_e} \approx 0.69\tau_e. \tag{10.59}$$

10.5.3 *Energy exchange times*

$$\tau_2 \equiv \frac{E^2}{\langle(\Delta E)^2\rangle} = \frac{w^3}{4A_D G(w/\mathcal{C}_s)}, \tag{10.60}$$

where the second expression follows from (10.51) and the approximation

$$(\Delta E)^2 = \left[\tfrac{1}{2}m\{(w + \Delta w_\parallel)^2 + (\Delta w_\perp)^2\} - \tfrac{1}{2}mw^2\right]^2 \approx (mw\Delta w_\parallel)^2.$$

At large velocities, $\tau_2 = w^2x^2/2A_D = (w^2/2\mathcal{C}_s^2)\tau_D \gg \tau_D$, showing that deflection is the dominant process. From Table 10.1 $G\{(1.5)^{\frac{1}{2}}\} = 0.203$, so for the conditions in which (10.59) applies, we get $\tau_2 \approx 0.8\tau_e Z$.

10.5.4 *Self-collision time*

When the test particles and scatterers are identical, a 'self-collision' time τ_c can be defined by $\tau_c = \tau_D$ at $w = (3kT/m)^{\frac{1}{2}}$ (Spitzer 1962). At $x = (1.5)^{\frac{1}{2}}$, Table 10.1 gives $\Phi - G = 0.714$, $G = 0.203$. Hence from (10.71),

$$\tau_c = 45.73 \frac{\epsilon_0^2}{e^4} \frac{(m_p A)^{1/2}}{\ln \Lambda} \frac{(kT)^{3/2}}{Z^4 n}, \tag{10.61}$$

where A is the particle mass number ($A_e = 1/1836$) and m_p is the mass of the proton. For this case (10.73) gives $\tau_c = 1.14\tau_2$, so τ_c is a measure of the relaxation times, both for the velocity distribution to become isotropic and for the energy distribution to thermalize. Notice that the electrons have a self-collision time $\frac{1}{43}$ of that for protons.

10.5.5 *Equipartition time*

Suppose that each of two groups of particles are in internal thermal equilibrium, i.e. have Maxwellian velocity distributions, but have different temperatures, say T and T_s. Then the equipartition time, τ_{eq} is implicitly defined by

$$\frac{dT}{dt} = \frac{T_s - T}{\tau_{eq}}. \tag{10.62}$$

An expression for τ_{eq} can be found as follows.

A particle of the species at temperature T changes its energy at the rate

$$\begin{aligned}\langle \Delta E\rangle &= \tfrac{1}{2}m\langle (w+\Delta w_{\parallel})^2 + (\Delta w_{\perp})^2 - w^2\rangle \\ &= \tfrac{1}{2}m\{2w\langle \Delta w_{\parallel}\rangle + \langle (\Delta w_{\parallel})^2\rangle + \langle (\Delta w_{\perp})^2\rangle\}.\end{aligned}$$

From (10.45), (10.51), and (10.52) this is

$$\langle \Delta E\rangle = \tfrac{1}{2}mA_D\left\{-\frac{2w}{\mathcal{C}_s^2}\left(1+\frac{m}{M_s}\right)G\left(\frac{w}{\mathcal{C}_s}\right) + \frac{1}{w}\Phi\left(\frac{w}{\mathcal{C}_s}\right)\right\}. \tag{10.63}$$

Hence by (10.39) this species increases its internal energy ρu per unit volume at the rate

$$\frac{4}{\sqrt{\pi}}\int_0^\infty \langle \Delta E\rangle \frac{n}{\mathcal{C}^3}\exp(-w^2/\mathcal{C}^2)w^2\,dw = \tfrac{3}{2}kn\frac{dT}{dt}. \tag{10.64}$$

By (10.63) the left-hand side of this equation is

$$\frac{mnA_D}{\mathcal{C}_s}\frac{2}{\sqrt{\pi}}\int_0^\infty \left\{-2\alpha\left(1+\frac{m}{m_s}\right)xG(\alpha x) + \frac{1}{\alpha x}\Phi(\alpha x)\right\}e^{-x^2}x^2\,dx,$$

where $\alpha = \mathcal{C}/\mathcal{C}_s = (Tm_s/T_s m)^{\frac{1}{2}}$. Using (10.47) and integration by parts we find that

$$4\int_0^\infty x^2 G(\alpha x)e^{-x^2}\,dx = \frac{\alpha}{(1+\alpha^2)^{3/2}}$$

and

$$2\int_0^\infty x\Phi(\alpha x)e^{-x^2}\,dx = \frac{\alpha}{(1+\alpha^2)^{1/2}}.$$

We can now evaluate the integral in (10.64) and rearrange that equation in the form of (10.62) to deduce that

$$\tau_{eq} = 3\sqrt{2}\pi^{3/2}\frac{\epsilon_0^2 m m_s}{n_s e^4 Z^2 Z_s^2 \ln\Lambda}\left(\frac{kT}{m}+\frac{kT_s}{m_s}\right)^{\frac{3}{2}}. \tag{10.65}$$

When the species involved are electrons and ions at comparable temperatures, the last term in (10.65) reduces to $(kT_e/m_e)^{\frac{1}{2}}$. Comparing this result with (9.54), we obtain the equipartition time $\tau_{eq} = \tau_{ei}^{\varepsilon}$ say, where

$$\tau_{ei}^{\varepsilon} = \tau_e\frac{m_i}{2m_e}. \tag{10.66}$$

Let τ_{ee}, τ_{ii} denote the self-collision times for electrons and ions, then from (10.61) and (10.66),

$$\tau_{ee} : \tau_{ii} : \tau_{ei}^{\varepsilon} = 1 : (m_i/m_e)^{\frac{1}{2}} : m_i/2m_e. \tag{10.67}$$

For example, in a hydrogen plasma the electrons reach thermal equilibrium 43 times more rapidly than the ions, whereas thermal equilibrium *between* the ions and electrons is 918 times more slowly attained then equilibrium in the electrons alone.

Let φ_e, φ_i denote the rates at which energy is transferred into the electron fluid from the ion fluid and visa versa, per unit volume (see (9.39) and (9.40)). Then $\varphi_e = \frac{3}{2}kn_e dT_e/dt = -\varphi_i$. Hence from (10.62) and (10.66),

$$\varphi_e = \frac{3m_e}{m_i}\frac{kn_e}{\tau_e}(T_i - T_e) = -\varphi_i. \tag{10.68}$$

There is also electron heating due to ohmic dissipation.

10.5.6 *Escape time for magnetically trapped particles*

In §10.5.2 we calculated the 90° deflection time τ_D. For a deflection through a smaller angle α the time required is τ_α, where

$$\frac{\tau_\alpha}{\tau_D} = \frac{\alpha^2}{(\pi/2)^2}. \tag{10.69}$$

The diffusivity, which by (1.34) is proportional to (displacement)^2/(time), is approximately constant and with a sequence of grazing collisions, the accumulated deflection angle is proportional to the displacement. Whence (10.69) follows.

By (9.88) magnetically trapped particles escape when the pitch angle of their trajectories exceeds $\alpha_c = \sin^{-1}(1 - R_m^{-1})^{\frac{1}{2}}$, where R_m is the mirror ratio. Hence in this case (10.69) gives the estimate for the escape time

$$\tau_{es} = \frac{4}{\pi^2}\alpha_c^2\,\tau_D = \frac{4}{\pi^2}\left[\sin^{-1}(1 - R_m^{-1})^{\frac{1}{2}}\right]^2 \tau_D. \tag{10.70}$$

10.6 The effect of magnetic fields

10.6.1 *The streaming derivative*

The magnetic field enters the theory by modifying the streaming derivative $\mathbb{D}$. In §5.2.3 (also see Mathematical note 7.5) we showed that

$$\mathbb{D} = \mathrm{D}^* + \mathbf{c}\cdot\nabla + (\dot{\mathbf{c}} - \mathbf{P})\cdot\frac{\partial}{\partial\mathbf{c}}, \tag{10.71}$$

where $\dot{\mathbf{c}}$ is the particle acceleration in a frame convected with the fluid, which in a neutral gas is given by

$$\dot{\mathbf{c}} = -\mathbf{c}\cdot\mathbf{e} \qquad (\mathbf{e} \equiv \tfrac{1}{2}(\nabla\mathbf{v} + \widetilde{\nabla\mathbf{v}})). \tag{10.72}$$

The presence of a magnetic field adds the Lorentz force per unit mass to the right-hand side of (10.72). In §9.5.1 it was shown that, in the absence of collisions,

$$\dot{\mathbf{c}} = \frac{Q}{m}\mathbf{c}\times\mathbf{B} - \mathbf{c}\cdot\mathbf{e}. \tag{10.73}$$

We shall incorporate this change by retaining the neutral gas definition of $\mathbb{D}$, viz.

$$\mathbb{D} \equiv \mathrm{D}^* + \mathbf{c}\cdot\nabla + \mathbf{a}\cdot\frac{\partial}{\partial\mathbf{c}} \qquad \big(\mathbf{a} \equiv -\mathbf{P} - \mathbf{c}\cdot\mathbf{e}\big), \tag{10.74}$$

and adding the Lorentz contribution separately via the operator

$$\delta \equiv \frac{Q}{m}\mathbf{c}\times\mathbf{B}\cdot\frac{\partial}{\partial\mathbf{c}} = -\omega_c\mathbf{b}\times\mathbf{c}\cdot\frac{\partial}{\partial\mathbf{c}}. \tag{10.75}$$

The total streaming derivative is now $\mathbb{D} + \delta$.

Thus in a magnetoplasma, the BGK kinetic equation, (5.34), becomes

$$\mathbb{D}f + \delta f = \frac{1}{\tau}(f_0 - f) + \mathrm{O}(\epsilon^2). \tag{10.76}$$

A similar change applies to the Boltzmann kinetic equation and of course also to the special Fokker–Planck form this equation takes when grazing collisions dominate the collision integral. However, the physical effects of magnetic fields are more readily explained using the BGK model. These theories are not accurate beyond first order in the Knudsen number. Second-order theory, which is much more important with strong magnetic fields, will be explained in Chapter 11.

10.6.2 *Physical basis of the BGK kinetic equation*

As $\tau\mathbb{D}f = \tau\mathbb{D}f_0 + \mathrm{O}(\epsilon)$, (10.76) can be written

$$f + \tau\delta f = f_0 - \tau\mathbb{D}f_0 + \mathrm{O}(\epsilon^2). \tag{10.77}$$

In §5.4.1 we gave a mean-free-path derivation of this equation in the case of a neutral gas, i.e. without the term $\tau\delta f$. In that case the right-hand side is the Taylor expansion of f_0 in a stream of molecules $\mathcal{B}$ at a position, speed and time corresponding to one collision interval earlier than the present.

That interpretation is not immediately applicable to (10.77) because of the presence of the Lorentz term $\tau\delta f$. In strong magnetic fields, the velocity $\mathbf{c}$ is rapidly changing in direction many times per collision interval, so a Taylor expansion involving $\mathbf{c}$ qua *particle* velocity is not valid. But there is another way of interpreting the right-hand side of (10.77). Because (10.76) is a local form, the velocity $\mathbf{c}$ in (10.77) has a fixed instantaneous value. (Recall that $\mathbf{c}$ is an independent variable, and its identification with the velocity of any particular bunch $\mathcal{B}$ is a matter of choice.) Hence the right-hand side of (10.77) is the Taylor expansion of $f_0\big(\mathbf{r} - \tau\mathbf{c},\, \mathbf{c} - \tau\mathbf{a},\, t - \tau\big)$, an expansion taken along a near-straight trajectory $\mathcal{T}$ that is tangentially to the orbits of the intervening particles. Let $d\boldsymbol{\nu}$ be a phase-space element following this tangential path, with velocity $\mathbf{c}$ and acceleration $\mathbf{a}$, then $\mathbb{D}$

defined in (10.71) is the streaming derivative following $d\boldsymbol{\nu}$ (see §5.2.3).*

The kinetic equation is a relation giving the balance between the streaming rate of change of particles within $d\boldsymbol{\nu}$ and the rate of collisional and other losses from $d\boldsymbol{\nu}$. Equation (10.76) can be rearranged as

$$\mathbb{D}f = \frac{1}{\tau}\Big(f_0 - [f + \tau\delta f]\Big),$$

which identifies δf as a loss term from $d\boldsymbol{\nu}$. It resembles the streaming losses due to the pressure gradient (see §5.8.2), but does not involve collisions.

As we shall show in §§10.7 to 10.9, there is a kinetic equation equivalent to (10.77), with the merit that it has no Lorentz term on the left-hand side. This is

$$f = f_0 - \tau\overline{\mathbb{D}f_0} + \mathrm{O}(\epsilon^2), \tag{10.78}$$

where the bar denotes weighted averages of functions of the peculiar velocity of the particles, extending from the distant past to the present. This is not a simple 'gyro-average', i.e. an average over *one* complete gyration, but an average over a great number of gyrations, that represents as closely as necessary the collisional *history* of the bunch of particles in question.

Equation (10.78) allows the following interpretation. The left-hand side is the phase space density of the particles $\mathcal{B}$ in a phase space element $d\overline{\boldsymbol{\nu}}$. For the same reasons as given in §5.4.1, it is equal to the equilibrium density of the bunch $\mathcal{B}$ a collision interval earlier, which is approximated by the right-hand side of (10.78). In strong magnetic fields $\mathcal{B}$ will never be far from its guiding centre G, so the right-hand side of (10.78) could be interpreted in terms of a streaming derivative $\overline{\mathbb{D}}$ following G through phase space. But it is important to average the *whole* term, and not the differential operator alone (see exercise 10.7). This point will become clearer when we have explained the averaging process itself.

10.7 Particle velocity gyro-averages

Our objective is to find those values $\overline{\mathbf{c}}$ and $\overline{\mathbf{cc}}$, which applied at a specific point Q_c on the orbit of a typical particle P, will correctly represent the collisional history of the bunch of particles $\mathcal{B}$ of which P is a member.

10.7.1 *Expression for* $\overline{\mathbf{c}}$

Suppose that in position Q_c' on the orbit, P's velocity is

$$\mathbf{c}(\theta) = \mathbf{c}_{\|} + c_{\perp}(\hat{\mathbf{c}}_{\perp}\, \cos\theta + \mathbf{b}\times\hat{\mathbf{c}}\, \sin\theta), \tag{10.79}$$

where $\hat{\mathbf{c}}_{\perp}$ and $\hat{\mathbf{a}} = \mathbf{b}\times\hat{\mathbf{c}}$ are unit vectors at Q_c, as indicated in Fig. 9.3. If P is at Q_c' at the time $t' = t - \theta/\omega_c$, then it will arrive at Q_c at time t. Because

*In §5.9.3 we described the case in a neutral gas when it is necessary to distinguish between particle and phase-space accelerations; here it is necessary to distinguish between velocities as well.

of collisions, the probability that P actually reaches Q_c is $\exp\{-(t-t')/\tau\} = \exp(-\theta/\varpi)$, where $\varpi \equiv \omega_c\tau$. The probability density over the variable θ is therefore $\exp(-\theta/\varpi)/\varpi$. With this exponential decay factor, the average value of $\mathbf{c}(\theta)$ is $\overline{\mathbf{c}} = \mathbf{c}_\parallel + \overline{\mathbf{c}}_\perp$, where

$$\overline{\mathbf{c}}_\perp = \int_0^\infty (\mathbf{c}_\perp \cos\theta + \mathbf{b}\times\mathbf{c}\sin\theta)\exp(-\theta/\varpi)\,d\theta/\varpi,$$

i.e.

$$\overline{\mathbf{c}}_\perp = \frac{1}{1+\varpi^2}\mathbf{c}_\perp + \frac{\varpi}{1+\varpi^2}\mathbf{b}\times\mathbf{c}. \tag{10.80}$$

In this expression $\mathbf{c} = \mathbf{c}_\perp + \mathbf{c}_\parallel$ is the *present* peculiar velocity at Q_c, whereas $\overline{\mathbf{c}}_\perp$ is the effective perpendicular velocity at Q_c, when allowance is made for the history of P.

Hence

$$\overline{\mathbf{c}} = \mathbf{c}\cdot\mathsf{k}, \tag{10.81}$$

where k is the second-order tensor defined in (9.98), viz.

$$\mathsf{k} = \mathbf{bb} - \frac{\varpi}{1+\varpi^2}\mathbf{b}\times\mathbf{1} + \frac{1}{1+\varpi^2}(\mathbf{1} - \mathbf{bb}). \tag{10.82}$$

Notice from (9.97) that $\tau\overline{\mathbf{c}}$ is the generalized mean free path.

10.7.2 *Expression for* $\overline{\mathbf{cc}}$

Similarly, from (10.79),

$$\begin{aligned}\overline{\mathbf{c}_\perp\mathbf{c}_\perp} &= \int_0^\infty \Big\{\mathbf{c}_\perp\mathbf{c}_\perp\cos^2\theta + (\mathbf{c}_\perp\mathbf{b}\times\mathbf{c} + \mathbf{b}\times\mathbf{c}\mathbf{c}_\perp)\sin\theta\cos\theta \\ &\qquad + \mathbf{b}\times\mathbf{c}\mathbf{b}\times\mathbf{c}\sin^2\theta\Big\}\exp(-\theta/\varpi)\,d\theta/\varpi,\end{aligned}$$

which yields

$$\begin{aligned}\overline{\mathbf{c}_\perp\mathbf{c}_\perp} &= \frac{1}{2}\left(1 + \frac{1}{1+4\varpi^2}\right)\mathbf{c}_\perp\mathbf{c}_\perp + \frac{\varpi}{1+4\varpi^2}(\mathbf{c}_\perp\mathbf{b}\times\mathbf{c} + \mathbf{b}\times\mathbf{c}\mathbf{c}_\perp) \\ &\qquad + \frac{1}{2}\left(1 - \frac{1}{1+4\varpi^2}\right)\mathbf{b}\times\mathbf{c}\mathbf{b}\times\mathbf{c}.\end{aligned}$$

Thence from $\overline{\mathbf{cc}} = \mathbf{c}_\parallel\mathbf{c}_\parallel + \mathbf{c}_\parallel\overline{\mathbf{c}}_\perp + \overline{\mathbf{c}}_\perp\mathbf{c}_\parallel + \overline{\mathbf{c}_\perp\mathbf{c}_\perp}$ and (10.80) we obtain

$$\begin{aligned}\overline{\mathbf{cc}} &= \{\mathbf{c}_\parallel\mathbf{c}_\parallel + \tfrac{1}{2}\mathbf{c}_\perp\mathbf{c}_\perp + \tfrac{1}{2}\mathbf{b}\times\mathbf{c}\mathbf{b}\times\mathbf{c}\} + \frac{1}{2(1+4\varpi^2)}\{\mathbf{c}_\perp\mathbf{c}_\perp - \mathbf{b}\times\mathbf{c}\,\mathbf{b}\times\mathbf{c}\} \\ &\quad + \frac{1}{1+\varpi^2}\{\mathbf{c}_\parallel\mathbf{c}_\perp + \mathbf{c}_\perp\mathbf{c}_\parallel\} + \frac{\varpi}{1+4\varpi^2}\{\mathbf{c}_\perp\mathbf{b}\times\mathbf{c} + \mathbf{b}\times\mathbf{c}\mathbf{c}_\perp\} \\ &\quad + \frac{\varpi}{1+\varpi^2}\{\mathbf{c}_\parallel\mathbf{b}\times\mathbf{c} + \mathbf{b}\times\mathbf{c}\mathbf{c}_\parallel\}.\end{aligned} \tag{10.83}$$

We can express this relation with the help of a fourth-order tensor W as follows:

$$\overline{\mathbf{cc}} = \mathbf{cc}:\mathsf{W}, \tag{10.84}$$

where

$$\mathbf{W} = \mathbf{W}_1 + \frac{1}{1+4\varpi^2}\mathbf{W}_2 + \frac{1}{1+\varpi^2}\mathbf{W}_3 + \frac{2\varpi}{1+4\varpi^2}\mathbf{W}_4 + \frac{\varpi}{1+\varpi^2}\mathbf{W}_5. \quad (10.85)$$

The tensors $\mathbf{W}_i$, $i = 1, 2, \ldots, 5$, can be represented as products of the following second-order tensors,

$$\mathbf{1}_\parallel \equiv \mathbf{bb}, \qquad \mathbf{1}_\wedge \equiv \mathbf{b}\times\mathbf{1}, \qquad \mathbf{1}_\perp \equiv \mathbf{1} - \mathbf{bb}, \quad (10.86)$$

with the convention that $\mathbf{1}_i{}^\diamond\mathbf{1}_j : \mathbf{A} = \mathbf{1}_i \cdot \mathbf{A} \cdot \mathbf{1}_j$ for any second-order tensor $\mathbf{A}$. We find from (10.83) that

$$\left.\begin{aligned} \mathbf{W}_1 &= \mathbf{1}_\parallel\mathbf{1}_\parallel + \tfrac{1}{2}(\mathbf{1}_\perp{}^\diamond\mathbf{1}_\perp - \mathbf{1}_\wedge{}^\diamond\mathbf{1}_\wedge), & \mathbf{W}_1 &= \mathbf{1}_\parallel\mathbf{1}_\parallel - \tfrac{1}{2}\mathbf{1}_\perp\mathbf{1}_\parallel, \\ \mathbf{W}_2 &= \tfrac{1}{2}(\mathbf{1}_\parallel{}^\diamond\mathbf{1}_\parallel + \mathbf{1}_\wedge{}^\diamond\mathbf{1}_\wedge), & \mathbf{W}_3 &= \mathbf{1}_\parallel{}^\diamond\mathbf{1}_\perp + \mathbf{1}_\perp{}^\diamond\mathbf{1}_\parallel, \\ \mathbf{W}_4 &= \tfrac{1}{2}(\mathbf{1}_\wedge{}^\diamond\mathbf{1}_\perp - \mathbf{1}_\perp{}^\diamond\mathbf{1}_\wedge), & \mathbf{W}_5 &= \mathbf{1}_\wedge{}^\diamond\mathbf{1}_\parallel - \mathbf{1}_\parallel{}^\diamond\mathbf{1}_\wedge, \end{aligned}\right\} \quad (10.87)$$

where the second form for $\mathbf{W}_1$ holds only for its contraction with a deviator, it being readily deduced that

$$\mathbf{1}_\perp \cdot \overset{\circ}{\mathbf{e}} \cdot \mathbf{1}_\perp - \mathbf{1}_\wedge \cdot \overset{\circ}{\mathbf{e}} \cdot \mathbf{1}_\wedge = \mathbf{1}_\perp\mathbf{1}_\perp : \overset{\circ}{\mathbf{e}} = -\mathbf{1}_\perp\mathbf{1}_\parallel : \overset{\circ}{\mathbf{e}}. \quad (10.88)$$

10.8 First-order transport in magnetoplasmas

10.8.1 *The distribution function*

From (5.46) and $\boldsymbol{\nu} = \mathbf{c}/\mathcal{C}$,

$$\mathbb{D}\ln f_0 = (\nu^2 - \tfrac{5}{2})\mathbf{c}\cdot\nabla\ln T + 2\mathcal{C}^{-2}\,\mathbf{cc} : \overset{\circ}{\mathbf{e}}, \quad (10.89)$$

whence by (10.78)

$$f = f_0\Big\{1 - \tau(\nu^2 - \tfrac{5}{2})\overline{\mathbf{c}}\cdot\nabla\ln T - 2\tau\mathcal{C}^{-2}\overline{\mathbf{cc}} : \overset{\circ}{\mathbf{e}}\Big\}. \quad (10.90)$$

With the averages given in (10.81) and (10.84) this reads

$$f = f_0\Big\{1 - \tau_2(\nu^2 - \tfrac{5}{2})\mathbf{c}\cdot\mathbf{lk}\cdot\nabla\ln T - 2\tau_1\mathcal{C}^{-2}\mathbf{cc} : \mathbf{W} : \overset{\circ}{\mathbf{e}}\Big\}, \quad (10.91)$$

where we have also introduced the two collision times discussed in §5.5.2.

10.8.2 *The heat flux vector and the pressure tensor*

We can either follow the method of §5.6 and integrate (10.91) directly to determine $\mathbf{q}$ and $\mathbf{p}$, or we can transfer the magnetic field modification from the peculiar velocity to the macroscopic gradients. Comparing (5.47) with (10.91), we conclude that the presence of the magnetic introduces changes to the gradients represented symbolically by

$$\nabla T \to \mathbf{lk}\cdot\nabla T, \qquad \overset{\circ}{\mathbf{e}} \to \mathbf{W} : \overset{\circ}{\mathbf{e}}. \quad (10.92)$$

It therefore follows from (5.61) that the heat flux vector in a magnetoplasma is

$$\mathbf{q} = -\,\mathbf{K}\cdot\nabla T, \qquad (\,\mathbf{K} = \kappa\mathbf{lk},\ \kappa = \tfrac{5}{2}Rp\tau_2), \quad (10.93)$$

where $\mathbf{K}$ is the thermal conductivity tensor. This is the expression previously obtained in §9.8.2 by elementary kinetic theory.

Similarly from (5.59) and (10.92) we deduce that the pressure tensor in a magnetoplasma is

$$\mathbf{p} = p\mathbf{1} - 2\mu\mathbf{W} : \overset{\circ}{\nabla}\mathbf{v} \qquad (\mu = p\tau_1). \tag{10.94}$$

The formulae for $\mathbf{p}$ is more intelligible in component form. Defining five coefficients of viscosity, μ_i, $i = 1, 2, \ldots 5$, by

$$\left.\begin{aligned} \mu_1 = \mu, \qquad \mu_3 &= \frac{\mu}{1+\varpi^2}, \qquad \mu_5 = \frac{\varpi\mu}{1+\varpi^2}, \\ \mu_2 &= \frac{\mu}{1+4\varpi^2}, \qquad \mu_4 = \frac{2\varpi\mu}{1+4\varpi^2}, \end{aligned}\right\}, \tag{10.95}$$

taking $\mathbf{b}$ along OZ and introducing the tensor $\mathbf{S} \equiv \overset{\circ}{\nabla}\mathbf{v}$, we find that

$$\left.\begin{aligned} p_{xx} &= p - \{2\mu_2 S_{xx} + (\mu - \mu_2)(S_{xx} + S_{yy}) + \mu_4 S_{xy}\}, \\ p_{xx} &= p - \{2\mu_2 S_{xx} + (\mu - \mu_2)(S_{xx} + S_{yy}) - \mu_4 S_{xy}\}, \\ p_{zz} &= p - 2\mu S_{zz}, \\ p_{xy} &= p_{yx} = -2\mu_2 S_{xy} - \mu_4(S_{yy} - S_{xx}), \\ p_{yz} &= p_{zy} = -2\mu_3 S_{yz} + 2\mu_5 S_{zx}, \\ p_{zx} &= p_{xz} = -2\mu_3 S_{zx} - 2\mu_5 S_{yz}. \end{aligned}\right\} \tag{10.96}$$

Notice that as $B \to 0$, $\varpi \to 0$, and $\mu_2, \mu_3 \to \mu$, $\mu_4, \mu_5 \to 0$, which reduces (10.96) to the isotropic pressure tensor.

10.9 Equivalence of BGK and gyro-averaged equations

It remains to show that equations (10.77) and (10.78) describe the same physical phenomenon. Introducing $\varphi \equiv (f - f_0)/f_0$, and noting that as $\partial f_0/\partial\mathbf{c}$ is parallel to $\mathbf{c}$, $\delta f_0 = 0$, we can write the first of these equations as

$$\varphi - \varpi\mathbf{b}\times\mathbf{c}\cdot\frac{\partial\varphi}{\partial\mathbf{c}} = -\tau\mathbb{D}\ln f_0 + \mathrm{O}(\epsilon^2). \tag{10.97}$$

Thus by (10.89), and two collision times, to first order in the Knudsen number,

$$\varphi_1 \quad \varpi\mathbf{b}\times\mathbf{c}\cdot\frac{\partial\varphi_1}{\partial\mathbf{c}} - -\tau_2(\nu^2 - \tfrac{5}{2})\mathbf{c}\cdot\nabla\ln T - 2\tau_1\boldsymbol{\nu}\boldsymbol{\nu} : \overset{\circ}{\mathbf{e}}\,. \tag{10.98}$$

10.9.1 *Solving the BGK equation*

To solve this equation, we assume that φ_1 can be expressed as

$$\varphi_1 = \boldsymbol{\nu}\cdot\mathbf{C} + \boldsymbol{\nu}\cdot\mathbf{A}\cdot\boldsymbol{\nu}, \tag{10.99}$$

where the vector $\mathbf{C}$ and tensor $\mathbf{A}$ depend on $\mathbf{r}$, t, ν^2, $\mathbf{b}\cdot\boldsymbol{\nu}$, and $\mathbf{B}$, but not on $\boldsymbol{\nu}$. Then, substituting

$$\mathbf{b}\times\mathbf{c}\cdot\frac{\partial\varphi_1}{\partial\mathbf{c}} = \mathbf{b}\times\boldsymbol{\nu}\cdot\frac{\partial\varphi_1}{\partial\boldsymbol{\nu}} = \mathbf{b}\times\boldsymbol{\nu}\cdot\mathbf{C} + \mathbf{b}\times\boldsymbol{\nu}\cdot(\mathbf{A}\cdot\boldsymbol{\nu} - \boldsymbol{\nu}\cdot\mathbf{A})$$

into (10.98), we find

$$\boldsymbol{\nu}\cdot\Big\{\mathbf{C}+\varpi\mathbf{b}\times\mathbf{C}+\tau_2(\nu^2-\tfrac{5}{2})\mathcal{C}\nabla\ln T\Big\} \\ +\boldsymbol{\nu}\cdot\Big\{\mathsf{A}-\varpi\mathbf{b}\times\mathsf{A}+\varpi\mathsf{A}\times\mathbf{b}+2\tau_1\,\overset{\circ}{\mathsf{e}}\Big\}\cdot\boldsymbol{\nu}=0. \quad (10.100)$$

To be consistent with the convention introduced in §10.7.2, we have written

$$\boldsymbol{\nu}\cdot(\mathbf{b}\times\mathsf{A})\cdot\boldsymbol{\nu}=\boldsymbol{\nu}\cdot(\mathbf{b}\times\mathbf{1}\cdot\mathsf{A})\cdot\boldsymbol{\nu}=(\mathbf{b}\times\mathbf{1}\cdot\boldsymbol{\nu})\cdot\mathsf{A}\cdot\boldsymbol{\nu}=\mathbf{b}\times\boldsymbol{\nu}\cdot\mathsf{A}\cdot\boldsymbol{\nu}.$$

As the expressions in curly braces in (10.100) are independent of $\boldsymbol{\nu}$, it follows that $\mathbf{C}$ and A are the solutions of

$$\mathbf{C}+\varpi\mathbf{b}\times\mathbf{C}+\tau_2(\nu^2-\tfrac{5}{2})\mathcal{C}\nabla\ln T=0, \quad (10.101)$$

and

$$\mathsf{A}-\varpi\mathbf{b}\times\mathsf{A}+\varpi\mathsf{A}\times\mathbf{b}+2\tau_1\,\overset{\circ}{\mathsf{e}}=0. \quad (10.102)$$

Mathematical note 10.9 *Solution of two vector equations*

Let vectors $\mathbf{G}$ and $\mathbf{H}$ satisfy

$$\left[\alpha_{\parallel}\mathbf{bb}+\alpha_{\wedge}\mathbf{b}\times\mathbf{1}+\alpha_{\perp}(\mathbf{1}-\mathbf{bb})\right]\cdot\mathbf{G}=\mathbf{H}, \quad (10.103)$$

where $\mathbf{b}$ is unit vector and $\alpha_{\parallel}$, $\alpha_{\wedge}$, $\alpha_{\perp}$ are scalar constants, then the solution of this equation is

$$\mathbf{G}=\left[\beta_{\parallel}\mathbf{bb}+\beta_{\wedge}\mathbf{b}\times\mathbf{1}+\beta_{\perp}(\mathbf{1}-\mathbf{bb})\right]\cdot\mathbf{H}, \quad (10.104)$$

where

$$\beta_{\parallel}=\frac{1}{\alpha_{\parallel}}, \qquad \beta_{\wedge}=-\frac{\alpha_{\wedge}}{\alpha_{\perp}^2+\alpha_{\wedge}^2}, \qquad \beta_{\perp}=\frac{\alpha_{\perp}}{\alpha_{\perp}^2+\alpha_{\wedge}^2}.$$

This is readily verified by substitution.

Let tensors A and D satisfy

$$\mathsf{A}-\varpi\mathbf{b}\times\mathsf{A}+\varpi\mathsf{A}\times\mathbf{b}=\mathsf{D}, \quad (10.105)$$

where D is given. Then

$$\mathsf{A}=\mathsf{W}:\mathsf{D}, \quad (10.106)$$

where W is the fourth-order tensor defined in (10.85) and (10.87). This solution can be derived directly from (10.105) (Woods 1987); we shall be content to confirm it by substitution.

From

$$\mathbf{1}_{\wedge}\cdot\mathbf{1}_{\parallel}=0, \quad \mathbf{1}_{\wedge}\cdot\mathbf{1}_{\perp}=\mathbf{1}_{\wedge}, \quad \mathbf{1}_{\wedge}\cdot\mathbf{1}_{\wedge}=-\mathbf{1}_{\perp},$$

which are readily verified, and (10.86) and (10.87), we can show that

$$\mathbf{b}\times\mathsf{W}_1-\mathsf{W}_1\times\mathbf{b}=\mathbf{1}_{\wedge}\cdot\mathsf{W}_1-\mathsf{W}_1\cdot\mathbf{1}_{\wedge}=0,$$
$$\mathbf{b}\times\mathsf{W}_2-\mathsf{W}_2\times\mathbf{b}=2\mathsf{W}_4, \qquad \mathbf{b}\times\mathsf{W}_3-\mathsf{W}_3\times\mathbf{b}=\mathsf{W}_5,$$
$$\mathbf{b}\times\mathsf{W}_4-\mathsf{W}_4\times\mathbf{b}=-2\mathsf{W}_2, \quad \mathbf{b}\times\mathsf{W}_5-\mathsf{W}_5\times\mathbf{b}=-\mathsf{W}_3.$$

Hence by (10.85) and (10.87),

$$\begin{aligned}\mathbf{W} - \varpi \mathrm{b}\times\mathbf{W} + \varpi\mathbf{W}\times\mathrm{b} &= \mathbf{W}_1 + \mathbf{W}_2 + \mathbf{W}_3 \\ &= \mathbf{1}_{\parallel}{}^{\diamond}\mathbf{1}_{\parallel} + \mathbf{1}_{\perp}{}^{\diamond}\mathbf{1}_{\perp} + \mathbf{1}_{\parallel}{}^{\diamond}\mathbf{1}_{\perp} + \mathbf{1}_{\perp}{}^{\diamond}\mathbf{1}_{\parallel} = \mathbf{1}^{\diamond}\mathbf{1},\end{aligned}$$

and as this is the unit fourth-order tensor (see (5.51)), it follows that when (10.106) is substituted in (10.105), we get

$$\mathbf{W}:\mathbf{D} - \varpi\mathrm{b}\times\mathbf{W}:\mathbf{D} + \varpi\mathbf{W}\times\mathrm{b}:\mathbf{D} = \mathbf{1}^{\diamond}\mathbf{1}:\mathbf{D} = \mathbf{D},$$

and the solution in (10.106) is verified.

Returning to (10.101), we have the case $\alpha_{\parallel} = \alpha_{\perp} = 1$, $\alpha_{\wedge} = \varpi$ of (10.103). Hence by (10.104) and (10.82) its solution is

$$\mathbf{C} = -\tau_2(\nu^2 - \tfrac{5}{2})\mathcal{C}\,\mathbf{k}\cdot\nabla\ln T.$$

Similarly the solution of (10.102) is $\mathbf{A} = -2\tau_1\mathbf{W}:\overset{\circ}{\mathbf{e}}$, and (10.99) becomes

$$\varphi_1 = -\tau_2(\nu^2 - \tfrac{5}{2})\mathbf{c}\cdot\mathbf{k}\cdot\nabla\ln T - \tau_1\mathcal{C}^{-2}\mathbf{cc}:\mathbf{W}:\overset{\circ}{\mathbf{e}},$$

in agreement with (10.91).

10.10 List of first-order transport formulae

10.10.1 *The two-fluid transport equations*

As remarked at the end of §9.3.2, to complete the set of two-fluid equations in a magnetoplasma, constitutive equations are required for $\mathbf{q}_e$, $\mathbf{q}_i$, $\boldsymbol{\pi}_e$, $\boldsymbol{\pi}_i$, $\mathbf{R}_e$, and φ_e. We have explained the physical principles determining these transport properties and obtained approximate expressions for most of the constitutive coefficients. It is not our purpose to pursue further details here, but the reader may find a complete list of equations helpful.

To complete the set (9.35)–(9.44) we required the following:

$$\mathbf{R}_e = en_e\{\boldsymbol{\eta}\cdot\mathbf{j} - \boldsymbol{\delta}\cdot\nabla T_e\}, \tag{10.107}$$

$$\mathbf{q}_e = -\mathbf{K}^e\cdot\nabla T_e - T_e\boldsymbol{\delta}\cdot\mathbf{j}, \quad \mathbf{q}_i = -\mathbf{K}^i\cdot\nabla T_i, \tag{10.108}$$

$$\boldsymbol{\pi}_e = -2\sum_{r=1}^{5}\mu_r^e\mathbf{W}_r:\overset{\circ}{\nabla\mathbf{v}_e}, \quad \boldsymbol{\pi}_i = -2\sum_{r=1}^{5}\mu_r^i\mathbf{W}_r:\overset{\circ}{\nabla\mathbf{v}_i}, \tag{10.109}$$

$$\varphi_e = -\varphi_i = 3\frac{m_e}{m_i}\frac{kn_e}{\tau_e}(T_i - T_e), \tag{10.110}$$

$$\xi_e = \mathbf{R}_e\cdot\mathbf{j}/en_e, \qquad \xi_i = 0, \tag{10.111}$$

where each second-order tensor coefficient is of the type

$$\boldsymbol{\delta} = \delta_{\parallel}\mathbf{bb} + \delta_{\wedge}\mathbf{1}\times\mathbf{b} + \delta_{\perp}(\mathbf{1} - \mathbf{bb}), \tag{10.112}$$

and the fourth-order tensors $\mathbf{W}_r$, $r = 1, 2, \ldots 5$, are defined in (10.87).

Three points about the thermoelectric tensor $\boldsymbol{\delta}$ should be noted. First its appearance in both $\mathbf{R}_e$ and $\mathbf{q}_e$ is a more general manifestation of the reciprocal theorem mentioned in §4.6.2. Secondly, the absence of a corresponding term from $\mathbf{q}_i$ is a good approximation, since thermal diffusion is almost entirely due to the electrons (see exercise 4.8). Thirdly, $\boldsymbol{\delta}$ satisfies $\mathbf{b} \times \boldsymbol{\delta} = \boldsymbol{\delta} \times \mathbf{b}$, this being the condition for lateral isotropy about $\mathbf{b}$ and (10.112) is the most general form of $\boldsymbol{\delta}$ possessing this property. Lateral isotropy requires all the second-order tensors in the constitutive equations to have this form, and it likewise constrains the form of the fourth-order tensor $\mathbf{W}$ (Woods 1986).

There are 23 scalar coefficients in these constitutive equations; values for 19 of them can be found in (10.93), (10.95) (written separately for the ion and electron fluids), (9.48), and (10.68). Some of the values are accurate, while others are satisfactory approximations. One source of error is the use of constant collision intervals in the evaluation of (10.81) and (10.84). According to (10.33), an expression for τ proportional to the cube of the particle speed would have given more accurate coefficients. A detailed account of the effect of this and other collision laws can be found in Chapter 8 of the work by Shkarofsky et al. (1966).

10.10.2 *Braginskii's transport coefficients*

Braginskii's (1965) frequently used formulae for the coefficients are reproduced below. They are based on approximate solutions of the two-component Fokker–Planck equations. Braginskii states that the largest errors in his coefficients are between 10 and 20 per cent and occur in the intermediate range $\varpi \sim 1$, but larger errors have been found in this range (Epperlein 1984). In very strong magnetic fields the errors are small, but if electric currents are passing through the magnetoplasma, transport is dominated by the second-order effects, described for heat flux in §9.7.3 and more generally in the next chapter.

The collision times used by Braginskii are τ_e for the electrons and $\tau_i = (2m_i/m_e)^{\frac{1}{2}}(T_i/T_e)^{\frac{3}{2}}Z^{-2}\tau_e$, (cf. (10.61)) for the ions. Thus from (9.57),

$$\tau_e = \frac{3.44 \times 10^5}{\ln \Lambda}\frac{\tilde{T}_e^{3/2}}{Z\tilde{n}_e}, \qquad \tau_i = \frac{2.09 \times 10^7}{\ln \Lambda}\frac{A^{1/2}\tilde{T}_i^{3/2}}{Z^3\tilde{n}_e}, \tag{10.113}$$

where $\tilde{T}_e$, $\tilde{T}_i$ are in electron volts, $\tilde{n}_e$ is in particles per cubic centimetre, $\ln \Lambda$ is defined in §9.4.3, and $A = m_i/m_p$ is the particle mass number.

The transport coefficients are specified below.

Resistivity

$$\left.\begin{aligned} \eta_{\parallel} &= \eta_0\alpha_0, & \eta_0 &\equiv \frac{m_e}{e^2 n_e \tau_e}, \\ \eta_{\wedge} &= -\eta_0\frac{\varpi_e(\alpha_1''\varpi_e^2+\alpha_0'')}{\Delta_e}, & \eta_{\perp} &= \eta_0\left\{1-\frac{\alpha_1'\varpi_e^2+\alpha_0'}{\Delta_e}\right\}, \end{aligned}\right\} \quad (10.114)$$

Thermoelectricity

$$\left.\begin{aligned} \delta_{\parallel} &= \frac{k}{e}\beta_0, \\ \delta_{\wedge} &= \frac{k}{e}\frac{\varpi_e(\beta_1''\varpi_e^2+\beta_0'')}{\Delta_e}, & \delta_{\perp} &= \frac{k}{e}\frac{\beta_1'\varpi_e^2+\beta_0'}{\Delta_e}, \end{aligned}\right\} \quad (10.115)$$

Electron thermal conductivity

$$\left.\begin{aligned} \kappa_{\parallel}^e &= \kappa_0^e\gamma_0, & \kappa_0^e &\equiv \frac{k}{m_e}p_e\tau_e, \\ \kappa_{\wedge}^e &= \kappa_0^e\frac{\varpi_e(\gamma_1''\varpi_e^2+\gamma_0'')}{\Delta_e}, & \kappa_{\perp} &= \kappa_0^e\frac{\gamma_1'\varpi_e^2+\gamma_0'}{\Delta_e}, \end{aligned}\right\} \quad (10.116)$$

where $\qquad \varpi_e \equiv |\omega_{ce}\tau_e|, \qquad \Delta_e \equiv \varpi^4 + \delta_1\varpi^2 + \delta_0,$

and values of the coefficients α, β, γ, δ for various choices of Z are given in Table 10.2.

Ion thermal conductivity

$$\left.\begin{aligned} \kappa_{\parallel}^i &= 3.91\kappa_0^i, & \kappa_0^i &\equiv \frac{k}{m_i}p_i\tau_i, \\ \kappa_{\wedge}^i &= \kappa_0^i\frac{\varpi_i(2.5\varpi_i^2+4.65)}{\Delta_i}, & \kappa_{\perp} &= \kappa_0^i\frac{2\varpi_i^2+2.65}{\Delta_i}, \end{aligned}\right\} \quad (10.117)$$

where $\qquad \varpi_i \equiv \omega_{ci}\tau_i \quad \text{and} \quad \Delta_i \equiv \varpi_i^4 + 2.70\varpi_i^2 + 0.677.$

Electron viscosity $(Z = 1)$

$$\left.\begin{aligned} \mu_1^e &= 0.73p_e\tau_e, \\ \mu_3^e &= p_e\tau_e\frac{2.05\varpi_e^2+8.50}{\Delta}, & \mu_2^e &= \mu_3^e(2\varpi_e), \\ \mu_5^e &= -p_e\tau_e\frac{\varpi_e(\varpi_e^2+7.91)}{\Delta}, & \mu_4^e &= \mu_5^e(2\varpi_e), \end{aligned}\right\} \quad (10.118)$$

where $\qquad \Delta = \varpi_e^4 + 13.8\varpi_e^2 + 11.6.$

Ion viscosity

$$\left.\begin{aligned}
\mu_1^i &= 0.96 p_i \tau_i, \\
\mu_3^i &= p_i \tau_i \frac{1.2\varpi_i^2 + 2.23}{\Delta}, & \mu_2^i &= \mu_3^i(2\varpi_i), \\
\mu_5^i &= -p_i \tau_i \frac{\varpi_i(\varpi_i^2 + 2.38)}{\Delta}, & \mu_4^i &= \mu_5^i(2\varpi_i),
\end{aligned}\right\} \quad (10.119)$$

where $$\Delta = \varpi_i^4 + 4.03\varpi_i^2 + 2.33.$$

For a physical explanation of the doubling of the argument in ν_2^e, μ_4^e, μ_2^i and μ_4^i, see the derivation in §10.7.2.

Table 10.2 *Dependence of transport coefficients on Z*

Z	1	2	3	4	∞		1	2	3	4	∞
α_0	0.51	0.44	0.40	0.38	0.29	β_1'	5.10	4.45	4.23	4.12	3.80
β_0	0.71	0.91	1.02	1.09	1.52	β_0'	2.68	0.95	0.59	0.45	0.15
γ_0	3.16	4.89	6.06	6.92	12.5	β_1''	1.5	1.5	1.5	1.5	1.5
δ_0	3.77	1.05	0.58	0.41	0.10	β_0''	3.05	1.78	1.44	1.29	0.88
δ_1	14.8	10.8	9.62	9.06	7.48	γ_1'	4.66	3.96	3.72	3.60	3.25
α_1'	6.42	5.52	5.23	5.08	4.63	γ_0'	11.9	5.12	3.53	2.84	1.20
α_0'	1.84	0.60	0.35	0.26	0.07	γ_1''	2.5	2.5	2.5	2.5	2.5
α_1''	1.70	1.70	1.70	1.70	1.70	γ_0''	21.7	15.4	13.5	12.7	10.2
α_0''	0.78	0.34	0.24	0.20	0.09						

10.10.3 *Strong magnetic fields*

It follows from the above expressions that when ϖ_e and ϖ_i are much larger than unity, good approximations for the case $Z = 1$ are:

$$\left.\begin{aligned}
\delta_\parallel &= 0.71\frac{k}{e}, \\
\eta_\parallel &= 0.51\eta_0, & \eta_\wedge &= 0, & \eta_\perp &= \frac{m_e}{e^2 n_e \tau_e}, \\
\kappa_\parallel^e &= 3.16\kappa_0^e, & \kappa_\wedge^i &= -\frac{5kp_e}{2eB}, & \kappa_\perp^e &= 4.66\frac{\kappa_0^e}{(\omega_{ce}\tau_e)^2}, \\
\kappa_\parallel^i &= 3.9\kappa_0^i, & \kappa_\wedge^i &= \frac{5kp_i}{2eB}, & \kappa_\perp^i &= 2\frac{\kappa_0^i}{(\omega_{ci}\tau_i)^2}.
\end{aligned}\right\} \quad (10.120)$$

Exercises 10

10.1 Prove that for heavy Maxwellian scatterers,

$$2\mathbf{A} = \frac{\partial}{\partial \mathbf{w}} \cdot \mathsf{B} - \frac{2}{\mathcal{C}^2} \mathsf{B} \cdot \mathbf{w},$$

where $\mathbf{A}$ and B are defined in (10.3).

10.2 Use (10.57) to obtain the estimate $\sigma = e^2 n_e \tau_e / m_e$ for the conductivity.

10.3 A beam of fast ions of mass m_b and speed v_b is being slowed down by collisions with thermal ions of mass m_i. Show that this takes a time

$$\tau_{si} = \frac{m_b}{m_i + m_b} \frac{4\pi\epsilon_0^2 m_b m_i v_b^3}{n_i e^4 \ln \Lambda}.$$

10.4 Show that plasma flowing parallel to a magnetic field $B\hat{\mathbf{z}}$ and sheared in the OX-direction, is subject to a retarding force of magnitude $-2\mu v_z''/(1+\varpi^2)$. For what physical reason does this force vanish in the limit $\varpi \to \infty$?

10.5 A velocity field $v(x)\hat{\mathbf{x}}$ changes only in the direction of flow; the magnetic field lies along OZ. In the notation of §10.8.2, show that

$$\mathsf{S} = \tfrac{2}{3}v'(\hat{\mathbf{x}}\hat{\mathbf{x}} - \tfrac{1}{2}\hat{\mathbf{y}}\hat{\mathbf{y}} - \tfrac{1}{2}\hat{\mathbf{z}}\hat{\mathbf{z}}), \quad \mathsf{W}_1 = \tfrac{1}{6}v'(\hat{\mathbf{x}}\hat{\mathbf{x}} + \hat{\mathbf{y}}\hat{\mathbf{y}} - 2\hat{\mathbf{z}}\hat{\mathbf{z}}),$$
$$\mathsf{W}_2 = \tfrac{1}{2}(\hat{\mathbf{x}}\hat{\mathbf{x}} - \hat{\mathbf{y}}\hat{\mathbf{y}}), \quad \mathsf{W}_3 = \mathsf{W}_5 = 0, \quad \mathsf{W}_4 = -\tfrac{1}{2}v'(\hat{\mathbf{x}}\hat{\mathbf{y}} + \hat{\mathbf{y}}\hat{\mathbf{x}}),$$

and hence

$$\boldsymbol{\pi} = -p\tau_1 v' \left\{ \tfrac{1}{3}(\hat{\mathbf{x}}\hat{\mathbf{x}} + \hat{\mathbf{y}}\hat{\mathbf{y}} - 2\hat{\mathbf{z}}\hat{\mathbf{z}}) + \frac{1}{1+4\varpi^2}(\hat{\mathbf{x}}\hat{\mathbf{x}} - \hat{\mathbf{y}}\hat{\mathbf{y}}) - \frac{2\varpi}{1+4\varpi^2}(\hat{\mathbf{x}}\hat{\mathbf{y}} + \hat{\mathbf{y}}\hat{\mathbf{x}}) \right\}.$$

10.6 Expressing $\mathbf{c}$ in cylindrical coordinates $(c_\perp, \vartheta, c_\parallel)$, show that (10.76) can be written

$$\mathbb{D}f - \omega_c \frac{\partial f}{\partial \vartheta} = \frac{1}{\tau}(f_0 - f) + \mathrm{O}(\epsilon^2),$$

where $$\mathbb{D} = \mathrm{D}^* + \mathbf{c}_\perp \cdot \nabla + \mathbf{c}_\parallel \cdot \nabla + \hat{\mathbf{c}} \cdot \mathbf{H} \frac{\partial}{\partial c_\perp} + c_\parallel^{-1} \hat{\boldsymbol{\theta}} \cdot \mathbf{H} \frac{\partial}{\partial \vartheta} + \mathbf{b} \cdot \mathbf{H} \frac{\partial}{\partial c_\parallel},$$

with $$\mathbf{H} \equiv \mathbf{F} - \mathrm{D}\mathbf{v} - \mathbf{c} \cdot \mathsf{e}.$$

By ignoring the effect of collisions, show that the average over ϑ of the streaming derivative alone is

$$\overline{\mathbb{D}} = \mathrm{D}^* + \mathbf{c}_\parallel \cdot \nabla - \mathbf{b} \cdot (\mathbf{P} + \mathbf{c} \cdot \mathsf{e}) \frac{\partial}{\partial c_\parallel},$$

and that when referred to the laboratory reference frame (cf. (7.34) and (9.73)), $\overline{\mathbb{D}}$ becomes

$$\overline{\mathbb{D}}_L = \frac{\partial}{\partial t} + \mathbf{u} \cdot \nabla + \frac{Q}{m} E_\parallel \frac{\partial}{\partial c_\parallel}.$$

10.7 With the definition of $\overline{\mathbb{D}}_L$ just given and $\overline{f}$ denoting the gyro-averaged distribution function, an equation of the type $\overline{\mathbb{D}}_L \overline{f} = \mathbb{C}$ is known as a 'drift' kinetic equation. Show that this method of averaging cannot correctly represent diffusion.

10.8 Referring to the result contained in exercise 7.7, show that by making use of the first relation in the exercise, Q_{12} can be expressed in the form

$$Q_{12} = f_1 \langle \Delta \mathbf{c}_1 \rangle - \tfrac{1}{2} \frac{\partial}{\partial \mathbf{c}} \cdot (f_1 \langle \Delta \mathbf{c}_1 \Delta \mathbf{c}_1 \rangle),$$

where the notation is that of (10.5). Deduce the Fokker–Planck equation.

11

TRANSPORT ACROSS STRONG MAGNETIC FIELDS

11.1 Kinetic equation correct to second-order

The first requirement in calculating the heat and momentum transport across strong magnetic fields is a kinetic equation accurate to second-order in the Knudsen number ϵ. By a 'strong' magnetic field is meant one for which $\varpi \equiv \omega_c \tau \gg 1$. The product $\varpi\epsilon$ is an important parameter in the theory. We shall show that if $\varpi\epsilon \gg 1$, second order transport dominates the process. This case, commonly obtaining in laboratory plasmas, yields some surprising results, including the possibility of (transient) heat flux *up* temperature gradients.

11.1.1 *Recapitulation of first-order theory*

In §10.6.2 we wrote the BGK equation as

$$f - \varpi \mathbf{b}\times\mathbf{c}\cdot\frac{\partial f}{\partial \mathbf{c}} = f_0 - \tau \mathbb{D} f_0 + \mathrm{O}(\epsilon^2) \qquad (\varpi \equiv \omega_c \tau), \tag{11.1}$$

where $\mathbb{D}$ has the same form as in a neutral gas, i.e.

$$\mathbb{D} = \mathrm{D}^* + \mathbf{c}\cdot\nabla + \mathbf{a}\cdot\frac{\partial}{\partial \mathbf{c}} \qquad \big(\mathbf{a} \equiv -\mathbf{P} - \mathbf{c}\cdot\mathbf{e}\big). \tag{11.2}$$

We then removed the Lorentz term by averaging over the history of the particle velocity, to obtain

$$f = f_0 - \tau\overline{\mathbb{D} f_0} + \mathrm{O}(\epsilon^2), \tag{11.3}$$

which lead to the first-order solution

$$f = f_0\Big\{1 - \tau_2(\nu^2 - \tfrac{5}{2})\mathbf{c}\cdot\mathbf{k}\cdot\nabla\ln T - 2\tau_1 \mathcal{C}^{-2}\mathbf{cc}:\mathbf{W}:\mathring{\mathbf{e}}\Big\}, \tag{11.4}$$

given in (10.91). This solution was also derived directly from (11.1).

The tensors $\mathbf{k}$ and $\mathbf{W}$ are defined by

$$\mathbf{k} = \mathbf{bb} - \frac{\varpi}{1+\varpi^2}\mathbf{b}\times\mathbf{1} + \frac{1}{1+\varpi^2}(\mathbf{1} - \mathbf{bb}), \tag{11.5}$$

and

$$\mathbf{W} = \mathbf{W}_1 + \frac{1}{1+4\varpi^2}\mathbf{W}_2 + \frac{1}{1+\varpi^2}\mathbf{W}_3 + \frac{2\varpi}{1+4\varpi^2}\mathbf{W}_4 + \frac{\varpi}{1+\varpi^2}\mathbf{W}_5, \tag{11.6}$$

where

$$\left.\begin{aligned} \mathbf{W}_1 &= \mathbf{1}_\parallel \mathbf{1}_\parallel - \tfrac{1}{2}\mathbf{1}_\perp \mathbf{1}_\parallel, \\ \mathbf{W}_2 &= \tfrac{1}{2}(\mathbf{1}_\parallel {}^\diamond \mathbf{1}_\parallel + \mathbf{1}_\wedge {}^\diamond \mathbf{1}_\wedge), \qquad & \mathbf{W}_3 &= \mathbf{1}_\parallel {}^\diamond \mathbf{1}_\perp + \mathbf{1}_\perp {}^\diamond \mathbf{1}_\parallel, \\ \mathbf{W}_4 &= \tfrac{1}{2}(\mathbf{1}_\wedge {}^\diamond \mathbf{1}_\perp - \mathbf{1}_\perp {}^\diamond \mathbf{1}_\wedge), \qquad & \mathbf{W}_5 &= \mathbf{1}_\wedge {}^\diamond \mathbf{1}_\parallel - \mathbf{1}_\parallel {}^\diamond \mathbf{1}_\wedge. \end{aligned}\right\} \tag{11.7}$$

From (11.4) it was shown in §10.8.2 that the first-order heat flux and viscous stress tensor are given by

$$\mathbf{q}_1 = -\tfrac{5}{2} Rp\tau_2 \mathsf{k} \cdot \nabla T, \qquad \boldsymbol{\pi}_1 = -2p\tau_1 \mathbf{W} : \overset{\circ}{\mathbf{e}} \,. \tag{11.8}$$

11.1.2 *A physical model*

In §10.6.2 we interpreted the right-hand side of (11.1) as the Taylor expansion for f_0 taken along a trajectory $\mathcal{T}$ that is tangential to the particle orbits. The streaming derivative follows an element $d\boldsymbol{\nu}$ moving along $\mathcal{T}$ with velocity $\mathbf{c}$ and acceleration $\mathbf{a}$, where $\mathbf{c}$ is arbitrarily chosen at $(\mathbf{r}, t)$ and has almost the same value at all points on $\mathcal{T}$. In this model the Lorentz term, $\delta f = -\omega_c \mathbf{b}\times\mathbf{c} \cdot \partial f/\partial \mathbf{c}$, becomes the rate at which the particles are being removed from $d\boldsymbol{\nu}$ by the magnetic field.

To reduce the error term in (11.1), we need to find the appropriate $O(\epsilon^2)$ term to add to its right-hand side. Comparison of (11.1) with (8.5) shows that were the Lorentz term absent, the required term would be

$$\mathcal{L} \equiv -\tau \mathbb{D}(f - f_0) + \tau \mathbf{P} \cdot \frac{\partial}{\partial \mathbf{c}}(f_0 - f) + \tau \boldsymbol{\alpha} \cdot \frac{\partial f_0}{\partial \mathbf{c}}, \tag{11.9}$$

where $\mathbf{P}$ and $\boldsymbol{\alpha}$ are defined in (8.6).

The first right-hand term in $\mathcal{L}$ merely replaces $\mathbb{D}f_0$ in (11.1) by $\mathbb{D}f$, and the second arises via the pressure gradient force, the reasoning for which (see §5.8) is independent of whether or not a magnetic field is present. The only term requiring attention is that involving the acceleration $\boldsymbol{\alpha}$. As explained in §5.9, $\boldsymbol{\alpha}$ is the small relative acceleration of molecules and phase-space elements just prior to the inverse scattering collision and this mismatch of accelerations generates the streaming gain (or loss) represented by the term in question. To calculate $\boldsymbol{\alpha}$ we need to take the set of penultimate mean free paths into account (cf. Fig. 5.2). As the phase-space element $d\boldsymbol{\nu}$ adopted here is the same as in a neutral gas, arguments similar to those used in §5.9 to determine $\boldsymbol{\alpha}$ can be applied in the present case. The first step is to modify Fig. 5.2.

It is simpler to deal first with the case of a strong magnetic field, in which a typical penultimate mean free path will be a helix, lying on a Larmor cylinder, extending a distance $\tau c_\parallel^*$ along $\mathbf{B}$ from the penultimate collision at P_c^* to the inverse collision at P_c' (see Fig. 11.1). Just as in §5.9.2, it is necessary to average over all trajectories like $P_c^* P_c'$ in order to determine the net effect of earlier collisions on the final (inverse) collision. In Fig. 5.2 is shown the trajectory of a pseudo-molecule $\bar{\mathsf{m}}$, representing the average dynamic behaviour of all the molecules converging on P_c'. We shall adopt a similar model for the magnetoplasma.

We first need to determine the trajectory of a pseudo-particle $\bar{\mathsf{p}}$ that represents the average dynamic effect of the particles that are scattered at P_c' into $d\boldsymbol{\nu}$. Evidently $\bar{\mathsf{p}}$ must have nearly the same acceleration as $d\boldsymbol{\nu}$, since the acceleration difference corresponding to $\boldsymbol{\alpha}$, say $\bar{\boldsymbol{\alpha}}$, is an $O(\epsilon)$ term.

Thus, like $d\boldsymbol{\nu}$, $\bar{\mathsf{p}}$ does not gyrate about the field lines, and as it represents the behaviour of real particles, it must move parallel to $\mathbf{B}$. If the beginning of $\bar{\mathsf{p}}$'s path is at P_c'', on the assumption that scattering into $d\boldsymbol{\nu}$ is equally likely from helical orbits with $\mathbf{c}_{\|}^{*} > \mathbf{c}_{\|}'$ as those with $\mathbf{c}_{\|}^{*} < \mathbf{c}_{\|}'$, we find that P_c'' must lie a vector distance $\tau\mathbf{c}_{\|}'$ from P_c' (cf. the similar, but more obvious situation in Fig. 5.2). Thus $\bar{\mathsf{p}}$ has a velocity $\mathbf{c}_{\|}'$ and an acceleration $-\mathbf{c}\cdot\mathsf{e}$, i.e. it moves parallel to the guiding centres, but with the same acceleration as the phase-space elements (like $d\boldsymbol{\nu}$) through which it passes.

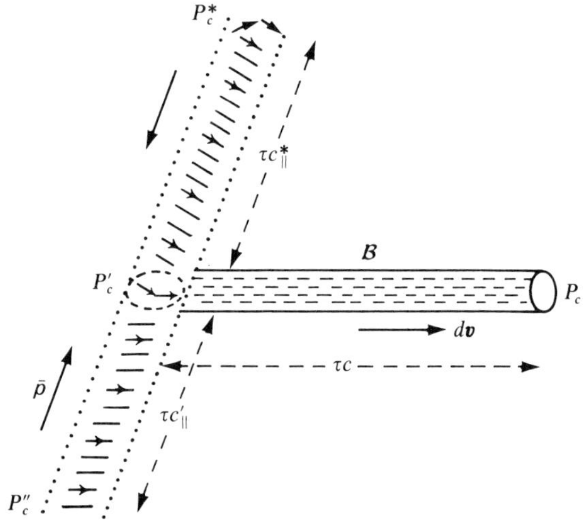

FIG. 11.1. Inverse scattering in a strong magnetic field

The acceleration difference $\bar{\boldsymbol{\alpha}}$ is responsible for streaming losses from $d\boldsymbol{\nu}$, in a manner similar to that described in §5.9.1 for a neutral gas. Then with $\boldsymbol{\alpha}$ replaced by $\bar{\boldsymbol{\alpha}}$, equations (11.1) and (11.9) yield the kinetic equation

$$f - \varpi\mathbf{b}\times\mathbf{c}\cdot\frac{\partial f}{\partial \mathbf{c}} = f_0 - \tau\mathbb{D}f + \tau\mathbf{P}\cdot\frac{\partial}{\partial \mathbf{c}}(f_0 - f) + \tau\bar{\boldsymbol{\alpha}}\cdot\frac{\partial f_0}{\partial \mathbf{c}}, \tag{11.10}$$

which is accurate to second-order in the Knudsen number, ϵ.

11.2 The acceleration term

11.2.1 *The pseudo-particle streaming derivative*

We follow the method of §5.9.3, adapting it to the trajectories shown in Fig. 11.1. To determine $\bar{\boldsymbol{\alpha}}$, we need to calculate the difference between the values of the acceleration $-\mathbf{c}\cdot\mathsf{e}$ at P_c'' and P_c'.

Let $\phi(\mathbf{r}, \mathbf{c}, t)$ denote a scalar property of $\bar{\mathsf{p}}$, whose change between P_c'' and P_c' we wish to determine. In the case of a strong field, the spatial change in ϕ is $\tau\mathbf{c}_{\|}\cdot\nabla\phi$. In the general case we adopt the gyro-averaging of §10.7.1 and replace $\mathbf{c}_{\|}$ by $\bar{\mathbf{c}} = \mathbf{c}\cdot\mathbb{k}$. This allows us to cover the complete range of magnetic field strength, which varies from the strong-field limit,

$\bar{\mathbf{c}} = \mathbf{c}_{\parallel}$, to the zero field limit, $\bar{\mathbf{c}} = \mathbf{c}$. Hence the total spatial change in ϕ is $\tau(\mathrm{D}^* + \mathbf{c}\cdot\mathbb{k}\cdot\nabla)\phi$, where we have added the convective change due to the (spinning) fluid motion (in the present application $\mathrm{D}^* = \mathrm{D}$).

Although $\bar{\mathsf{p}}$'s acceleration varies, the change over $P_c^* P_c'$ is small. Hence for the purpose of finding the velocity-space change in ϕ, say $\Delta\phi$, we may assume the acceleration to be constant and equal to $-\mathbf{c}\cdot\mathbf{e}$. Thus $\Delta\phi = -\tau\mathbf{c}\cdot\mathbf{e}\cdot\partial\phi/\partial\mathbf{c}$. Adding the spatial changes to $\Delta\phi$, we arrive at the total change in ϕ:

$$\tau\overline{\mathbb{D}}\phi = \tau\big(\mathrm{D}^*\phi + \mathbf{c}\cdot\mathbb{k}\cdot\nabla\phi - \mathbf{c}\cdot\mathbf{e}\cdot\frac{\partial\phi}{\partial\mathbf{c}}\big). \tag{11.11}$$

Evidently $\overline{\mathbb{D}}$ is not a streaming derivative following $d\boldsymbol{\nu}$, for this would have $\mathbf{c}$ in place of $\mathbf{c}\cdot\mathbb{k}$. It is a mixed 'streaming' derivative, obtained by tracking the pseudo-particles, which follow the guiding centres through physical space and the phase-space elements through velocity space.

In evaluating the spatial change of a vector of the type $\boldsymbol{\Phi} = c\cdot\mathbf{A}$, where the tensor $\mathbf{A}$ is independent of $\mathbf{c}$, since the gyro-average of $\mathbf{c}\cdot\nabla(\mathbf{c}\cdot\mathbf{A})$ is $\overline{\mathbf{cc}}:\nabla\mathbf{A}$, by (10.84) we obtain

$$\tau\overline{\mathbb{D}}\boldsymbol{\Phi} = \tau\big(\mathrm{D}^*\boldsymbol{\Phi} + \mathbf{c}\cdot\mathbf{W}:\nabla\boldsymbol{\Phi} - \mathbf{c}\cdot\mathbf{e}\cdot\frac{\partial\boldsymbol{\Phi}}{\partial\mathbf{c}}\big). \tag{11.12}$$

It should be remarked that the acceleration $-\mathbf{c}\cdot\mathbf{e}$ appearing in the last terms of (11.11) and (11.12) is *not* the acceleration that would be experienced by a real particle moving parallel to the magnetic field, since this would equal $-\mathbf{c}_{\parallel}\cdot\mathbf{e}$. For the purpose of calculating streaming losses from $d\boldsymbol{\nu}$, we remove the gyratory motion, i.e. freeze the peculiar velocity to the almost constant value it has along the trajectory $\mathcal{T}$ followed by $d\boldsymbol{\nu}$. And as $d\boldsymbol{\nu}$ has an acceleration $-\mathbf{c}\cdot\mathbf{e}$, the pseudo-particle $\bar{\mathsf{p}}$, which is scattered into $d\boldsymbol{\nu}$, must have almost the same acceleration.

11.2.2 *The change in acceleration*

In the particular case that $\boldsymbol{\Phi}$ in (11.12) equals the acceleration itself, care is required not to force the gyro-constraint on $\mathbf{c}\cdot\mathbf{e}$ in calculating its streaming derivative. To apply (11.11) and (11.12), which amounts to replacing $\mathbf{c}\cdot\nabla T$ and $\mathbf{cc}:\nabla\overset{\circ}{\mathbf{e}}$ in the neutral gas formula for $\boldsymbol{\alpha}$ by $\mathbf{c}\cdot\mathbb{k}\cdot\nabla T$ and $\mathbf{cc}:\mathbf{W}:\nabla\overset{\circ}{\mathbf{e}}$, it is necessary first to evaluate $\mathbb{D}_S(\mathbf{c}\cdot\mathbf{e})$ (see (5.82)) and *then* to make these replacements. Our starting point is therefore the formula for $\boldsymbol{\alpha}$ in the neutral gas. From (5.92) and (8.30) this is

$$\begin{aligned}\boldsymbol{\alpha} = \tau_1\Big\{&\mathbf{c}\cdot\mathrm{D}^*\overset{\circ}{\mathbf{e}} + \mathbf{cc}:\nabla\overset{\circ}{\mathbf{e}} - \mathcal{C}^2\nabla\cdot\overset{\circ}{\mathbf{e}} - \mathbf{c}\cdot\overset{\circ}{\overline{\overset{\circ}{\mathbf{e}}\cdot\overset{\circ}{\mathbf{e}}}} - \tfrac{2}{3}\nabla\cdot\mathbf{v}\,\mathbf{c}\cdot\overset{\circ}{\mathbf{e}}\Big\}\\ &-\tfrac{1}{2}\tau_2 T^{-1}(\mathbf{cc} - \tfrac{5}{2}\mathcal{C}^2\mathbf{1})\cdot\big\{\mathrm{D}^*\nabla T + \overset{\circ}{\mathbf{e}}\cdot\nabla T + \nabla\cdot\mathbf{v}\,\nabla T\big\}\\ &-\tau_1\mathbf{c}\cdot\overset{\star}{\nabla}\mathbf{F} + \mathrm{O}(\epsilon^2).\end{aligned}$$

In addition to the replacements just described, we need to ensure that the constraint on $\boldsymbol{\alpha}$ given in (5.83) is also satisfied by $\bar{\boldsymbol{\alpha}}$. For example, on changing the term $\tau_1 \mathbf{cc} : \nabla \overset{\circ}{\mathbf{e}}$ to $\tau_1 \mathbf{cc} : \mathbf{W} : \nabla \overset{\circ}{\mathbf{e}}$, we must also change $\tau_1 \mathcal{C}^2 \nabla \cdot \overset{\circ}{\mathbf{e}}$ to $\tau_1 \mathcal{C}^2 \nabla \cdot \mathbf{W} : \overset{\circ}{\mathbf{e}}$ (cf. (5.90)). These modifications lead to

$$\begin{aligned} \bar{\boldsymbol{\alpha}} = \tau_1 \Big\{ & \mathbf{c} \cdot \mathrm{D}^* \overset{\circ}{\mathbf{e}} + \mathbf{cc} : \mathbf{W} : \nabla \overset{\circ}{\mathbf{e}} - \mathcal{C}^2 \nabla \cdot \mathbf{W} : \overset{\circ}{\mathbf{e}} - \mathbf{c} \cdot \overline{\overset{\circ}{\mathbf{e}} \cdot \overset{\circ}{\mathbf{e}}} - \tfrac{2}{3} \nabla \cdot \mathbf{v}\, \mathbf{c} \cdot \overset{\circ}{\mathbf{e}} \Big\} \\ & - \tfrac{1}{2} \tau_2 T^{-1} (\mathbf{cc} - \tfrac{5}{2} \mathcal{C}^2 \mathbf{1}) \cdot \Big\{ \mathsf{lk} \cdot \mathrm{D}^* \nabla T + \overset{\circ}{\mathbf{e}} \cdot \nabla T + \nabla \cdot \mathbf{v}\, \mathsf{lk} \cdot \nabla T \Big\} \\ & - \tau_1 \mathbf{c} \cdot \mathsf{lk} \cdot \overset{\star}{\nabla} \mathbf{F} + \mathrm{O}(\epsilon^2), \end{aligned} \tag{11.13}$$

for a magnetoplasma.

Notice that the term containing $\overset{\circ}{\mathbf{e}} \cdot \nabla \ln T$ is not changed. This fact proves to be so important for the transport of energy in a magnetoplasma, that a physical supporting argument is desirable. A useful comparison can be made with (3.40) for the peculiar velocity in elementary kinetic theory. For small values of t/τ_1, this equation shows that the change in the peculiar velocity over a mean free path $\boldsymbol{\lambda} = \tau_2 \mathbf{c}$, due to a temperature gradient, is

$$\Delta \mathbf{c} = -\tfrac{1}{2}(\nu^2 - \tfrac{5}{2}) \mathbf{c} \boldsymbol{\lambda} \cdot \nabla \ln T.$$

At large values of τ_1, (3.28) shows that the acceleration is $\dot{\mathbf{c}} = -\mathbf{c} \cdot \overset{\circ}{\mathbf{e}}$, and therefore in a time interval τ_2, $\boldsymbol{\Lambda} = -\tau_2 \mathbf{c} \cdot \overset{\circ}{\mathbf{e}}$ is the displacement in velocity space. Hence, corresponding to $\Delta \mathbf{c}$, there will be an acceleration change $\Delta \dot{\mathbf{c}}$ proportional to $\frac{1}{2} \mathbf{c} \boldsymbol{\Lambda} \cdot \nabla \ln T$. The resulting mismatch in acceleration is $-\Delta \dot{\mathbf{c}}$ (cf. (5.82)), and so we expect a contribution to $\bar{\boldsymbol{\alpha}}$ proportional to $-\frac{1}{2} \tau_2 \mathbf{cc} \cdot \overset{\circ}{\mathbf{e}} \cdot \nabla \ln T$. Thus *acceleration* is the origin of the term in question, for which reason it is not suppressed by the magnetic field.

11.3 Cross-field transport

11.3.1 *The relative distribution function*

As in §8.1.2, we introduce the relative distribution function

$$\varphi \equiv (f - f_0)/f_0, \qquad \varphi = \varphi_1 + \varphi_2 + \cdots, \tag{11.14}$$

and rewrite (11.10) in the form (cf. (8.8))

$$\varphi - \varpi \mathbf{b} \times \mathbf{c} \cdot \frac{\partial \varphi}{\partial \mathbf{c}} = -\tau f_0^{-1} \mathbb{D}[f_0(1+\varphi)] + \varphi_2^{(\alpha)} - \tau f_0^{-1} \mathbf{P} \cdot \frac{\partial}{\partial \mathbf{c}}(f_0 \varphi), \tag{11.15}$$

where here

$$\varphi_2^{(\alpha)} = \tau \bar{\boldsymbol{\alpha}} \cdot \frac{\partial \ln f_0}{\partial \mathbf{c}} = -2\tau \mathcal{C}^{-2} \mathbf{c} \cdot \bar{\boldsymbol{\alpha}}, \tag{11.16}$$

and the error term is $\mathrm{O}(\epsilon^3)$. Proceeding as in §8.1.2, we obtain the first- and second-order kinetic equations:

$$\varphi_1 - \varpi \mathbf{b} \times \mathbf{c} \cdot \frac{\partial \varphi_1}{\partial \mathbf{c}} = -\tau \mathbb{D}_0 \ln f_0, \tag{11.17}$$

and

$$\varphi_2 - \varpi \mathbf{b} \times \mathbf{c} \cdot \frac{\partial \varphi_2}{\partial \mathbf{c}} = \Phi_2, \tag{11.18}$$

where

$$\Phi_2 = -\tau f_0^{-1}\mathbb{D}_0(f_0\varphi_1) - \tau\mathbb{D}_1 \ln f_0 + \varphi_2^{(\alpha)} - \tau f_0^{-1}\mathbf{P}\cdot\frac{\partial}{\partial\mathbf{c}}(f_0\varphi_1). \tag{11.19}$$

The solution of (11.17) is given in (11.4)

$$\varphi_1 = -\tau_2(\nu^2 - \tfrac{5}{2})\mathbf{c}\cdot\mathbf{k}\cdot\nabla\ln T - 2\tau_1 C^{-2}\mathbf{cc}:\mathbf{W}:\mathring{\mathbf{e}}; \tag{11.20}$$

the solution of (11.18) is found similarly, either by taking its gyro-average to remove the Lorentz term or directly (see §10.9).

By (11.13) and (11.14) we find

$$\begin{aligned}\varphi_2^{(\alpha)} &= \tau_1\tau_2\Big\{-2\mathbf{c}\boldsymbol{\nu\nu}\,\vdots\,\mathbf{k}\cdot\nabla\mathring{\mathbf{e}} + 2\mathbf{k}\cdot\nabla\cdot\mathring{\mathbf{e}}\cdot\mathbf{c}\Big\}\\ &\quad+\tau_2^2T^{-1}(\nu^2-\tfrac{5}{2})\mathbf{c}\cdot\Big(\mathbf{k}\cdot\mathrm{D}^*\nabla T + \mathring{\mathbf{e}}\cdot\nabla T + \nabla\cdot\mathbf{v}\,\mathbf{k}\cdot\nabla T\Big)\\ &\quad+2\tau_1^2\Big\{-\boldsymbol{\nu}\cdot\mathrm{D}^*\mathring{\mathbf{e}}\cdot\boldsymbol{\nu} + \boldsymbol{\nu}\cdot\overset{\circ}{\overline{\mathring{\mathbf{e}}\cdot\mathring{\mathbf{e}}}}\cdot\boldsymbol{\nu} + \tfrac{2}{3}\nabla\cdot\mathbf{v}\,\boldsymbol{\nu\nu}:\mathring{\mathbf{e}}\Big\},\end{aligned} \tag{11.21}$$

where we have written $\tau = \tau_2$ for energy transport and $\tau = \tau_1 = \frac{2}{3}\tau_2$ for momentum transport.

11.3.2 *Strong magnetic fields*

The function Φ_2 defined in (11.18) is algebraically complicated, containing a score of terms; and for a theory covering the complete range of ϖ it cannot be simplified. We shall deal with the important special case

$$\varpi \gg 1, \qquad \mathbf{b}\cdot\nabla = 0, \tag{11.22}$$

i.e. the field is strong and parallel gradients can be ignored. As *parallel* transport is always dominated by the first-order terms, there is little point in calculating the second-order terms for this direction.

We shall retain only the dominant term in Φ_2. From (11.6), (11.7), (11.8), (11.22), and (8.22) we find that $\mathbf{k}\cdot\nabla T$, $\mathbf{W}:\mathring{\mathbf{e}}$, $\mathbf{q}_1$, $\boldsymbol{\pi}_1$ and $\mathbb{D}_1\ln f_0$ are all $\mathrm{O}(\varpi^{-1})$ quantities. It follows that the dominant terms in Φ_2 are those in $\varphi_2^{(\alpha)}$ that do not contain $\mathbf{k}$ or $\mathbf{W}$. Thus from (11.18) and (11.21),

$$\varphi_2 - \varpi\mathbf{b}\times\mathbf{c}\cdot\frac{\partial\varphi_2}{\partial\mathbf{c}} = \Phi_2^* + \mathrm{O}(\varpi^{-1}), \tag{11.23}$$

where

$$\begin{aligned}\Phi_2^* &= \tau_2{}^2T^{-1}(\nu^2-\tfrac{5}{2})\mathbf{c}\cdot\mathring{\mathbf{e}}\cdot\nabla T\\ &\quad+2\tau_1{}^2\Big\{-\boldsymbol{\nu}\cdot\mathrm{D}^*\mathring{\mathbf{e}}\cdot\boldsymbol{\nu} + \boldsymbol{\nu}\cdot\overset{\circ}{\overline{\mathring{\mathbf{e}}\cdot\mathring{\mathbf{e}}}}\cdot\boldsymbol{\nu} + \tfrac{2}{3}\nabla\cdot\mathbf{v}\,\boldsymbol{\nu\nu}:\mathring{\mathbf{e}}\Big\}.\end{aligned} \tag{11.24}$$

11.3.3 *Second-order transport*

The first-order heat flux is given in (11.8). It can be deduced from equation (10.98) for φ_1. A comparison of this equation with (11.23) for φ_2 shows that the replacement

$$\nabla T \longrightarrow -\tau_2 \overset{\circ}{\mathbf{e}} \cdot \nabla T$$

in (11.8) is all that is necessary to change $\mathbf{q}_1$ into $\mathbf{q}_2$. Thus

$$\mathbf{q}_2 = \tfrac{5}{2}\tau_2^2 Rp\,\mathbf{lk} \cdot \overset{\circ}{\mathbf{e}} \cdot \nabla T. \tag{11.25}$$

Hence by (11.5), (11.22), and $\varpi = \omega_c \tau_2$,

$$\mathbf{q}_2 = -\frac{5kp}{2QB}\tau_2 \mathbf{b}\times \overset{\circ}{\mathbf{e}} \cdot \nabla T. \tag{11.26}$$

The first-order viscous stress tensor in (11.8) derives from the term $-2\tau_1 \boldsymbol{\nu\nu} : \overset{\circ}{\mathbf{e}}$ on the right of (10.98). Thus the calculation of $\boldsymbol{\pi}_1$ amounts to replacing $\boldsymbol{\nu\nu}$ in this term by $p\mathbf{W}$. Applying the same rule to (11.23), we obtain

$$\boldsymbol{\pi}_2 = 2p\tau_1^2 \mathbf{W} : \mathbf{S} \qquad \left(\mathbf{S} \equiv \overset{\circ}{\overline{\overset{\circ}{\mathbf{e}} \cdot \overset{\circ}{\mathbf{e}} - \mathrm{D}^* \overset{\circ}{\mathbf{e}}}}\right), \tag{11.27}$$

where the term containing $\nabla \cdot \mathbf{v}$ has been omitted from $\mathbf{S}$ because it is proportional to $\boldsymbol{\pi}_1$ and dominated by it.

By (11.6) the $\mathrm{O}(\varpi^{-1})$ term in $\mathbf{W}$ is $\frac{1}{2}\mathbf{W}_4 + \mathbf{W}_5$. Hence with $\varpi = \omega_c \tau_1$, it follows from (11.22) that, to sufficient accuracy

$$\boldsymbol{\pi}_2 = -\frac{p\tau_1}{2\omega_c}\left\{\mathbf{S} \times \mathbf{b} - \mathbf{b} \times \mathbf{S} + 3\mathbf{bb} \cdot \mathbf{S} \times \mathbf{b} - 3\mathbf{b} \times \mathbf{S} \cdot \mathbf{bb}\right\}. \tag{11.28}$$

11.3.4 *Physical principles in cross-field transport*

By 'tying' particles to field lines, for given temperature and fluid velocity gradients, strong magnetic fields suppress the amplitude of the relative distribution $\varphi = (f - f_0)/f_0$. Particles are no longer free—as in a neutral gas—to transport 'information' about the macroscopic variables through vector distances $\tau \mathbf{c}$ in all directions, e.g. for heat flux the vector mean free path is replaced by $\tau \mathbf{c} \cdot \mathbf{lk}$, the largest non-parallel term in which is $\mathrm{O}(\varpi^{-1})$. What a *uniform* magnetic field does not alter—excepting the change in $\boldsymbol{\alpha}$—is the expression for the rate at which particles are removed from the typical phase-plane element $d\boldsymbol{\nu}$ by collisions. We stress 'uniform' here, because as we shall show in §11.5, in some circumstances magnetic mirrors make a very substantial reduction in this rate.

As will be shown in §11.4.2, in the cross field direction, the magnetic field strongly suppresses first-order transport but greatly enhances second-order transport. In the electron gas in a tokamak (see §11.6) the latter can be several orders of magnitude *larger* than the former, reversing one's usual expectation.

The vital importance of the 'acceleration' term $\varphi_2^{(\alpha)}$, for the second-order theory should be noticed. Were it absent from Φ_2, the dominance of the left-hand Lorentz term in (11.18) would force φ_2 to be $\mathrm{O}(\varpi^{-2})$. The first-order term φ_1 would then dominate φ_2 for all directions of transport, allowing it to be omitted. This is the case with the standard kinetic theories, e.g. the BGK and Fokker–Planck models, so that they give entirely wrong results for plasmas in strong magnetic fields, an assertion that we shall later support with experimental evidence.

The expression for $\mathbf{q}_2$ in (11.26) was deduced by a mean-free-path approach in §9.7.3 and a special case of it was verified by an independent physical argument in §9.9.3. As it leads to results that are quite unexpected, it is important to have these independent derivations, which may be regarded as providing support for the more general kinetic theory from which (11.25) is derived.

The relation between $\boldsymbol{\pi}_2$ and $\boldsymbol{\pi}_1$ is of the same type (albeit rather more algebraically complicated) as in a neutral gas (see (8.44)), allowing $\boldsymbol{\pi}_2$ in a magnetoplasma to be explained by an argument like that given in §8.4.3.

11.4 Heat flux from a cylindrical magnetoplasma

11.4.1 *The second-order heat flux*

We shall consider the case of a cylindrical magnetoplasma, with a strong, helical magnetic field, $\mathbf{B} = B_z\hat{\mathbf{z}} + B_\theta\hat{\boldsymbol{\theta}}$, where $(\hat{\mathbf{r}}, \hat{\boldsymbol{\theta}}, \hat{\mathbf{z}})$ is the triad of unit vectors. It will be assumed that conditions are independent of the axial and azimuthal variables. Then $\nabla T = \hat{\mathbf{r}}T'$, where the dash denotes the radial derivative. The radial component of the second-order heat flux follows from (11.26):

$$Q_r \equiv \hat{\mathbf{r}} \cdot \mathbf{q}_2 = -\frac{5kp}{2QB}\tau_2 HT', \tag{11.29}$$

where

$$H \equiv \mathbf{b}\times\hat{\mathbf{r}} \cdot \overset{\circ}{\nabla}\mathbf{v} \cdot \hat{\mathbf{r}}. \tag{11.30}$$

For the unit vector parallel to $\mathbf{B}$ we have

$$\mathbf{b} = b_z\hat{\mathbf{z}} + b_\theta\hat{\boldsymbol{\theta}} \qquad (b_z = B_z/B,\ b_\theta = B_\theta/B), \tag{11.31}$$

where B is the field strength.

The radial velocity v_r of either the ion or electron fluid is suppressed by the strong field to values much less than either the azimuthal component, v_θ, or the axial component v_z. With axial symmetry and uniform conditions along the axis, we have

$$\nabla\mathbf{v} = v_\theta'\hat{\mathbf{r}}\hat{\boldsymbol{\theta}} - \frac{v_\theta}{r}\hat{\boldsymbol{\theta}}\hat{\mathbf{r}} + v_z'\hat{\mathbf{r}}\hat{\mathbf{z}},$$

therefore

$$\overset{\circ}{\nabla}\mathbf{v} = \tfrac{1}{2}(v_\theta' - v_\theta/r)(\hat{\mathbf{r}}\hat{\boldsymbol{\theta}} + \hat{\boldsymbol{\theta}}\hat{\mathbf{r}}) + \tfrac{1}{2}v_z'(\hat{\mathbf{z}}\hat{\mathbf{r}} + \hat{\mathbf{r}}\hat{\mathbf{z}}). \tag{11.32}$$

The scalar defined in (11.30) becomes

$$H = \tfrac{1}{2}r(v_\theta/r)'b_z - \tfrac{1}{2}v_z'b_\theta. \tag{11.33}$$

11.4.2 *Ratio of second to first-order heat fluxes*

From (11.5), (11.8), and (11.22), $\hat{\mathbf{r}} \cdot \mathbf{q}_1 = -(5Rp/2\varpi^2)\tau_2 T'$,

so that
$$|\hat{\mathbf{r}} \cdot \mathbf{q}_2/\hat{\mathbf{r}} \cdot \mathbf{q}_1| = \varpi\epsilon, \tag{11.34}$$

where
$$\epsilon = \tau_2 H \tag{11.35}$$

is the appropriate Knudsen number. Thus the second-order heat flux is dominant when
$$|\varpi\epsilon| \gg 1, \qquad (\epsilon \ll 1), \tag{11.36}$$

where of course the constraint on ϵ is required for the validity of the theory generally.

11.5 Influence of magnetic mirrors on transport

11.5.1 *Real collisions*

The collision interval τ_1 is the average time that a particle P can remain in the phase-space volume element $d\boldsymbol{\nu}$ before it, or rather its excess momentum, is removed by a *sequence* of collisions (see §3.3). The convenient shorthand of terming τ_1 the time that P remains within $d\boldsymbol{\nu}$ before being removed by 'a' collision is usually adopted. Similarly, while τ_2 is the relaxation time for P's excess energy, we can attach this time to P itself, without ambiguity.

In intermediate kinetic theory, we initially adopt a single collision time τ (see (11.15)), and later, when the terms relating to energy and momentum transport are distinguished, τ is assigned the value τ_2 or τ_1 accordingly.

The effect of any force able to remove P from $d\boldsymbol{\nu}$ must be accounted for in the kinetic equation. But forces acting on the fluid element, i.e. on the whole collection of particles, are not of interest here, since the kinetic equation is formulated (or should be formulated) in the convected fluid frame, relative to which such forces are absent. Our interest here is in forces that act *selectively*, affecting some particles, but not others.

The model most convenient for our purposes is that described in the last paragraph of §10.6.2, where $d\bar{\boldsymbol{\nu}}$ is a phase-space element closely following the guiding centre G, i.e. it has a velocity approximately equal to $\mathbf{c}_{\parallel}$ (see (11.2) and ex. 10.6). The particles repeatedly pass through $d\bar{\boldsymbol{\nu}}$, as it moves parallel to G, and τ is the average length of time that P can continue to pass through $d\bar{\boldsymbol{\nu}}$ on each gyration about the field lines. A real collision (as defined above) will certainly stop P from cycling through $d\bar{\boldsymbol{\nu}}$, but magnetic mirrors can also do this.

11.5.2 *Pseudo-collisions with magnetic mirrors*

Suppose that particles are cycling through $d\bar{\boldsymbol{\nu}}$ with velocity components $(c'_{\parallel}, c'_{\perp})$, where $c_{\parallel} - dc_{\parallel} < c'_{\parallel} < c_{\parallel} + dc_{\parallel}$. If the local magnetic field has a strength B that increases steadily with distance to a maximum value B_m

(located at a magnetic mirror), then as explained in §9.6, to conserve the particle magnetic moment, $mc_\perp^2/2B$, energy is transferred from the parallel to the perpendicular motion, which thus reduces $c'_\parallel$ below the range covered by $d\bar{\boldsymbol{\nu}}$: in other words, particles are removed from $d\bar{\boldsymbol{\nu}}$. And similarly $d\bar{\boldsymbol{\nu}}$ will acquire particles whose initial speeds exceeded $c_\parallel$, but which are slowed up into $d\bar{\boldsymbol{\nu}}$ by the increasing magnetic field. Hence in a direction parallel to **b**, changing fields have the same effect as soft, head-on collisions, removing and introducing particles into $d\bar{\boldsymbol{\nu}}$, but not displacing them in a direction perpendicular to **b**. These psuedo-collisions are 'soft' because of the gradual changes in B, but in the following, to simplify the account, we shall treat them as being 'hard collisions', occurring suddenly at the mirror locations, and involving only the particles actually reflected from the mirrors.

We shall term the small time interval that a particle continues to cycle through $d\bar{\boldsymbol{\nu}}$ between average psuedo-collisions, the *residence time*, and denote it by $\tau_\parallel^*$. For particles trapped between magnetic mirrors, $\tau_\parallel^*$ is proportional to the time it takes particles to travel between these mirrors, a time known as the 'bounce' time. In the toroidal confinement machine, known as a 'tokamak', there is a continuum of magnetic mirrors, and as we shall shortly establish, the bounce time τ_b is a function of the angular displacement θ of the mirrors, measured around the minor axis of the torus. In this case it is necessary to find an average bounce time by summing their reciprocals (cf. the addition of collision frequencies), weighted by the particle numbers involved at each point in the continuum. More precisely, it is the sum of the weighted *diffusivities* that yields the appropriate residence time (see (1.35)).

11.6 Tokamak magnetic fields

11.6.1 *Coordinate systems*

A tokamak is an axi-symmetric toroidal device used in fusion power research, the aim of which is to confine a high temperature plasma by a strong toroidal magnetic field B_φ. A toroidal current flows in the plasma and produces a relatively weak poloidal field, B_θ. A fuller description will be given in §11.9; for the present our interest is in how the magnetic field in a torus of circular cross-section is able to trap the charged particles comprising the plasma.

Figure 11.2 shows the coordinate systems usually adopted to describe the toroidal plasma, namely the usual cylindrical coordinates $(R,\ Z,\ \varphi)$ and the 'local' coordinates $(r,\ \theta,\ \varphi)$. This choice gives the azimuthal angle φ the opposite sign from that usual with cylindrical coordinates, but it gives the 'correct' sign with the more important local coordinate triad. The major radius is R_0. A point $P(r,\ \theta,\ \varphi)$ is a distance r from the minor axis, at a poloidal angle θ measured from the plane containing this axis, $\theta = 0$ being on the outside of the torus, and at an angle φ measured from some reference plane containing the major axis. Symmetry about this axis makes the location of this plane unimportant. In the cylindrical coordinates P is at the point $R = R_0 + r\cos\theta,\ Z = r\sin\theta$.

Let $\mathbf{A}$ denote a vector with local components $(A_r,\ A_\theta,\ A_\varphi)$; then its divergence and curl are given by

$$\nabla \cdot \mathbf{A} = \frac{1}{rR}\frac{\partial}{\partial r}(rRA_r) + \frac{1}{rR}\frac{\partial}{\partial \theta}(RA_\theta), \tag{11.37}$$

and

$$\nabla \times \mathbf{A} = \frac{1}{rR}\frac{\partial}{\partial \theta}(RA_\varphi)\hat{\mathbf{r}} - \frac{1}{R}\frac{\partial}{\partial r}(RA_\varphi)\hat{\boldsymbol{\theta}} + \left\{\frac{1}{r}\frac{\partial}{\partial r}(rA_\theta) - \frac{1}{r}\frac{\partial A_r}{\partial \theta}\right\}\hat{\boldsymbol{\varphi}}, \tag{11.38}$$

where $\hat{\mathbf{r}},\ \hat{\boldsymbol{\theta}},\ \hat{\boldsymbol{\varphi}}$ is the triad of unit orthogonal vectors associated with $r,\ \theta,\ \varphi$.

To simplify the theory, it will be assumed that in the local coordinate system, the radial components of the current density and the magnetic field are zero. Then from $\nabla \cdot \mathbf{B} = 0,\ \nabla \cdot \mathbf{j} = 0$ and (11.37), it follows that RB_θ and Rj_θ are functions of r only, i.e.

$$B_\theta(r,\ \theta) = \frac{B_{\theta 0}(r)}{1 + \varepsilon\cos\theta}, \qquad j_\theta(r,\ \theta) = \frac{j_{\theta 0}(r)}{1 + \varepsilon\cos\theta}, \tag{11.39}$$

where
$$\varepsilon \equiv r/R_0, \qquad R = R_0(1 + \varepsilon\cos\theta), \tag{11.40}$$

and $B_{\theta 0},\ j_{\theta 0}$ are the distributions in the cylindrical limit, $\varepsilon = 0$. These

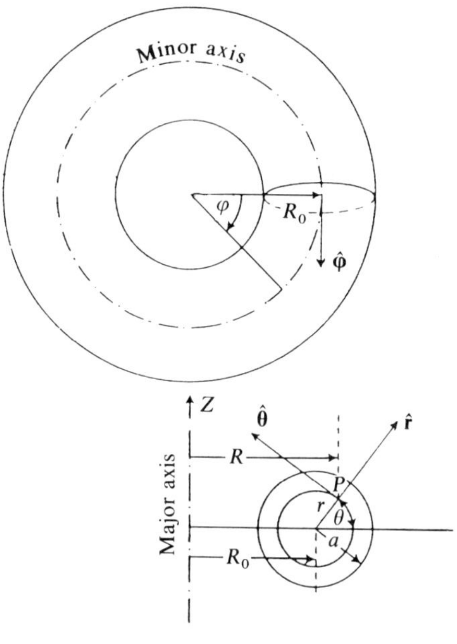

FIG. 11.2. Cylindrical and local coordinates

expressions are not exact, since plasma pressure and internal inductance slightly displace the circles $B_{\theta 0} =$ const. towards the outside of the torus. But this is a small effect that can be ignored here.

As B_r and j_r are assumed to be zero, the $\hat{\boldsymbol{\theta}}$-component of $\nabla \times \mathbf{B} = \mu_0 \mathbf{j}$ (see (9.6)) is

$$-\frac{1}{R}\frac{\partial}{\partial r}(RB_\varphi) = \mu_0 j_\theta = \frac{R_0}{R}\mu_0\, j_{\theta 0},$$

hence

$$B_\varphi(r,\,\theta) = \frac{B_{\varphi 0}(r)}{1+\varepsilon\cos\theta}, \tag{11.41}$$

and

$$\mu_0 j_{\theta 0} = -\frac{\partial}{\partial r}B_{\varphi 0}. \tag{11.42}$$

11.6.2 *Magnetic mirrors*

A magnetic field line is determined by the equations $dr = 0$, $R\,d\varphi/B_\varphi = r\,d\theta/B_\theta$,

or

$$r = \text{const.}, \qquad d\varphi = \frac{q\,d\theta}{1+\varepsilon\cos\theta}, \tag{11.43}$$

where

$$q \equiv \frac{rB_\varphi}{R_0 B_\theta} = \varepsilon\,\frac{B_{\varphi 0}}{B_{\theta 0}}. \tag{11.44}$$

If $\varepsilon \ll 1$, $\varphi \approx q\theta$, so q is the number of rotations of a field line about the major axis, per rotation about the minor axis. For reasons to do with plasma stability, q is termed the 'safety factor'; typical values lie between one and five. The total distance travelled along a field line per 360° rotation about the minor axis of the torus is approximately $2\pi R_0 q$; this is termed the 'connection length'.

The magnetic field lies in the direction of the unit vector

$$\mathbf{b} = \frac{B_\theta}{B}\hat{\boldsymbol{\theta}} + \frac{B_\varphi}{B}\hat{\boldsymbol{\varphi}} = \frac{1}{(1+ß^2)^{\frac{1}{2}}}(\hat{\boldsymbol{\varphi}} + ß\hat{\boldsymbol{\theta}}),$$

where by (11.44),

$$ß \equiv \frac{B_\theta}{B_\varphi} = \frac{B_{\theta 0}}{B_{\varphi 0}} = \frac{\varepsilon}{q}. \tag{11.45}$$

This ratio is typically less than 1/10, so terms $\mathrm{O}(ß^2)$ can usually be ignored, in which case

$$\mathbf{b} = \hat{\boldsymbol{\varphi}} + ß\hat{\boldsymbol{\theta}}. \tag{11.46}$$

From (11.39) and (11.41) the magnetic field strength is

$$B = \frac{B_0}{1+\varepsilon\cos\theta}, \qquad B_0 = (B_{\varphi 0}^2 + B_{\theta 0}^2)^{\frac{1}{2}}, \tag{11.47}$$

and hence:

$$\text{at } \theta = 0,\ \ B = \frac{B_0}{1+\varepsilon} = B_{min}; \quad \text{at } \theta = \pi,\ \ B = \frac{B_0}{1-\varepsilon} = B_{max}. \tag{11.48}$$

Along a given field line, a particle meets these extrema a distance $\pi R_0 q$

apart. Thus the tokamak field presents a continuum of magnetic mirrors along the inside of the torus.

11.7 Trapping in tokamak fields

11.7.1 *The fraction of trapped particles*

The mirror ratio defined in §9.6.3 is $(1+\varepsilon)/(1-\varepsilon)$, and hence by (9.89) the fraction of trapped particles is

$$f_T = \left(\frac{2\varepsilon}{1+\varepsilon}\right)^{\frac{1}{2}} \approx (2\varepsilon)^{\frac{1}{2}}, \tag{11.49}$$

the approximation serving for ε sufficiently small.

Let the plasma boundary be at $r = a$, then R_0/a is termed the 'aspect ratio'. A typical value for this ratio is about three. At $\varepsilon = 1/3$ (11.49) gives $f_T \approx 0.71$, while at $\varepsilon = 1/6$, $f_T \approx 0.53$. Hence a large fraction of the particles in a tokamak is trapped in its magnetic field.

Of course, a particle need not travel the whole distance to the mirror at $\theta = \pi$ before being reflected. If its parallel velocity is sufficient to carry it only to $\theta = \pm\theta_0$ before reflection, then in the derivation of f_T, the extrema in B are replaced by $B_0(1 \pm \varepsilon\cos\theta_0)^{-1}$. The fraction trapped in $-\theta_0 < \theta < \theta_0$ is therefore

$$f_T(\theta_0) = \left(\frac{2\varepsilon}{1+\varepsilon}\right)^{\frac{1}{2}} \sin\tfrac{1}{2}\theta_0. \tag{11.50}$$

It follows that the fraction of particles reflected in $(\theta_0,\, \theta_0 + d\theta_0)$ is

$$df_T(\theta_0) = \left(\frac{2\varepsilon}{1+\varepsilon}\right)^{\frac{1}{2}} \tfrac{1}{2}\cos\tfrac{1}{2}\theta_0\, d\theta_0 \qquad (0 < \theta_0 < \pi). \tag{11.51}$$

11.7.2 *Bounce time in a tokamak field*

The next step is to find an expression for the time $\tau_b(\theta_0)$ that it takes a particle P to travel between the reflection points at $\theta = \pm\theta_0$. The velocity with which the guiding centre drifts away from a field line will be assumed to be small enough to be ignored in the calculation of $\tau_b(\theta_0)$.

The element of distance along a field line is

$$ds = \left\{(R\,d\varphi)^2 + (r\,d\theta)^2\right\}^{\frac{1}{2}},$$

and hence from (11.40) and (11.43) the parallel speed of P is

$$c_\parallel = \frac{ds}{dt} = \left\{1 + \left(\frac{R_0 q}{r}\right)^2\right\}^{\frac{1}{2}} \frac{r\,d\theta}{dt} \approx R_0 q \frac{d\theta}{dt},$$

since $(R_0 q/r)$ is typically about 10. By (11.47) and $M \equiv mc_\perp^2/2B$ (see (9.78)), we have

$$c_\parallel = c\left\{1 - \left(\frac{c_\perp}{c}\right)^2\right\}^{\frac{1}{2}} = c\left\{1 - \frac{1+\varepsilon}{1+\varepsilon\cos\theta}\sin^2\psi\right\}^{\frac{1}{2}},$$

where $$\sin^2\psi \equiv \frac{2B_0 M}{mc^2}\frac{1}{1+\varepsilon}$$

is the value of $(c_\perp/c)^2$ at $\theta = 0$. Thus

$$|c_\parallel| = (2\varepsilon)^{\frac{1}{2}}\, c\left(\frac{\kappa^2 - \sin^2\frac{1}{2}\theta}{1+\varepsilon\cos\theta}\right)^{\frac{1}{2}} \qquad \left(\kappa^2 \equiv \frac{1+\varepsilon}{2\varepsilon}\cos^2\psi\right), \tag{11.52}$$

and $$\frac{d\theta}{dt} = \frac{c_\parallel}{R_0 q} = \pm\frac{c(2\varepsilon)^{\frac{1}{2}}}{R_0 q}\left\{\frac{\kappa^2 - \sin^2\frac{1}{2}\theta}{1+\varepsilon\cos\theta}\right\}^{\frac{1}{2}}, \tag{11.53}$$

where '+' applies to θ increasing, and '-' to θ decreasing.

The reflection points are at $d\theta/dt = 0$, i.e. at $\theta = \pm\theta_0$, whence $\kappa = \sin\frac{1}{2}\theta_0$ and the required bounce time is

$$\tau_b(\theta_0) = \frac{2R_0 q}{c(2\varepsilon)^{\frac{1}{2}}}\int_0^{\theta_0}\left\{\frac{1+\varepsilon\cos\theta}{\sin^2\frac{1}{2}\theta_0 - \sin^2\frac{1}{2}\theta}\right\}^{\frac{1}{2}} d\theta.$$

The denominator in the integrand varies from 0 to ~ 1, while the numerator varies between $(1+\varepsilon)$ and $(1-\varepsilon)$, with an average ~ 1. With ε small, little error is made by replacing the numerator by unity. The integral can then be evaluated with help of the transformation $\sin\phi = \sin\frac{1}{2}\theta/\sin\frac{1}{2}\theta_0$. We find that

$$\tau_b(\theta_0) = \frac{2\sqrt{2}R_0 q}{c\sqrt{\varepsilon}}K(\sin\tfrac{1}{2}\theta_0), \tag{11.54}$$

where K is the complete elliptic integral of the first kind; this result is due to Kadomtsev and Pogutse (1967).

In typical tokamaks collision times are usually somewhat larger than an average bounce time, $\tau/\tau_b \sim 10$ or so being common. Thus particles will move between reflection points many times before escaping to new orbits.

11.8 Transport in tokamaks

11.8.1 *The diffusivity due to trapped particles*

The next step is to calculate the parallel diffusivity due to the trapped particles. We start with the increment $d\chi_\parallel^{(T)}$ to the parallel diffusivity due to the fraction df_T of particles trapped between magnetic mirrors a distance $\lambda_m = R_0 q\theta_0$ apart, and taking a time $\tau_b(\theta_0)$ to move from one mirror to the other. This is (see (1.34)) $d\chi_\parallel^{(T)} = (\lambda_m^2/\tau_b)df_T$ and therefore the total diffusivity is

$$\chi_\parallel^{(T)} = \int_{-\pi}^{\pi}\frac{(R_0 q\theta_0)^2}{\langle\tau_b\rangle}\, df_T(\theta_0),$$

where we have replaced τ_b by its average over all particle speeds. By (11.51),

(11.54) and $\varepsilon = r/R_0$, this gives

$$\chi_{\parallel}^{(T)} = \frac{4qr\bar{c}}{(1+\varepsilon)^{\frac{1}{2}}} \int_0^{\pi/2} \frac{x^2 \cos x}{K(\sin x)}\, dx. \tag{11.55}$$

As ε typically lies in the range (0, 0.3), $(1+\varepsilon)^{\frac{1}{2}} \approx 1$ is a reasonable approximation. The integral is approximately 0.219, and with $\bar{c}$ given in (2.42), with acceptable error, we arrive at the simple expression

$$\chi_{\parallel}^{(T)} = rq\mathcal{C}. \tag{11.56}$$

11.8.2 *The parallel diffusivity due to untrapped particles*

The diffusivity of untrapped particles is determined by the usual relation,

$$\chi_{\parallel}^{(U)} = \lambda_{\parallel}^2/\tau_2 = \overline{c_{\parallel}^2}\tau_2 = \tfrac{1}{2}\mathcal{C}^2\tau_2, \tag{11.57}$$

where $\lambda_{\parallel}$ is the parallel mean free path, $\overline{c_{\parallel}^2}$ has been evaluated from the parallel component of (2.43), and for the present we are restricting attention to energy transport. Except near the minor axis ($r = 0$), $\chi_{\parallel}^{(U)}$ is typically about 10^2 times larger than $\chi_{\parallel}^{(T)}$, so its contribution to the total diffusivity could be enormous. What prevents this is the axial symmetry about the major axis, at least in steady conditions. The untrapped or 'passing' particles experience conditions that repeat after each connection length, $2\pi R_0 q$ (see §11.6.2). These conditions and the particle orbits are generically independent of the toroidal coordinate φ. It follows that the concept of parallel thermal diffusivity is not relevant for these particles, allowing them to be ignored so far as radial energy transport is concerned.

By destroying the axial symmetry about the major axis, particles that either *become* trapped or detrapped by fluctuations in the magnetic field strength, are able to contribute to the parallel diffusivity. Let f_{TU} and f_{UT} denote the fraction of trapped particles about to become untrapped and the fraction of untrapped particles about to become trapped. The particles just untrapped suddenly acquire a diffusivity $\chi_{\parallel}^{(U)} f_{TU}$, while those just trapped lose a diffusivity $\chi_{\parallel}^{(U)} f_{UT}$. The net effect is $\chi_{\parallel}^{(U)}(f_{TU} - f_{UT})$, giving a total parallel diffusivity of

$$\chi_{\parallel} = rq\mathcal{C} + \tfrac{1}{2}\mathcal{C}^2\tau_2\, \delta f_T, \qquad (\delta f_T \equiv f_{TU} - f_{UT}). \tag{11.58}$$

11.8.3 *Residence time for energy transport*

The purpose of calculating $\chi_{\parallel}$ is *not* to determine the diffusion of heat parallel to the magnetic field—this process is dominated by the terms first-order in ϵ. Our aim (see §11.5.2) is to obtain an expression for the residence time $\tau_{\parallel}^*$ to replace the collision interval τ_2 in the formula (11.26) for the second-order heat flux. This time is related to $\chi_{\parallel}$ by $\frac{1}{2}\mathcal{C}^2\tau_{\parallel}^* = \chi_{\parallel}$, (cf. (11.57)) and therefore from (11.58),

$$\tau_{\|}^{*} = \frac{2rq}{\mathcal{C}} + \tau_2 \delta f_T \qquad \left(\mathcal{C} = (2kT/m)^{\frac{1}{2}}\right). \tag{11.59}$$

In steady conditions δf_T is zero.

The physical significance of the first term in $\tau_{\|}^{*}$ is that of a weighted bounce time, equal to the average time that particles are 'in' a phase-space element $d\bar{\boldsymbol{\nu}}$ until removed by a head-on 'collision' with a magnetic mirror. Here 'in' is shorthand for 'cycling through'. The element $d\bar{\boldsymbol{\nu}}$ tracks a bunch of guiding centres at $(\mathbf{X},\, t)$ say, and adopting the mean-free-path account of §9.7.2, we describe the particles in $d\bar{\boldsymbol{\nu}}$ as being labelled by the temperature at $(\mathbf{X}_s^*,\, t - 2rq/\mathcal{C})$. By confining the guiding centres to a restricted range, repeatedly traversed, the mirrors reduce the scope for deviations from the equilibrium temperature, and so reduce the energy flux from the trapped particles.

It follows from (11.26) that the second-order heat flux from a tokamak is given by

$$\mathbf{q}_2 = -\frac{5kp}{2_Q B}\tau_{\|}^{*}\mathbf{b}\times \overset{\circ}{\mathbf{e}} \cdot \nabla T. \tag{11.60}$$

In unsteady conditions, when particles are trapped or detrapped by changing magnetic fields, the second term in $\tau_{\|}^{*}$ becomes dominant, giving rise to transport instabilities that are sometimes sufficient to completely disrupt the plasma discharge. We shall return to this feature in §11.10.3.

11.8.4 *Collision time for momentum flux*

It should be emphasized that $\tau_{\|}^{*}$ replaces the usual collision interval τ only for particle displacements from $d\bar{\boldsymbol{\nu}}$ that are parallel to $\mathbf{B}$. For the cross-field transfer of viscous forces, particle displacements perpendicular to $\mathbf{B}$ are necessary, and the corresponding time interval between events causing such displacements is unaffected by the 'head-on' loss and gain of particles due to reflection from magnetic mirrors. Particles can be removed from $d\bar{\boldsymbol{\nu}}$—in the sense that their orbits no longer traverse $d\bar{\boldsymbol{\nu}}$ each gyration about the field lines—in a direction perpendicular to $\mathbf{B}$, only by the usual collisional process, i.e. the second-order viscous stress tensor in a tokamak is given by (11.28), viz.

$$\boldsymbol{\pi}_2 = -\frac{p\tau_1}{2\omega_c}\left\{\mathsf{S}\times\mathbf{b} - \mathbf{b}\times\mathsf{S} + 3\mathbf{b}\mathbf{b}\cdot\mathsf{S}\times\mathbf{b} - 3\mathbf{b}\times\mathsf{S}\cdot\mathbf{b}\mathbf{b}\right\}, \tag{11.61}$$

where

$$\mathsf{S} \equiv \overline{\overset{\circ}{\mathbf{e}}\cdot\overset{\circ}{\mathbf{e}}}^{\circ} - \mathrm{D}^{*}\overset{\circ}{\mathbf{e}}. \tag{11.62}$$

The reflection of particles from magnetic mirrors may be regarded as being ideal or reversible collisions between the impinging particles and a 'wall', producing a force normal to it, without transverse components. Such a force cannot generate or transmit viscous forces. On the other hand the energy flux that leads to (11.60) is delivered by a term that has its origin in the component $\mathbf{cc} \cdot \nabla\nabla \cdot \mathbf{v} \propto \mathbf{cc} \cdot \nabla \mathrm{D}p$ of $\mathbb{D}_S(\mathbf{c} \cdot \mathbf{e})$ (see (5.87), (8.30), (11.21), and remarks in §5.10.3). Hence it depends on the *scalar* pressure. Such a flux can be interrupted, or limited, by the ideal mirror collisions. This argument supports our choice of $\tau_{\|}^*$ for energy transport, and its rejection for momentum transport.

11.8.5 *The Knudsen number constraint*

Equations (11.60) and (11.61) are correct to terms of order ϵ^2 and ϖ^{-1}. For terms of higher order to be negligible, it is necessary that $\varpi^{-1} \ll 1$ and $\epsilon \ll 1$. The first constraint offers no difficulty in tokamaks, where ϖ is typically 10^5 or larger, but the second may present a problem. For both transport phenomena the restriction amounts to

$$\epsilon = \tau^* \| \overset{\circ}{\nabla} \mathbf{v} \| \ll 1, \tag{11.63}$$

where τ^* is either $\tau_{\|}^*$ for heat transport or τ_1 for fluid momentum transport. To determine ϵ we need an estimate of the angular velocity H defined in (11.31), but to find this we need to develop the theory further. We shall return to this important question in §§11.9.3 and 11.11.3.

11.9 Energy losses from tokamaks

It is not our purpose here to give a full account of the application of second-order theory to tokamak physics; readers interested in this can consult Woods (1987). The rather different approach to second-order theory adopted in that work will be outlined in Appendix A2. Nor shall we describe the other theories that have attempted to explain the remarkably high rates at which energy escapes from tokamak fields. The main resort has been to a variety of turbulence generating instabilities (Liewer 1985), but no theoretical model of *turbulence* transport has yet passed the test of giving reliable predictions over a wide range of tokamak parameters. For a general survey of tokamak physics, see Wesson (1987).

Our present aim is to develop tokamak transport theory only to the point where comparisons with experiment can be made for the most important phenomena. Besides the high rates of energy loss, there is also a curious oscillation in the temperature (and other variables) known as a 'sawtooth' instability. A steady climb in temperature is followed by an abrupt collapse and this occurs periodically under normal operating conditions. When the plasma density exceeds a certain critical value, the oscillation disappears and a sudden irreversible collapse of the temperature may occur. This is termed a 'major disruption'. The second-order theory explains these phenomena and gives reasonably accurate estimates for such parameters as the sawtooth period and the critical number density beyond which tokamaks are not viable.

11.9.1 *Some tokamak parameters*

Let $r = a$ specify the edge of the plasma* in a tokamak of circular cross section, then the *aspect ratio* is R/a, where R is the major radius.† As the $\hat{\varphi}$-component of $\nabla \times \mathbf{B} = \mu_0 \mathbf{j}$ gives

$$\mu_0 j_\varphi = \frac{1}{r}\frac{\partial}{\partial r}(rB_\theta), \tag{11.64}$$

the total plasma current is

$$I_p = 2\pi \int_0^a j_\varphi r\, dr = \frac{2\pi a}{\mu_0} B_{\theta a}. \tag{11.65}$$

Tokamak magnetic fields have $B_\theta \ll B_\varphi$, i.e.

$$s \equiv \frac{B_\theta}{B_\varphi} = \frac{\mu_0}{B_\varphi r}\int_0^r j_\varphi(r')r'\, dr' \ll 1; \tag{11.66}$$

also the variation in $B_\varphi \approx B$ across the cross-section is negligible.

The safety factor is defined in (11.46):

$$q(r) = \frac{rB_\varphi}{RB_\theta} = \frac{\varepsilon}{s}. \tag{11.67}$$

At the boundary, by (11.65) and (11.67), it has the value

$$q_a = \frac{2\pi a^2}{\mu_0 I_p}\frac{B_\varphi}{R}. \tag{11.68}$$

The average current density, $\bar{j}_\varphi$, is given by

$$\mu_0 \bar{j}_\varphi = \frac{2B_\varphi}{Rq_a}. \tag{11.69}$$

Let $j_{\varphi 0}$ be the current density on the minor axis, then it follows from (11.67) and (11.66) that on $r = 0$ the safety factor has the value q_0, where

$$q_0 = \frac{2B_\varphi}{\mu_0 j_{\varphi 0} R}. \tag{11.70}$$

Notice that $q_a = q_0 j_{\varphi 0}/\bar{j}_\varphi$, so with fixed q_0, peaked current profiles correspond to large values of q_a.

Let

$$p = \frac{2}{a^2}\int_0^a p(r) r\, dr \tag{11.71}$$

defined the volume-averaged pressure, then

$$\beta_t \equiv \frac{2\mu_0 \bar{p}}{B_\varphi^2}, \qquad \beta_p \equiv \frac{2\mu_0 \bar{p}}{B_{\theta a}^2} = \left(\frac{Rq_a}{a}\right)^2 \beta_t, \tag{11.72}$$

*In the (usual) absence of a magnetic field structure known as a 'diverter' to reduce the influx of ionized impurities, this boundary is termed a 'limiter'.

†No confusion will arise if in the remainder of this chapter we omit the subscript from R_0, since the distinction between R and R_0 is no longer required.

define important parameters known as the toroidal and poloidal 'betas'.

In the electron fluid we shall need the following averages:

$$\bar{n}_e \equiv \frac{1}{a}\int_0^a n_e(r)\,dr, \qquad \langle n_e\rangle \equiv \frac{2}{a^2}\int_0^a n_e(r)r\,dr, \tag{11.73}$$

and

$$\langle T_e\rangle \equiv \frac{2}{\langle n_e\rangle a^2}\int_0^a n_e(r)T_e(r)r\,dr. \tag{11.74}$$

11.9.2 *The electron thermal diffusivity*

The experimental evidence is that high thermal diffusivity in the electron fluid is the main cause for the energy losses from tokamaks. Conduction in the ion fluid makes a relatively small contribution. Radiation is important, but mainly from the plasma periphery, where energy is delivered from the central regions by a combination of conduction and convection. Although plasma flow losses are high, they are usually insufficient to convect energy from tokamaks at more than a small fraction of the observed rates.

From (11.59), (11.60), and (11.31), in steady operating conditions

$$Q_{er} \equiv \hat{\mathbf{r}}\cdot\mathbf{q}_2 = -\frac{5kp_e}{2eB}\frac{2rq}{\mathcal{C}_e}H_eT_e'. \tag{11.75}$$

For our purpose it is sufficiently accurate to evaluate H_e in cylindrical coordinates—the toroidal geometry enters through q. Thus from (11.33) and $b_z \approx 1$, $b_\theta \approx s$, we get

$$H_e = \tfrac{1}{2}r\left(v_{e\theta}/r\right)' - \tfrac{1}{2}sv_{e\varphi}' \approx -\tfrac{1}{2}\left\{r\left(\frac{j_\theta}{ren_e}\right)' - s\left(\frac{j_\varphi}{en_e}\right)'\right\},$$

where the approximation follows from $\mathbf{j} = en_e(\mathbf{v}_i - \mathbf{v}_e)$, and the assumption that $|v_{i\theta}| \ll |v_{e\theta}|$ and $|v_{i\varphi}| \ll |v_{e\varphi}|$, i.e. that the electrons are moving much more rapidly than the relatively heavy ions.

We eliminate j_θ by adopting the equilibrium condition $\mathbf{j}\times\mathbf{B} = \nabla p$, which follows from (9.33) plus neglect of viscosity. Thus

$$j_\theta = sj_\varphi + p'/B_\varphi \qquad (p = p_e + p_i), \tag{11.76}$$

and the expression for H_e becomes

$$H_e = \frac{-1}{2en_e}\left\{j_\varphi r\left(\frac{s}{r}\right)' + rn_e\left(\frac{p'}{rn_eB_\varphi}\right)'\right\}.$$

By (11.66) this can be written

$$H_e = \frac{1}{2en_e}\left\{\frac{\mu_0 j_\varphi}{B_\varphi}\frac{2}{r^2}\int_0^r \{j_\varphi(r^*) - j_\varphi(r)\}r^*\,dr^* - rn_e\left(\frac{p'}{rn_eB_\varphi}\right)'\right\}. \tag{11.77}$$

By (11.75) the thermal conductivity is

$$\kappa_e = \frac{5kp_e}{eB_\varphi \mathcal{C}_e} rqH_e, \tag{11.78}$$

so ignoring the small variations in B_φ across the plasma cross-section we can write the thermal diffusivity, $\chi_e = \kappa_e/(\frac{3}{2}kn)$, in the form

$$\chi_e = \frac{5m_e rq\mathcal{C}_e}{6e^2 n_e B_\varphi^2}\left\{\mu_0 j_\varphi \frac{2}{r^2}\int_0^r \{j_\varphi(r^*) - j_\varphi(r)\}r^*\,dr^* - rn_e\left(\frac{p'}{rn_e}\right)'\right\}. \tag{11.79}$$

The appearance of this expression can be simplified by introducing the relative magnitudes

$$J = j_\varphi/\bar{j}_\varphi, \quad P = p/\bar{p}, \quad N = n_e/\langle n_e\rangle, \tag{11.80}$$

expressed in terms of the volume averages defined in §11.9.1. Then changing the variable to

$$y = r^2/a^2, \tag{11.81}$$

and using a dot to denote d/dy, (11.79) becomes

$$\chi_e = \frac{10m_e}{3\mu_0 e^2 n_e}\frac{a\mathcal{C}_e}{R^2 q_a}\frac{q}{q_a}y^{\frac{1}{2}}\left\{\frac{J}{y}\int_0^y [J(y') - J(y)]\,dy' - \tfrac{1}{2}\beta_p y(\ddot{P} - \dot{P}\dot{N}/N)\right\}. \tag{11.82}$$

11.9.3 *Validity of the theory*

We can now attend to the Knudsen number constraint given in (11.63). The appropriate component of the velocity gradient tensor is defined in (11.30). Hence in the electron gas, by (11.59), the constraint becomes

$$\epsilon = \tau_\parallel^* H_e \ll 1 \qquad (\tau_\parallel^* = 2rq/\mathcal{C} + \tau_2\delta f_T). \tag{11.83}$$

Of the two terms in H_e, the second in (11.77) is usually the smaller. So to obtain an estimate, we adopt the integral term, replace j_φ by $\bar{j}_\varphi$ and $\{j_\varphi(r^*) - j_\varphi(r)\}$ by $\frac{1}{2}\bar{j}_\varphi$ (which gives an overestimate), then by (11.69) arrive at

$$H_e \sim B_\varphi/(en_e\mu_0 R^2 q_a^2).$$

For the Jet tokamak typical values for the variables in H_e ($B_\varphi = 3$, $en_e = 5$, $Rq_a = 10$ in MKS units) give $H_e \sim 4.8 \times 10^3$. For $\tau_\parallel^*$ we have $\tau_\parallel^* \sim aq_a/\mathcal{C}_e + \tau_e\delta f_T$. In JET $a \sim 1, q_a \sim 3$ and $\langle T_e\rangle \sim 6$keV give $\tau_\parallel^* \sim 6.5 \times 10^{-8} + 1.8 \times 10^{-4}\delta f_T$. Thus in steady conditions ($\delta f_T = 0$), $\tau_\parallel^* H_e \sim 3 \times 10^{-4}$, leaving ample scope for (11.83) to be satisfied, even with substantial changes in the values of the parameters involved.

In unsteady conditions, the magnitude of δf_T is typically only about one or two per cent (Woods 1987), giving $\tau_\parallel^* \sim 3 \times 10^{-6}$ and $\epsilon \sim 1.4 \times 10^{-2}$. The constraint is still satisfied, even though the change in $\tau_\parallel^*$ increases the heat flux by a factor of 50 or so.

For the ions, since $\mathcal{C}_i = (m_e/m_i)^{1/2}\mathcal{C}_e$ and the fluid velocities are similarly related (see exercise 11.7), the value of ϵ is much the same for both species, i.e. so far as ion heat transfer is concerned, the Knudsen number constraint is satisfied.

11.9.4 *The electron-energy confinement time*

We start from the simplest form of energy conservation in the electron gas, namely

$$\frac{1}{r}\frac{\partial}{\partial r}(rQ_{er}) \approx j_\varphi E_\varphi. \tag{11.84}$$

Here it is assumed that ohmic heating alone supplies the electron thermal energy and that conduction is the only loss mechanism. Integrating (11.84) over the minor radius, and taking E_φ as constant across the torus cross-section, we get

$$\big[rQ_{er}\big]_{r=a} = \int_0^a j_\varphi E_\varphi \, r\, dr = \tfrac{1}{2}a^2 \int_0^1 j_\varphi E_\varphi \, dy \tag{11.85}$$

and

$$\big[rQ_{er}\big]_{r=a} = a\langle Q_{er}\rangle / 2A_\epsilon, \tag{11.86}$$

where

$$\langle Q_{er}\rangle \equiv \frac{2}{a^2}\int_0^a Q_{er}\, r\, dr = \int_0^1 Q_{er}\, dy \tag{11.87}$$

and

$$A_\epsilon = \tfrac{1}{2}\int_0^a y^{-1/2}\, dy \int_0^y j_\varphi(y')\, dy' \Big/ \int_0^1 j_\varphi \, dy. \tag{11.88}$$

The electron thermal energy in the torus is proportional to

$$W_e = \int_0^a \tfrac{3}{2}p_e \, r\, dr = \tfrac{3}{4}a^2 k\langle T_e\rangle\langle n_e\rangle = \tfrac{3}{4}a^2\bar{p},$$

using the definitions given in §11.9.1. A measure of the time that the electron energy in the plasma remains confined is

$$\tau_{Ee} = W_e / \big[Q_{er}\big]_{r=a} = 1.5A_\epsilon \bar{p}a/\langle Q_{er}\rangle, \tag{11.89}$$

the advantage of the second form being that the confinement time is not sensitively dependent on the value of Q_{er} at a single surface. A similar measure of the time it takes to replace this energy by ohmic heating is

$$\tau^*_{Ee} = W_e \Big/ \int_0^a j_\varphi E_\varphi \, r\, dr. \tag{11.90}$$

With the replacement time readily inferred from observations on tokamaks, the equilibrium condition $\tau_{Ee} = \tau^*_{Ee}$ provides a means of checking a theory for Q_{er}.

11.9.5 *The use of empirical profiles*

The complexity of (11.75) is such that to obtain a manageable formula for τ_{E_e}, it is expedient to adopt empirical profiles for T_e and n_e. Of course in a complete theory, these profiles should be deduced, but as the dominant term in (11.79)—the integral term—is only weakly dependent on the profile shapes, the method has the merit of providing a first check on the theory.

One-parameter profiles that resemble those obtaining in experiment are

$$n_e = n_{e0}(1-y)^{\alpha_n}, \quad T_e = T_{e0}(1-y)^{\alpha_t}, \quad j_\varphi = j_{\varphi 0}(1-y)^{\delta}, \tag{11.91}$$

and as $j_\varphi = \eta_\| E_\varphi$ is approximately proportional $T_e^{3/2}$ (see §10.10.2), it is reasonable to assume that $\delta = \frac{3}{2}\alpha_t$. We also need the safety factor distribution

$$q = q_a y \left\{1 - (1-y)^{\delta+1}\right\}^{-1},$$

which follows from (11.66), (11.67), and (11.91).

At this stage we shall assume that the density is low enough to make the parameter β_p much less than unity, which allows us to omit the profile-sensitive derivative term from (11.79). Then using the empirical profiles in the above equations, after some straightforward algebra, we arrive at the formula

$$\tau_{E_e} = \mathcal{F}_\epsilon \frac{\mu_0 e^2}{(2m_e)^{\frac{1}{2}}} \frac{\langle n_e \rangle a R^2 q_a}{\langle kT_e \rangle^{\frac{1}{2}}}, \tag{11.92}$$

where $\mathcal{F}_\epsilon = \mathcal{F}_\epsilon(\alpha_n, \alpha_t)$ is the profile shape factor. As shown in Table 11.1, this factor is a slowly varying function of its arguments.

Table 11.1 *Values of the shape factor $\mathcal{F}_\epsilon$*

$\alpha_n \backslash \alpha_t$	1.0	1.5	2.0	2.5	3.0	3.5
0.5	0.51	0.40	0.34	0.31	0.27	0.24
1.0	0.60	0.49	0.43	0.39	0.36	0.32
1.5	0.66	0.56	0.51	0.47	0.43	0.39
2.0	0.71	0.61	0.57	0.54	0.50	0.46

11.10 Comparison of theory and experiment

11.10.1 *Some JET tokamak results at low β_p*

Experimental results are usually expressed in terms of the line-averaged densities $\bar{n}_e$. Using the density profile in (11.91), it is a simple matter to replace $\langle n_e \rangle$ by $\bar{n}_e$. Typical values for the profile indices are $\alpha_t = 2$ and $\alpha_n = 1$, giving $\mathcal{F}_\epsilon \approx 0.43$. Then (11.92) can be written

$$\tau_{E_e} = 1.93 \times 10^{-20} \frac{\bar{n}_e a R^2 q_a}{\langle \tilde{T}_e \rangle^{\frac{3}{2}}} \quad \text{s} \quad (\tilde{T}_e \text{ in eV}). \tag{11.93}$$

An alternative form follows from (11.68):

$$\tau_{E_e} = 0.97 \times 10^{-19} \frac{\bar{n}_e a^3 R B_\varphi}{\langle \tilde{T}_e \rangle^{\frac{3}{2}} \tilde{I}_p} \quad \mathrm{s} \quad (\tilde{I}_p \text{ in MA}). \tag{11.94}$$

Table 11.2 *Typical JET discharge parameters at low β_p*

I_p (MA)	B_φ (T)	A (m^2)	$10^{-19}\bar{n}_e$ (m^{-3})	$\langle T_e \rangle$ (keV)	Z^*	τ_E(exp) (s)	τ_{E_e}(th) (s)
2.3	2.5	4.3	1.62	1.78	5.1	0.19	0.19
3.0	2.5	4.6	1.80	2.00	5.0	0.18	0.17
2.3	2.5	4.5	1.41	2.80	7.2	0.22	0.14
2.1	2.5	4.4	2.30	2.00	5.1	0.33	0.29
2.1	2.5	4.4	2.11	1.94	3.2	0.28	0.27
1.7	2.6	7.0	0.69	3.35	5.8	0.21	0.17
2.7	2.1	6.4	2.19	1.64	4.2	0.28	0.35
2.4	2.6	6.2	1.81	2.30	4.0	0.31	0.32
3.1	2.6	6.4	1.93	2.50	4.2	0.25	0.27
3.0	3.4	5.0	2.84	2.46	4.1	0.40	0.37
3.2	3.4	5.0	3.01	2.86	6.0	0.62	0.34
2.1	2.5	4.8	2.12	2.33	2.5	0.42	0.28

In Table 11.2 is listed some early JET measurements (Rebut *et al.* 1985) at values of $\beta_p \sim 0.1$. The JET torus cross-section is elongated in the vertical direction, so for an equivalent radius we have taken $(A/\pi)^{1/2}$, where A is the area of the plasma cross-section. The major radius of the torus is 3m. The last column is the theoretical confinement time, calculated from (11.94). Bearing in mind that changes in the profile shapes can alter τ_{E_e} by 50 per cent or so away from the average, we see that the theoretical values are as close as can be expected to the experimental values in the penultimate column. We conclude that the theory is well supported by experiment at low values of β_p.

11.10.2 *Measurements in eight tokamaks under general conditions*

The theory given above has been extended to allow for a range of values of beta poloidal, for other forms of power injection besides ohmic heating and to include conduction losses in the ion gas.* It would take us too far from the main topic of this book to detail these extensions, but it is worth recording here that the agreement found above between theory and measurements made in the JET tokamak also holds in more general conditions.

*See the final chapter of *Principles of Magnetoplasma Dynamics*, OUP, 1987.

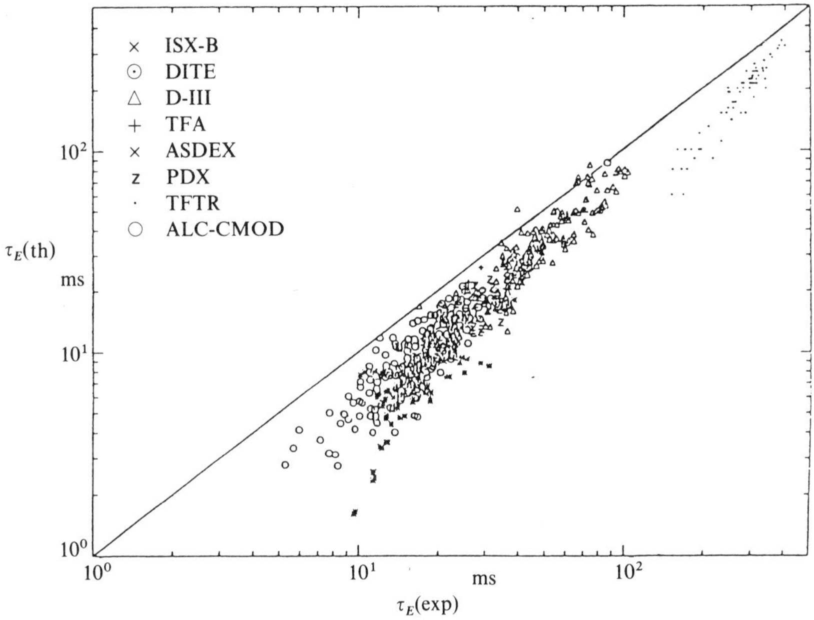

FIG. 11.3. Energy confinement times in a variety of tokamaks

Let τ_E, denote the *total* energy confinement time, then after considerable algebra based on (11.60), applied to each of the electron and ion gases, it can be shown that (cf. equation (18.65) of Woods (1987))

$$\tau_E = 1.09 \times 10^{-14} \frac{I_p^{4/5} R^{7/5} a^{4/5} \bar{n}_e^{3/5} q_a^{2/5}}{P^{3/5}(1 + 0.65\beta_p^{-1})} \left(\frac{8\mathcal{R}^{3/2}}{7+\mathcal{R}}\right)^{2/5}, \tag{11.95}$$

where besides the symbols already defined, we have introduced P for the total power input (ohmic plus auxiliary) and $\mathcal{R} = 1 + \langle n_i \rangle \langle T_i \rangle / \langle n_e \rangle \langle T_e \rangle$.

For comparison of (11.95) with experiment, the Kaye–Goldston (1985) tokamak data base, augmented by an ALCATOR-C data base, was adopted. This gave results from eight tokamaks in all. The experimental points are shown in Fig. 11.3, where the straight line represents theory. Bearing in mind the several approximations employed in applying the theory, e.g. the use of empirical temperature and density profiles, the neglect of radiation losses and of energy transfers between the electrons and ions, the agreement is good.

11.10.3 *Minor and major disruptions*

Besides the energy confinement time, there are a number of other phenomena related to the transport of energy and mass from tokamaks requiring explanation. We shall discuss two of these and describe in general terms how they are explained by second-order theory. For further details, the reader is referred to Woods (1987).

The low β_p regime is stable and reproducible, but as β_p is increased, making the derivative term in (11.82) more important, this stability is lost. A finite-amplitude oscillation with a sawtooth appearance occurs in several of the dependent variables, and is especially marked in the electron temperature. The mechanism for this is as follows.

It is a consequence of the p' term in χ_e that steeper temperature profiles *reduce* the thermal conductivity. Thus once a profile begins to steepen, it 'holds' more energy and hence steepens further, and so on. Finally the point is reached where at the limiter (plasma boundary) the outwards heat flux ceases and the energy starts to flow inwards. Of course this is prevented by the usual conservation restraint on energy, and the climb up the sawtooth ramp is checked. At this stage the safety factor on the axis, q_0, is too small for mhd stability to be maintained, and an oscillation of the magnetic axis sets in. This oscillation modulates the magnetic field and by altering the fraction of trapped particles, activates the second term in (11.59). As this term is usually one or two orders of magnitude larger than the steady state residence time, there is a very rapid increase in thermal diffusivity. The result is a sudden drop in the central temperature—termed a *minor disruption*—and a corresponding increase in the temperature outside what is known as the 'inversion radius'. This activity is the collapse phase of the sawtooth.

It is found from the theory that when the central region of the temperature profile has flattened a certain amount, heat begins to flow *inwards* from the inversion radius, which stage marks the end of the temperature collapse. Now the climb up the long temperature ramp of the sawtooth begins and the process repeats. With the aid of a one-parameter family of temperature profiles that exhibits the oscillatory behaviour described above, it is possible to deduce an estimate for the sawtooth period τ_s. Table 11.3 compares theory and experiment for typical sawtooth periods in a number of tokamaks. The theory is satisfactory, especially considering that the one-parameter family of profiles is a crude representation of the actual profiles.

For a given value of the plasma current, there is a critical average number density above which tokamak discharges are unstable. The instability develops very rapidly, with a sudden cooling of the central plasma and a flattening of the current profile, usually followed by a slower decay of the current to zero. This instability is known as a *major disruption*. It is a serious defect in tokamaks, since by limiting the attainable number density, it restricts their operation to an uneconomic regime. Of course understanding it *may* suggest a means of overcoming the limitation.

The explanation provided by the second-order theory is relatively simple. At the end of the collapse phase in a minor disruption the profile is

Table 11.3 *Sawtooth periods in a number of tokamaks*

Machine	R (m)	B_φ (T)	V (V)	$\tilde{T}_{e0}$ (eV)	$10^{-19}\bar{n}_e$ (m^{-3})	Z^*	τ_s(exp) (ms)	τ_s(th) (ms)
PLT	1.3	352	2	850	14	2	9	10.0
T10	1.5	3.5	2	800	8	1.5	13	5.7
TFR	0.98	2.5	2.5	1200	4.5	6	2.5	3.4
T-4	1.0	2.6	5.5	700	6	5	1	1.6
DITE	1.17	1.34	3	600	1.2	5	1	0.8
TOSCA	0.3	0.4	2	200	1.5	1.5	0.12	0.09
JET	3	3.4	0.91	2000	1.55	4.0	39	44
JET	3	3.4	0.77	2850	2.43	3.36	90	102
JET	3	3.4	0.76	3000	3.01	2.88	138	101

quite flat, and as remarked above such profiles lose energy more rapidly than steep profiles. If the energy supplied by ohmic heating or any other method is at too low a rate, the flat profile may lose energy more rapidly than can be replaced by the supply. In this case, what starts ostensibly as a minor disruption may continue into the extended collapse of a major disruption. Increasing the power supply will therefore extend the domain of stable operation; this is what is observed. The theory leads to a stability diagram in a (ξ, β_p) plane, where ξ is a non-dimensional measure of the rate at which energy is supplied to the plasma within the inversion radius. This is in general agreement with observations.

11.11 The flow of plasma from tokamaks

11.11.1 *The viscous stress tensor in a cylindrical plasma*

In steady conditions it follows from (5.8) that

$$\mathrm{D}^*\overset{\circ}{\mathbf{e}} = \mathbf{v}\cdot\nabla\overset{\circ}{\mathbf{e}} - \boldsymbol{\Omega}\times\overset{\circ}{\mathbf{e}} + \overset{\circ}{\mathbf{e}}\times\boldsymbol{\Omega} \qquad (\boldsymbol{\Omega} = \tfrac{1}{2}\nabla\times\mathbf{v}).$$

Hence the tensor S defined in (11.27) is

$$\mathsf{S} = \overline{\overset{\circ}{\mathbf{e}}\cdot\overset{\circ}{\mathbf{e}}}^{\circ} - \mathbf{v}\cdot\nabla\overset{\circ}{\mathbf{e}} + \boldsymbol{\Omega}\times\overset{\circ}{\mathbf{e}} - \overset{\circ}{\mathbf{e}}\times\boldsymbol{\Omega}. \tag{11.96}$$

In the case of the cylindrical magneto-plasma described in §11.4.1, we have

$$\overset{\circ}{\mathbf{e}} = a(\hat{\mathbf{r}}\hat{\boldsymbol{\theta}} + \hat{\boldsymbol{\theta}}\hat{\mathbf{r}}) + c(\hat{\boldsymbol{\varphi}}\hat{\mathbf{r}} + \hat{\mathbf{r}}\hat{\boldsymbol{\varphi}}), \tag{11.97}$$

with $$a = \tfrac{1}{2}(v_\theta' - v_\theta/r), \qquad c = \tfrac{1}{2}v_\varphi'.$$

The fluid spin is given by $$\boldsymbol{\Omega} = \tfrac{1}{2}(v_\theta' + v_\theta/r)\hat{\boldsymbol{\varphi}} - \tfrac{1}{2}v_\varphi'\hat{\boldsymbol{\theta}};$$

also $$\mathbf{v}\cdot\nabla\overset{\circ}{\mathbf{e}} = 2a\frac{v_\theta}{r}(\hat{\boldsymbol{\theta}}\hat{\boldsymbol{\theta}} - \hat{\mathbf{r}}\hat{\mathbf{r}}) + c\frac{v_\theta}{r}(\hat{\boldsymbol{\theta}}\hat{\boldsymbol{\varphi}} + \hat{\boldsymbol{\varphi}}\hat{\boldsymbol{\theta}}).$$

From these expressions we find that

$$\mathsf{S} = -\tfrac{5}{3}(a^2 + c^2)\hat{\mathbf{r}}\hat{\mathbf{r}} + \tfrac{1}{3}(7a^2 - 2c^2)\hat{\boldsymbol{\theta}}\hat{\boldsymbol{\theta}} + \tfrac{1}{3}(7c^2 - 2a^2)\hat{\boldsymbol{\varphi}}\hat{\boldsymbol{\varphi}} + 3ac(\hat{\boldsymbol{\theta}}\hat{\boldsymbol{\varphi}} + \hat{\boldsymbol{\varphi}}\hat{\boldsymbol{\theta}}). \tag{11.98}$$

With $\mathbf{b} = b_\varphi\hat{\boldsymbol{\varphi}} + b_\theta\hat{\boldsymbol{\theta}} \qquad (b_\varphi \equiv B_\varphi/B,\ b_\theta \equiv B_\theta/B)$,

we find from (11.28) that

$$\boldsymbol{\pi}_2 = -\frac{p\tau_1}{2\omega_c}\Big\{G(\hat{\mathbf{r}}\hat{\boldsymbol{\theta}} + \hat{\boldsymbol{\theta}}\hat{\mathbf{r}}) + H(\hat{\mathbf{r}}\hat{\boldsymbol{\varphi}} + \hat{\boldsymbol{\varphi}}\hat{\mathbf{r}})\Big\}, \tag{11.99}$$

where $G \equiv (4a^2 + c^2)b_\varphi - 3acb_\theta + 9b_\theta\{(a^2 - c^2)b_\varphi b_\theta + ac(b_\varphi^2 - b_\theta^2)\}$,

and $H \equiv -(a^2 + 4c^2)b_\theta + 3acb_\varphi + 9b_\varphi\{(a^2 - c^2)b_\varphi b_\theta + ac(b_\varphi^2 - b_\theta^2)\}$.

In tokamak fields we may assume that $b_\varphi \approx 1,\ b_\theta \approx 0$, reducing (11.99) to

$$\boldsymbol{\pi}_2 = -\frac{p\tau_1}{2\omega_c}\Big\{(4a^2 + c^2)(\hat{\mathbf{r}}\hat{\boldsymbol{\theta}} + \hat{\boldsymbol{\theta}}\hat{\mathbf{r}}) + 12ac(\hat{\mathbf{r}}\hat{\boldsymbol{\varphi}} + \hat{\boldsymbol{\varphi}}\hat{\mathbf{r}})\Big\}. \tag{11.100}$$

11.11.2 *The radial fluid velocity*

With the help of

$$\nabla = \hat{\mathbf{r}}\frac{\partial}{\partial r} + \frac{\hat{\boldsymbol{\theta}}}{r}\frac{\partial}{\partial \theta}, \qquad \frac{\partial \hat{\mathbf{r}}}{\partial \theta} = \hat{\boldsymbol{\theta}}, \qquad \frac{\partial \hat{\boldsymbol{\theta}}}{\partial \theta} = -\hat{\mathbf{r}},$$

we obtain the fluid stresses

$$-\nabla p = -p'\hat{\mathbf{r}}, \qquad -\nabla\cdot\boldsymbol{\pi}_1 = \frac{1}{r^2}\left\{\frac{r^2 p}{\omega_c}a\right\}'\hat{\mathbf{r}}, \tag{11.101}$$

and

$$-\nabla\cdot\boldsymbol{\pi}_2 = \frac{1}{r^2}\left\{\frac{r^2 p\tau_1}{2\omega_c}(4a^2 + c^2)\right\}'\hat{\boldsymbol{\theta}} - \frac{6}{r}\left\{\frac{rp\tau_1}{\omega_c}ac\right\}'\hat{\boldsymbol{\varphi}}, \tag{11.102}$$

where for completeness we have added the zeroth and first-order terms (see exercise 11.5). These formulae can be used to calculate the small radial fluid velocity as follows.

Omitting the external force $\mathbf{f}$ from (9.71) and noting that in the present application $\mathrm{D}\mathbf{v}$ is zero, we have

$$\mathbf{v}_\perp = v_r\hat{\mathbf{r}} + v_\theta\hat{\boldsymbol{\theta}} = \frac{\mathbf{E}}{B}\times\hat{\boldsymbol{\varphi}} + \frac{1}{\rho\omega_c}(\mathbf{R} - \nabla\cdot\mathbf{p})\times\hat{\boldsymbol{\varphi}}, \tag{11.103}$$

where $\mathbf{p} = p\mathbf{1} + \boldsymbol{\pi}_1 + \boldsymbol{\pi}_2 + \cdots$. With axial symmetry, E_θ is zero, so (11.103) yields

$$v_r = \frac{1}{\rho\omega_c}(\mathbf{R} - \nabla\cdot\mathbf{p})\cdot\hat{\boldsymbol{\theta}}. \tag{11.104}$$

By (10.107), $\mathbf{R} = en_e\boldsymbol{\eta}\cdot\mathbf{j} = -\mathbf{R}_i$, where we have dropped the relatively unimportant thermoelectric term. Hence from (11.100) and (11.102) we obtain

$$v_r = -B^{-1}\boldsymbol{\eta}\cdot\mathbf{j}\cdot\hat{\boldsymbol{\theta}} + \frac{1}{\rho\omega_c r^2}\left\{\frac{r^2 p\tau_1}{2\omega_c}\left[(v'_\theta - v_\theta/r)^2 + \tfrac{1}{4}{v'_\varphi}^2\right]\right\}'. \quad (11.105)$$

A more familiar form for the first term in (11.105) follows from the equilibrium condition $\mathbf{j}\times\mathbf{B} = \nabla p$. This gives

$$v_r^{(\eta)} = -\frac{\eta_\perp}{B^2}p'.$$

But when toroidal effects are taken into account, this term increases to the Pfirsch–Schülter (1968) value

$$v_r^{(\eta)} = -\left(1 + \frac{2\eta_\parallel}{\eta_\perp}q^2\right)\frac{\eta_\perp}{B^2}p'. \quad (11.106)$$

However, in normal operating conditions the second term in (11.105) can be shown to be orders of magnitude larger than the first.

The second-order expression in (11.105) differs from that given in the author's earlier work describing second-order theory for magnetoplasmas,* by having $\frac{1}{4}$ instead of unity as the coefficient of ${v'_\varphi}^2$. However, because v_r is so strongly dependent on the shape of the temperature profile, comparisons with experiment based on empirical profiles in the theory are unreliable to within an order of magnitude or so. With the 'correct' choice of profile, (11.105) gives results in agreement with experimental measurements of the loss rate of plasma from tokamaks.

11.11.3 *Validity of the theory*

For the theory to be valid for momentum transport, it appears from (11.63) that the condition

$$\epsilon = \tau_1 || \overset{\circ}{\nabla}\mathbf{v} || \ll 1 \quad (11.107)$$

should be satisfied. The corresponding expression for heat transfer was easily satisfied, because—as explained in §11.9.3—we had the much smaller time $\tau_\parallel^*$ in place of τ_2. Periodicity allows us to make a similar reduction to τ_1 as follows.

The connection length of the magnetic field lines is $2\pi Rq$ (see §11.6.2), hence the projections of the guiding centres on to a plane orthogonal to the minor axis traces a closed curve in a time $2\pi Rq/c_\parallel = \tau_g$ say. Let ϕ represent a macroscopic function, then by tokamak periodicity, following a bunch of guiding centres, $\phi(t-\tau) = \phi(t-[\tau - n\tau_g])$, where n is an integer. We may chose n such that $\tau - n\tau_g = \alpha\tau_g$, where, assuming symmetry about a plane perpendicular to the major axis, $0 \le \alpha \le 1/2$. Now we can replace $\phi(t-\tau) = \phi(t) - \tau\mathbb{D}\phi + \cdots$ in a non-periodic system by $\phi(t-\tau) =$

*See Appendix A2 for a brief account of this treatment.

$\phi(t) - \alpha\tau_g\mathbb{D}\phi + \cdots$ in the tokamak. Thus the constraint $\epsilon = |\tau\mathbb{D}| \ll 1$ (cf. (5.33)) can be replaced by $|\alpha\tau_g\mathbb{D}| \ll 1$. This is equivalent to writing (11.107) as

$$\epsilon = (\pi R q_a/\mathcal{C}) \| \overset{\circ}{\nabla} \mathbf{v} \| \ll 1, \tag{11.108}$$

where we have estimated $\frac{1}{2}\tau_g$ by $\pi R q_a/\mathcal{C}$.

Returning to the numbers given in §11.9.3 for the JET tokamak, in place of the time $aq_a/\mathcal{C}$, we now have $\pi R q_a/\mathcal{C}$, which is about ten times larger. This gives $\epsilon \sim 3 \times 10^{-3}$ for the electron fluid. For the ions $\mathcal{C}$ is reduced by a factor $(m_e/m_i)^{1/2}$, but as the ion fluid velocities are likewise reduced by ambipolarity (see exercise 11.6), the net effect is to leave ϵ much the same in both fluids. The conclusion is that the theory is valid for momentum transport in a tokamak.

11.12 Energy losses from reversed field pinches

A toroidal reversed field pinch (RFP) is distinguished from a tokamak in having its toroidal and poloidal magnetic fields of comparable magnitude, i.e. $s = B_\theta/B_\varphi$ is not small and $q-1 = (rB_\varphi/RB_\theta)-1$ is not positive. With some generalization, the diffusion theory developed in §§11.9 and 11.11 for tokamaks can be applied to RFPs.

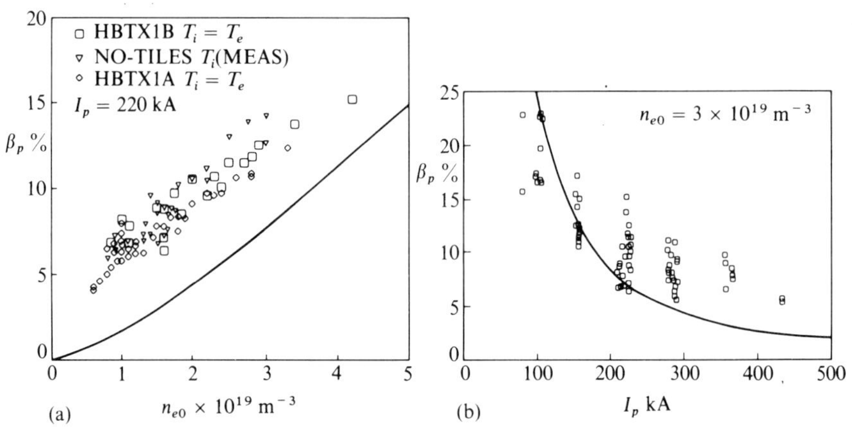

FIG. 11.4. Theory and experiment for RFPs

Deane (1989) found that in RFPs the dominant mechanism for energy transport is not conduction, but convection. As his theory was an extension of that presented in Woods (1987), it initially contained the flaw mentioned at the end of §11.11.2. This has now been removed. Using (11.99) and (11.104), Deane found an expression for the energy confinement time in RFPs, from which he developed several useful scaling laws. Each of these laws showed fair agreement with Bodin's (1987) experimental data for the

'HBTX' RFP. For example Deane obtained the relation

$$\beta_p = 1.5 \times 10^{-19}\, n_{e0}^{1.35}\, I_p^{-1.61} \tag{11.109}$$

between the poloidal beta, the electron number density on the minor axis and the total plasma current. Apart from assuming profile shapes for n_e and T_e (cf. §11.9.5), he introduced no other empirical elements into (11.109).

Figs. 11.4a and 11.4b show the dependence of β_p for (a) a fixed current of $I_p = 220$ kA and for (b) a fixed number density of $n_{e0} = 3 \times 10^{19}$ m^{-3}. The experimental points shown in the figures were obtained by Bodin (1987) from two versions of the HBTX RFP. The curves were obtained from (11.109). Considering the various simplifying assumptions made in the theory and the difficulties of making accurate measurements—as evidenced by the spread of the observational points—the agreement between theory and experiment is satisfactory. It provides some support for the complicated expression given in §11.3.3 for the second-order viscosity tensor.

Exercises 11

11.1 The axial current density profile in a cylindrical magnetoplasma has a shallow minimum on the axis, a maximum on a surface $r = r_0$ and falls to zero at $r = a > r_0$. The temperature profile is parabolic, with its maximum on the axis. Show that this configuration is thermally unstable and explain how it would evolve.

11.2 In §8.5.4 it was argued that it is not physically possible for heat flux to be generated in a neutral gas by fluid shear alone. Show that according to the theory of §11.3.1, this is also the case in a magnetoplasma, i.e. there is no heat flux in the absence of a temperature gradient.

11.3 In equation (11.105) v_θ occurs only in the combination $(v'_\theta - v_\theta/r)$. Why must this be so?

11.4 Show that toroidal geometry adds a term $\frac{1}{2} s v_\varphi \cos\theta\,/R$ to the value of H defined in (11.31), and that when this term is averaged over θ with $(1+\varepsilon\cos\theta)$ as a weighting factor, it vanishes. Deduce that the net heat flux from a torus is the same as that from a cylinder of the same minor radius provided particle trapping is accounted for.

11.5 Given that in a tokamak the axial fluid velocities are much larger than the azimuthal components, use (11.69) and the approximation $v_{e\varphi} \approx -j_\varphi/en_e$ to prove that v_r is independent of B_φ and hence that the confinement time for the plasma is independent of the strength of the magnetic field. What purpose does the field serve?

11.6 Ambipolar diffusion of plasma across field lines is defined by the condition $v_{er} = v_{ir}$. (This corresponds to the physical constraint of no radial electric current.) Show that ambipolarity imposes a constraint on the fluid velocities of the two species, and that in particular, when axial components are dominant and the temperatures are the same for the electrons and ions, this condition is approximated by

$$v_{i\varphi} = -G v_{e\varphi}, \qquad G = \left(\frac{m_e \tau_e}{m_i \tau_i}\right)^{1/2}.$$

APPENDIX

A1. Physical constants in MKS units

Electron mass	m_e	9.1096×10^{-31} kg
Proton mass	m_p	1.6726×10^{-27} kg
Electron charge	e	1.6022×10^{-19} C
Boltzmann constant	k	1.3806×10^{-23} J K^{-1}
Permeability (free space)	μ_0	$4\pi \times 10^{-7}$ H m^{-1}
Permittivity (free space)	ϵ_0	8.854×10^{-12} F m^{-1}
Velocity of light	c	2.9979×10^{8} m s^{-1}

A2. Departures from received theory

1. *Introduction*

That the present book departs from the usual or classical treatment of kinetic theory is noted in the Preface, and also at several points throughout the work. For the classical treatment, the general texts identified in the list of references following this appendix can be consulted. It may be helpful to readers new to the subject to have the significant points of departure collected together, so that the extent and nature of the changes will be immediately apparent.

A central parameter in the subject is the ratio of microscopic to macroscopic length (or time) scales. This is known as the Knudsen number, which we have denoted by ϵ. Throughout this book we have been concerned with the case when ϵ is much smaller than unity. By 'first-order transport' is meant the simplest theory, with terms of order ϵ^2 being neglected. None of the classical work on first-order transport is altered by our approach. Hence Chapters 1 (except §1.5), 2, 3 (except §§3.6, 3.9), 4, 6, 7 (except §7.11), 9 (except §§9.5, 9.7, 9.9), and 10, which cover this work, do not depart from received theory, except of course in the details of presentation.

The 'intermediate' kinetic theory in Chapter 5 starts with an account of first-order transport and then extends this step by step to second-order transport. This is the point at which the treatment in this book differs substantially from accepted theory, the second-order terms being obtained by *physical* rather than mathematical reasoning. And the resulting terms are different. The new material in §§5.7 to 5.12 forms the basis of the second-order transport studied and applied in Chapters 8 and 11.

For neutral gases second-order terms are usually dominated—as one would expect—by the first-order terms, so the changes we have introduced have only a small impact on transport. But this is measurable and enables us to distinguish between the standard theory and that advanced in this work. However, in magnetoplasmas the $O(\epsilon^2)$ terms can be orders of magnitude *larger* than the $O(\epsilon)$ terms, and have a dramatic effect on transport. The reason for this unexpected result is that in the case of strong magnetic fields, there are *two* small parameters in the theory. Besides ϵ, there is also $\eta = (\omega_c \tau)^{-1}$, where ω_c is the cyclotron frequency (see §9.2.1) and τ is the time interval between successive collisions. The stronger the field, the smaller is η: in tokamaks typical values of the parameters are $\epsilon \sim 10^{-2}$ and $\eta \sim 10^{-6}$. For radial heat flux, the leading terms are $O(\epsilon\eta^2)$ and $O(\epsilon^2\eta)$, so the one classified as being 'second-order' may be $\sim 10^4$ times larger than the first-order term.

2. *Pressure at points interior to the gas is due to collisions*

In the usual accounts, the pressure tensor is *defined* to be the total momentum flux tensor, $\rho\langle \mathbf{cc} \rangle$ (see §1.5). Particle collisions are not necessarily involved, so that, with this definition, a collection of non-colliding particles will have a pressure. But this 'pressure' is not the same as that adopted in fluid mechanics, where it is defined to be the *force* per unit area transmitted across a conceptual (or mathematical) surface S embedded in the fluid. This force is similar to that exerted by gas molecules on a boundary surface, i.e. it is due to molecular *interactions* across S. Newton's second law plays an essential role. Kinetic pressure cannot be identified with fluid pressure without a caveat about the presence of particle collisions, but this is ignored in the usual accounts. In a theory correct to $O(\epsilon)$ this lacuna does not matter, but it does generate errors in a $O(\epsilon^2)$ theory. Throughout this book we have used *fluid* pressure. See §§1.5, 3.6, 3.9, 5.3.2, 5.8, 7.11, and 8.5.

The simplest example is provided by an atmosphere; with collisions there is an equilibrium pressure gradient given by $\nabla p = \rho \mathbf{g}$, where $\rho \mathbf{g}$ is the gravitational force per unit volume. Without collisions there is no equilibrium and the molecules fall separately with acceleration $\mathbf{g}$. The usual kinetic definition of pressure allows no distinction between these cases (see §5.3.2 and exercise 7.1). A second example is cited in the footnote at the end of this appendix.

3. *The acceleration term in the kinetic equation*

The basic equation of kinetic theory expresses the conservation of particle numbers for an element $d\boldsymbol{\nu}$ of phase space. The phase-space density is f; the $f\,d\boldsymbol{\nu}$ particles in $d\boldsymbol{\nu}$ are changed by streaming (convection through phase space) and collisional scattering. The equation describing this—known as the kinetic equation—is

$$\mathrm{D}f + \mathbf{c} \cdot \nabla f + \mathbf{a} \cdot \frac{\partial f}{\partial \mathbf{c}} = \left(\frac{df}{dt} \right)_{col}, \tag{A1}$$

where D is the rate of change following a fluid element $\mathcal{F}$, $\mathbf{c}$ and $\mathbf{a}$ are the velocity and acceleration of $d\boldsymbol{\nu}$ relative to $\mathcal{F}$, and $(df/dt)_{col}$ is the rate of change of particles in unit volume of $d\boldsymbol{\nu}$ due to collisions.

The velocity $\mathbf{c}$ is an independent variable, so that, subject to the condition that its average $\langle\mathbf{c}\rangle$ is zero, its value may be freely chosen. We could also choose $\mathbf{a}$ freely, but if it does not match the average acceleration of the particles within $d\boldsymbol{\nu}$, we cannot expect the right-hand term in (A1) to be due only to collisions—there will also be a streaming term $\delta\mathbf{a}\cdot\partial f/\partial\mathbf{c}$ to balance the mismatch, $\delta\mathbf{a}$, in the accelerations of the 'container' $d\boldsymbol{\nu}$ and the particles within it (see §5.2.3). (The distinction between the motion of the volume element $d\boldsymbol{\nu}$ and that of the particles momentarily contained within it, is largely ignored in the received theory.)

Standard kinetic theory (SKT) is usually formulated in the laboratory frame; of course the physics represented is, or should be, independent of the choice of reference frame. In the laboratory frame SKT assumes the acceleration to be equal to the externally-imposed body force per unit mass, say $\mathbf{F}$. In the convected or fluid frame, this becomes

$$\mathbf{a} = \mathbf{a}_* \equiv \mathbf{F} - \mathrm{D}\mathbf{v} - \mathbf{c}\cdot\nabla\mathbf{v}, \qquad \text{(A2)}$$

where $\mathbf{v}$ is the fluid velocity. The term $-\mathbf{c}\cdot\nabla\mathbf{v}$ enters because the peculiar velocity $\mathbf{c}$ is measured from origins having different (fluid) velocities in the sheared flow. The equation of fluid motion allows us to introduce the pressure gradient force $\mathbf{P} = -\nabla\cdot\mathsf{p}/\rho$, where p is the pressure tensor. Thus

$$\mathbf{a}_* = -\mathbf{P} - \mathbf{c}\cdot\nabla\mathbf{v}.$$

It is then assumed, that with this acceleration, the right-hand side of (A1) is due entirely to collisions, i.e. no *streaming* corrections are required. Thus the loss rate from $d\boldsymbol{\nu}$ is taken to be f/τ, where τ is a suitable average of the time taken by a particle between successive collisions.

In this book we depart from (A2) by taking $\mathbf{a}$ to be

$$\mathbf{a} = -\mathbf{c}\cdot\mathsf{e}, \qquad \mathsf{e} \equiv \tfrac{1}{2}\left(\nabla\mathbf{v} + \widetilde{\nabla\mathbf{v}}\right), \qquad \text{(A3)}$$

where e is the symmetrical part of the velocity gradient tensor. The reasoning for this choice is that, for the collisional loss rate to be expressible in the isotropic form f/τ, the average of the accelerations of the containers $d\boldsymbol{\nu}$ must exactly match the average acceleration of the particles within these containers. By definition, the latter average is the same as the the acceleration of the fluid element. It follows that, since $\mathbf{a}$ is measured relative to this element, $\langle\mathbf{a}\rangle$ is zero, a condition that is satisfied by (A3), but not by (A2). It is also neccessary to remove the influence of fluid vorticity by determining $\mathbf{c}$ in a frame that *spins* with the fluid element, a modification that replaces $\mathbf{c}\cdot\nabla\mathbf{v}$ by $\mathbf{c}\cdot\mathsf{e}$ (see §§5.7, 5.8 and 5.9). Without this change, the second-order transport terms become dependent on the choice of reference frame (see §§8.5.2 and 5.12).

4. *Inverse collisions*

In SKT microscopic reversibility is deployed to infer a symmetry between the rates at which particles are collided in and out of an element of phase space (see §7.7.2). This is valid only for single, binary collisions, but in the approach of this book, for a theory accurate to $O(\epsilon^2)$, we find it necessay to trace particles back in time through *two* collisions. This destroys the symmetry just described. For a brief account of this divergence from standard theory, the reader is referred to §5.12.

5. *Second-order transport in a magnetoplasma*

A key equation in magnetoplasma transport is the formula for the second-order heat flux across a strong magnetic field, due to a temperature gradient ∇T. This is

$$\mathbf{q} = -\frac{5kp}{2QB}\,\tau_2\,\mathbf{b}\times\,\overset{\circ}{\nabla\mathbf{v}}\,\cdot\nabla T, \tag{A4}$$

where $\overset{\circ}{\nabla\mathbf{v}}$ is the traceless, symmetrical part of the velocity gradient tensor, Q is the particle charge, $B = |\mathbf{B}|$ is the magnetic field, $\mathbf{b}$ is unit vector parallel to $\mathbf{B}$ and τ_2 is an appropriate collision interval (see §§11.5.2 and 11.8.3). This formula cannot be derived from Boltzmann's equation or from any of the other standard kinetic equations, yet it is readily established by mean-free-path arguments (see §§9.7 and 9.9.3). It also follows from the new kinetic equation developed in §§11.1 and 11.2.

Another method of deducing (A4), and a similar formula for the second-order viscous stress tensor, was given in the author's 1987 text. But this method was insecure, being based on a kinetic equation derived by an extension of standard kinetic theory that, while plausible, lacked a detailed physical model. An outline of the argument adopted is as follows.

6. *Heuristic derivation of a magnetoplasma kinetic equation*

In a neutral gas the BGK equation reads (see §5.8)

$$\mathbb{D}f = (f_0 - f)/\tau \qquad \big(f = f(\mathbf{r},\,\mathbf{c},\,t)\big), \tag{A5}$$

where f_0 is the equilibrium velocity distribution and (see (A1))

$$\mathbb{D} \equiv \mathrm{D} + \mathbf{c}\cdot\nabla - (\mathbf{P} + \mathbf{c}\cdot\mathbf{e})\cdot\frac{\partial}{\partial\mathbf{c}}. \tag{A6}$$

This is not quite the standard theory, since that does not allow for the fluid particle spin, i.e. has $\nabla\mathbf{v}$ in place of $\mathbf{e}$.

In a magnetoplasma, for a component gas of particles of mass m and charge Q, the acceleration in (A6) is augmented by the Lorentz term $-\omega_c\mathbf{b}\times\mathbf{c}$, where $\omega_c \equiv QB/m$. Hence (A5) is generalized to

$$\mathbb{D}f = -\frac{1}{\tau}(f - f_0) - \delta(f - f_0), \tag{A7}$$

where
$$\delta \equiv -\omega_c \mathbf{b} \times \mathbf{c} \cdot \frac{\partial}{\partial \mathbf{c}}.$$

It is readily shown that $\delta f_0 = 0$. Relative to the streaming derivative $\mathbb{D}$, which is still defined by (A6), the term $-\delta(f - f_0)$ is the loss (or gain) rate of particles from the element $d\boldsymbol{\nu}$ due to interations with the magnetic field.

Notice that (A5) can be written (exactly)

$$(1 + \tau\, \mathbb{D})(f - f_0) = -\tau\, \mathbb{D} f_0,$$

so with the Knudsen number expansion,

$$f = f_0 + f_1 + f_2 + \cdots \qquad (f_r = \mathrm{O}(\epsilon^r)), \qquad \tau\mathbb{D} = \mathrm{O}(\epsilon),$$

we get
$$f - f_0 = -(1 - \tau\mathbb{D})(\tau\mathbb{D} f_0) + \mathrm{O}(\epsilon^3). \tag{A8}$$

Without the factor $(1 - \tau\mathbb{D})$ on the right of (A8), the error term is $\mathrm{O}(\epsilon^2)$. We may therefore interpret this factor, which changes the current value of $(-\tau\mathbb{D} f_0)$ to its value a collision time earlier, as being what is required to reduce the error term from $\mathrm{O}(\epsilon^2)$ to $\mathrm{O}(\epsilon^3)$.

In a magnetoplasma, equation (A7), plus the interpretation just given, suggest that, in a strong magnetic field, we should replace (A8) by

$$f - f_0 + \tau\, \delta(f - f_0) = -(1 - \tau\, \overline{\mathbb{D}})(\tau\, \mathbb{D} f_0) + O(\epsilon^3), \tag{A9}$$

where $\overline{\mathbb{D}}$ is a streaming derivative following an element $d\overline{\boldsymbol{\nu}}$ that moves parallel to the guiding centres of the gyrating particles, and has the same acceleration as these centres. Since guiding centres have an average motion with the *same* velocity and acceleration as the fluid element (another departure from prevailing opinion*; see §9.5 and Woods 1987, pp. 212–15), the pressure force makes no contribution to the additional acceleration in $\overline{\mathbb{D}}$. Hence

$$\overline{\mathbb{D}} = \mathrm{D} + \mathbf{c}_{\|} \cdot \nabla - \mathbf{c} \cdot \mathbf{e} \cdot \frac{\partial}{\partial \mathbf{c}}. \tag{A10},$$

where $\mathbf{c}_{\|}$ is the velocity of $\overline{\mathbb{D}}$ relative to the fluid particle. (For the Taylor expansion in (A9) to be valid, we cannot follow the rapid changes in velocity and acceleration of the gyrating particles, so it is necessary to replace $\mathbb{D}$ by $\overline{\mathbb{D}}$; cf. §11.2.1.) Notice that (A9) does not follows from (A7), which yields the quite different (and incorrect) equation

$$f - f_0 + \tau\, \delta(f - f_0) = -\tau\, \mathbb{D}(f_0 + f_1) + O(\epsilon^3). \tag{A11}$$

The evidence against (A11), known to the author in 1982, was that it did not lead to (A4), which relation followed from elementary kinetic the-

*It is generally accepted that, considered as individual gyrating *particles*, electrons and ions are not subject to pressure gradient forces, whereas taken collectively in elements of electron and ion *fluids*, the same particles do respond to these forces. This schism is a consequence of adopting two conflicting definitions of pressure—a *flux* for the particles, but a *force* for the fluid (see 2 above). The outcome is that the centres about which the particles gyrate—the so-called guiding centres—have an average motion relative to the fluid particle. But since the particles follow the guiding centres, their mass is also transported relative to the fluid particle, an evident contradiction.

ory without ambiguity. That (A4) was also in agreement with experiment (see §11.10) added conviction that the standard kinetic theory (SKT) for a magnetoplasma was seriously flawed at the $O(\epsilon^2)$ level. The argument leading to (A9)—which does yield (A4)—was devised to bypass the problem with SKT, until a more complete theory could be found. But it was discovered later (1990) that *even for a neutral gas*, the standard theory is not 'exact' for all Knudsen numbers, i.e. (A5) itself has an error term $O(\epsilon^2)$. It follows that the error term in (A8) is $O(\epsilon^2)$, not $O(\epsilon^3)$ as supposed; hence the argument leading to (A9) is unsound. Despite this, the earlier kinetic equation gave almost the same expressions as we have obtained in this book (see remarks at the end of §11.11.2.).

REFERENCES

[The pages on which the references are cited are given in square brackets. General texts that amplify parts of the subject matter of this book in various ways and which provide extensive lists of references, are marked by asterisks.]

Alsmeyer, H. (1976). *J. Fluid Mech.* **74**(3), 497–513. [187]

Beenakker, J.J.M. and Mc Court, F.R. (1970). *Ann. Rev. Phys. Chem.* **21**, 47. [180]

Bhatnager, P.L., Gross, E.P. and Krook, M. (1954). *Phys. Rev.* **94**, 511–25. [86]

*Bird, G.A. (1976). *Molecular gas dynamics*. Clarendon Press, Oxford. [161, 162, 187]

Bodin, H.A.B. (1987). *Proc. Conf. on Physics of Mirrors, Reverse Field Pinches and Compact Tori*, ISPP, Varenna. [270]

Boltzmann, L. (1872). *Sber. Akad. Wiss Wien* **66**, 275–379. [51, 115, 135, 144]

Braginskii, S.I. (1965). *Reviews of Plasma Physics*, Vol. 1 (ed. M.A. Leontovich) pp. 205–311. Consultants Bureau, New York. [210, 238]

Burnett, D. (1935). *Proc. London Math. Soc.* **39**, 385–430, and in **40**, 382–435. [136, 156]

*Cercignani, C. (1988). *The Boltzmann equation and its applications*. Springer–Verlag, New York. [189]

Candel, S.M. and Thivet, F.J.P. (1992) (Laboratoire d'Energétique Moléculaire et Macroscopique, Combustion, Ecole Centrale Paris.) Private communication. [187]

Chapman, S. and Cowling, T.G. (1941). *Proc. Roy. Soc. A.* **179**, 159. [178]

*Chapman, S. and Cowling, T.G. (1970). *The mathematical theory of non-uniform gases*. Cambridge University Press. [27, 66, 74, 80, 115, 124, 130, 136, 158, 185]

Chapman, S. and Dootson, F.W. (1917). *Phil. Mag.* **33**, 248. [77]

Chapman, S. (1916). *Phil. Trans. Roy. Soc. A* **217**, 115. [77, 114, 136]

Clausius, K. and Dickel, G. (1939). *Z. Phys. Chem. B* **44**, 397. [77]

Deane, G.B. (1989). *The transport of mass and energy in toroidal fusion machines*. Doctoral thesis, University of Oxford. [270]

Eggermont, G.E.J., Hermans, P.W., Hermans, L.J.F., Knaap, H.F.P. and Beenakker, J.J.M. (1978). *Z. Naturforsch.* **33a**, 749–60. [180]

Elliot, J.P. and Baganoff, D. (1974). *J. Fluid. Mech.* **65**, 603–24. [187]

Enskog, D. (1911). *Phys. Zeit.* **12**, 56 and 533. [77]

Enskog, D. (1912). *Ann. d. Physik* **38**, 731. [80]

Enskog, D. (1917). *Kinetische Theorie der Vorgänge in mässig verdünnten Gasen*, Diss., Uppsala. [136]

Epperlein, E.M. (1984). *J. Phys. D: Appl. Phys.* **17**, 1823–7. [238]

Euchen, A. (1913). *Phys. Z.* **14**, 324. [66]

*Ferziger, J.H. and Kaper, H.G. (1972). *Mathematical theory of transport in gases*. North Holland, Amsterdam. [131, 158, 175, 189, 199]

Fiscko, K.A. and Chapman, D.R. (1988). *Proc. AIAA Thermophysics, Plasmadynamics and Lasers Conference*, San Antonio, Texas, AIAA-88-2733. [187]

Grad, H. (1949). *Comm. Pure and Appl. Math.* **2**, 311–407. [136, 178]

Grad, H. (1952). *Comm. Pure and Appl. Math.* **5**, 257. [178]

Greenspan, M. (1956). *J. Acoust. Soc. Am.* **28**(4), 644–8. [181, 183]

Hermans, L.J.F., Schutte, A., Knapp, H.F.P., and Beenakker, J.J.M. (1970). *Physica* **46**, 491–506. [180]

Hilbert, D. (1912). *Grundzüge einer allgemeinen Theorie der linearen Integralgleichungen.* (Chelsea Pub. Co., New York, 1953). [136]

*Jeans, J. (1960). *An introduction to the kinetic theory of gases*. Cambridge University Press. [65]

Jones, R.C. and Furry W.H. (1946). *Rev. Mod. Phys.* **18**, 151. [77, 80]

Kadomtsev, B.B. and Pogutse, O.P. (1967). *Soviet Phys.* JETP **24**, 172. [255]

Kaye, S.M. and Goldston, R.J. (1985). *Nuclear Fusion* **25**, 65. [265]

*Kennard, E.H. (1938). *Kinetic theory of gases*. McGraw–Hill Book Co. Inc., New York and London.

Kihara, T. (1949). *Imperfect Gases* (Japanese). Asakusa Bookstore, Tokyo. (English Translation by US Office of Air Research, Wright–Paterson Air Base); also see *Rev. Mod. Phys.* **25**, 831 (1953). [158]

Leipmann, H.W., Narasimha, R. and Chakine, M.T. (1962). *Phys. Fluids.* **5**, 1313. [186]

Liewer, P.C. (1985). *Nuclear Fusion* **25**(5), 543. [258]

Lorentz, H.A. (1905 Lecture Notes) *Theory of electrons*, 2nd. ed. Stechert, New York (1923): Dover Reprint (1952). [136, 199]

Lumpkin, F.E. and Chapman, D.R. (1991). *Proc. 29th Aerospace Sciences Meeting*, Reno, Nevada, AIAA 91–0771. [187, 188]

Maxwell, J.C. (1860). *Phil. Mag.* **19**, 19–32; **20**, 21–37. [30, 49, 63]

Maxwell, J.C. (1867). *Phil. Trans. Roy. Soc.* **157**, 49–88; (See also *Scientific papers of James Clerk Maxwell*, W.D. Niven, ed. New York: Dover, (1965), vol. 2, 43.) [31, 49, 50, 59]

Maxwell, J.C. (1879). *Phil. Trans. Roy. Soc.* **170**, 231–56. [178]

Pfirsch, D. and Schlüter, A. (1968). *Rep. Max-Planck Inst.* MPA/PA/7/62. [269]

Pham-Van-Diep, G.C., and Erwin, D.A. (1989). *Proc. 16th. Int. Symp. on Rarified Gas Dynamics*, ed. by E.P. Muntz, *et al. Progress in Aeronautics and Astronautics*, **118**, 271. [187]

*Present, R.D. (1958). *Kinetic theory of gases*. McGraw-Hill Book Co., New York. [62, 126]

Rebut, P.H. *et al.*, (1985). *Proc. 12th. European Conf. on Controlled Fusion and Plasma Physics*, Budapest. Pt. 1, 11. [264]

Reese, J.M. (1991). (Mathematical Institute, University of Oxford.) Private communication. [186]

Rosenbluth, M.N., MacDonald, W.M., and Judd, D.L. (1957). *Phys. Rev.* **107**, 1. [219]

Spitzer, L. and Härm, R. (1953). *Phys. Rev.* **89**, 977. [199]

*Spitzer, L., (1962). *Physics of fully ionized gases*, 2nd. ed. Interscience, New York. [199, 227, 228]

*Shkarofsky, I.P., Johnston, T.W., and Bachynsky, M.P. (1966). *The particle kinetics of plasmas*. Addison–Wesley, Reading, Mass. [199, 222, 238]

Sherman, F.S., and Talbot, L. (1960). *Proc. First Int. Sym. on Rarefied Gas Dynamics*, ed. F.M. Devienne. Pergammon, New York, p. 161. [185]

*Vincenti, W.G. and Kruger, Jr. W.G. (1965). *Introduction to physical gas dynamics*. Krieg. [186]

Watson, G.N. (1952). *Theory of Bessel functions*, (2nd. ed.) Cambridge University Press. [153]

*Wesson, J. (1987). *Tokamaks*. Clarendon Press, Oxford. [258]

*Williams, M.M.R. (1971). *Mathematical methods in particle transport theory*. Butterworths, London. [189]

Woods, L.C. (1983a). *J. Fluid Mech.* **136**, 423–433. [90, 100]

Woods, L.C. (1983b). *J. Plasma Phys.* **29**(1), 143–154. [191, 208]

*Woods, L.C. (1986). *The thermodynamics of fluid systems*. Oxford University Press. [18, 22, 83, 238]

*Woods, L.C. (1987). *Principles of magnetoplasma dynamics*. Oxford University Press. [191, 202, 258, 261, 264, 265, 266, 270]

INDEX